To Milt
with
Dick

Museum Studies and Wildlife Management

Selected Papers

Compiled and Annotated by Richard C. Banks

Produced by
Smithsonian Institution Press
for
National Fish and Wildlife Laboratory
U.S. Fish and Wildlife Service
E-600, National Museum of Natural History
Washington, D.C. 20560

1979

It is with deep respect and appreciation
that I dedicate this book to
JOHN W. ALDRICH
who in his 33-year career with the
U.S. Fish and Wildlife Service provided a
constant example of the role of museum-related
studies in the field of wildlife management.

Printed in the United States of America

Cover: The stylized animal footprints illustrated
are used by the Smithsonian's National Zoological Park
in a trail system to guide the visitor
through the Zoo.

Library of Congress Catalogue Card Number: 79-16312

Contents

Introduction

There seems to be an expanding breach between the disciplines of museum biology and wildlife management. More properly, I should state that the apparent breach is between practitioners of these fields; as I hope to indicate in this compilation of research papers, the disciplines are and always have been closely related and interdependent. Indeed, the management orientation of the present U.S. Fish and Wildlife Service has a substantial proportion of its roots in the field and museum-oriented activities of its predecessor agency, the Bureau of Biological Survey. Early wildlife managers were expected to be competent preparators of specimen material that would find its way to museum collections (Davis, 1938: see reference in Section I).

Part of the reason for this schism may be the maturation of the relatively young management science, concurrent with a general decline in systematic biology. As the management field developed into a recognized science in its own right, emphasis was very properly directed toward expansion of its own techniques and capabilities. As managers became ecologists more and wardens less, their need for more information on living populations supplanted the need for information that could be obtained from preserved samples of those populations, much of which undoubtedly seemed to be readily available in cut-and-dried manual form. In the training of new managers, museum-oriented studies have gradually faded to something that other biologists do. Unfortunately, this ordering of priorities overlooked the fact that taxonomy and other museum-based studies are not static, that their results are not etched in the ivory of the museum towers. The museum worker delights in acquiring new material that will, on further study, disprove his (or his colleague's) earlier work, or refine a point for the sake of refining. The museum worker is, to a great extent, a purveyor of knowledge for its own sake. He produces information that can be used, by himself or anyone else; he seldom suggests in his writing who should use the information, or how it might be used.

This schema is, of course, both oversimplified and exaggerated. The twain do meet and even overlap occasionally, and the information produced does reach those who can use it effectively. Occasionally, too, the need for information reaches one who is particularly adept at producing information of just that kind.

Nonetheless, the breach persists. Of the present 45 Cooperative Fishery or Wildlife Research Units, only 15 are located at universities where there are major active research collections of specimens of the appropriate groups. Because most potential wildlife managers (including administrators) are trained in schools lacking large associated museums, they tend to be only slightly acquainted with what systematists do, and can do, for them—and with what they can do for systematists. Because future museum workers are trained in and around museums and in zoology rather than wildlife departments, they sometimes fail to recognize the applicability of their work to "practical" problems. Thus, early separation of workers by school and/or by department tends to establish communication gaps, if not barriers, that later are difficult to overcome.

Perhaps legislation and regulations designed to protect species threatened by drastic population declines have done more than anything to bring these fields back into contact in recent times. Museum workers, in field studies to gather material or investigate population differences, have frequently been the first to draw attention to these declines. Managers, in preparing lists of declining species and in developing and enforcing regulations, more routinely must turn to museum workers for current scientific names and identifications. However, the museum worker views some of the regulations as overprotective and restrictive of access to material necessary for the further study of biological problems, and some managers see the museum worker as a threat to the continued existence of certain resources.

The papers assembled here are selected examples that indicate the relevance of museum studies to various aspects of the very broadly defined field of wildlife management. They were gathered in the hope that together they would aid the understanding of workers in both disciplines in a way they could not while scattered through the literature. To workers in both fields, I hope to indicate that the fields are not isolated, that the breach is only an apparent one—although this should not suggest that they are identical. To the wildlife administrator and enforcement official, I hope to show that administrative and management decisions, including the formulation and enforcement of regulations, are and must be based on sound and solid taxonomic information and knowledge of the characteristics of populations. To the biologist in the field, I hope to suggest that he can contribute material that will be of use in furthering museum studies, and that both the specimens and studies based on them are valuable to him. For the student, I hope to open new vistas, to broaden horizons, and to give new ideas. To the museum worker, I want to indicate the need for the assessment of the practical application of his work, a reason for working on groups that often seem to be overworked and unglamorous, and an appreciation of the problems faced by managers at all levels.

The included papers were selected according to several criteria. First, of course, was their appropriateness to the concept of the volume. An effort was made to cover all four classes of vertebrates. Shorter papers were chosen over longer ones, where there was a choice, and papers were selected to represent as many journals and other publication outlets as possible. Finally, I attempted to find appropriate papers by a wide variety of authors. Papers considered but not used, and other papers that bear on the general theme, are given as references in the appropriate section. Neither the included papers nor the additional readings begin to exhaust the topic. More references will be found in the bibliographies of papers reprinted or cited.

The commentary that introduces each section was developed so that the included papers could be placed in perspective. As stated earlier, papers by museum workers often fail to note or suggest how they can be applied to management problems, even when done in response to a specific need, and of course the management result or effect based on the study is often delayed for some time and the action taken does not always cite the background information.

In a review of another anthology of papers (*Jour. Mammal.* 58:119-120, 1977), I suggested that such collections should not be used as a substitute for chasing down references in their original sources. Browsing in the earlier literature is a worthwhile expenditure of a portion of one's time, and I hope that those who choose to pursue any of the references given will avail themselves of the opportunity to browse.

It is obvious that a compilation of this nature and scope required help from others. I requested and received references for possible inclusion from John W. Aldrich, Robert L. Brownell, Howard W. Campbell, Bruce B. Collette, Robert Rush Miller, and Don E. Wilson. Other papers were suggested by Daniel W. Anderson, Elmer Birney, Robert H. Gibbs, Jr., William F. Perrin, and Robert G. Tuck. To all, I express my appreciation. Of course, errors of judgment in selecting or interpreting papers are my responsibility.

The entire manuscript, including the selection of papers and my own commentaries, was reviewed by Bruce B. Collette, Laurence R. Jahn, Henry M. Reeves, Henry W. Setzer, Rollin W. Sparrowe, and Robert E. Stevens; I am grateful to them for their suggestions and comments.

I appreciate the cooperation of the authors and publishers of the papers included in giving permission to use their material, and in providing original illustrations for halftone plates wherever possible.

Publication of this book was subsidized by the National Fish and Wildlife Laboratory of the U.S. Fish and Wildlife Service.

Richard C. Banks
National Fish and Wildlife Laboratory
U.S. Fish and Wildlife Service
E-600 National Museum of Natural History
Washington, D.C. 20560

SECTION I. Definitions

One must always define his terms, as difficult as that may be.

The first paper in this compilation is the introduction to another, *Readings in Wildlife Conservation*. The definition of conservation offered there can serve also for management as used in the title of the present volume, for I think of management in a very broad sense to include administrative and operational decision making as well as work with animals in the field. Management may be but a means to the end of conservation, but it is such a continuing means that it essentially becomes the end.

I tried to avoid the terms taxonomy and systematics in the Introduction to this survey, because museum work involves much more than these. It includes the amassing, storage, and consultation of bits of knowledge, much as a computer does it, the "bits" in this instance being specimen samples of free-living populations. As Hammer indicates in the second selection, modern systematics is the synthesis of these bits into new information, and in a sense all museum work is systematics just as all management is conservation. Of course, museum work is not confined by the walls of a building, but extends into the field.

Each paper presented beyond adds to these broad definitions. Having read the papers, one will have grasped the concepts. If not, the final paper in the book offers alternate definitions of both systematics and conservation.

I do not include a definition of wildlife. The term includes all living animals, but I have restricted my examples to the vertebrate classes. The traditional distinction between "fish" and "wildlife," in a management sense, is ignored here.

Additional Readings

Davis, William B. 1938. Letter to the editor: "What is a wildlife specialist?" *Jour. Wildl. Mgt.* 2: 272-273.

Miller, Alden H. 1963. The curator as a research worker. *Curator* 6: 282-286.

INTRODUCTION

These papers have been selected to present a broad perspective of wildlife conservation.

Wildlife conservation is the social process, including both professional and lay activities, which defines and seeks to attain wise use and perpetuation of wild animals and the biotic communities to which they belong.

Conservation is a social process because it includes the continuous melding of personal values and requires selections between salmon and electricity, elk and plywood, soybeans and prairie chickens. These value judgements are influenced by attitudes and practices of the past. Thus history provides insight for understanding today's conservation process. Public interest and participation in these value judgements may be stimulated by educational programs of public agencies and by activities of conservation organizations such as the Sierra Club and the great variety of sportsmen's clubs. Public agencies responsible for wildlife must seek awareness of the changing desires of people–for in the United States they own the wildlife.

The conservation process includes interactions among many kinds of people–trained professionals, politicians, and citizens, including agencies and organizations having a diversity of biases and interests. The wisdom of goals selected by this socio-political process will be debated as long as people differ in their backgrounds and interests. Thus, conservation remains a social process.

In addition, conservation is a social process because successful programs require communication and cooperation among professionals from many agencies–the State Division of Wildlife, the Forest Service, the Soil Conservation Service, the Bureau of Reclamation and many others. These professionals work in six activities–research, management, public relations, education, law enforcement and administration. Performance in all these activities and several agencies may be critical to success. Failure in one section of an agency often limits possibilities for achievement in others and affects morale and performance throughout the agency and even in other agencies.

Wildlife conservation is based upon knowledge from many sciences–ecology, chemistry, physics, mathematics, statistics, genetics, zoology, botany, economics and sociology. It employs the tools of many professions–agriculture, forestry, engineering, medicine, law, administration, public relations, accounting and planning. Wildlife conservation therefore has need for the specialist and the generalist. The specialist should have knowledge broad enough to comprehend the interrelated roles of the six conservation activities mentioned above. He should be aware of historical influences upon today's conservation and should have a sensitivity toward human values and the philosophical question of what constitutes *wise* use. The generalist should be adequately trained in the sciences so that he can communicate with specialists and can use scientific information to establish realistic goals for solving conservation problems.

Papers in this book either present important ideas or philosophies or are good examples of activities in wildlife conservation. We hope they contribute to the education of those who will, as professionals or as laymen, participate in the conservation process. We believe an increased awareness of all aspects of conservation will help participants to relate their wildlife interests or problems to the whole conservation process, and will provide incentive to each participant and encourage teamwork among them.

This book has been a cooperative effort of The Wildlife Society. More than forty members suggested over 250 titles for consideration. Others reviewed our selections, edited our introductory remarks and offered helpful suggestions. Our sincere thanks go to these, our advisors and critics. However, responsibility for any errors in reproducing selected papers is ours. Thanks are also due our respective departments at Colorado State University, the University of Missouri, and Oklahoma State University for funding our efforts and providing clerical assistance.

James A. Bailey
William Elder
Ted McKinney

Reprinted from *ASC Newsletter*

TAXONOMY & SYSTEMATICS: THEIR SIGNIFICANCE

Richard M. Hammer

As implied in the article by Welton L. Lee ("The San Francisco Bay Project: A New Approach to Using Systematics," *ACS Newsletter,* April 1977, Vol. V(2): 15-17), most taxonomists and systematists are appalled and irritated by the general feeling among much of the scientific community in regard to the values and contributions of taxonomy and systematics. It is not uncommon to hear the opinion that "taxonomy is a dead science" or that "systematics is not even a science but simply an outdated mechanistic hobby of 19th Century naturalists." This attitude prevails because of two problems—one relating to definitions and the other to lack of communication.

The first problem involves the fact that many people do not clearly distinguish between the meanings of the terms taxonomy and systematics (*sensu lato*). "Taxonomy is the theory and practice of naming and classifying organisms" (Mayr, 1969. Principles of Systematic Zoology). It should be emphasized that taxonomy provides the scientific community with a basic and common communication system of uniqueness, universality and stability. On the other hand, when one understands the meaning of the term systematics correctly, it implies a much broader scope. Simpson (1961. Principles of Animal Taxonomy) has properly redefined systematics as "the scientific study of the kinds and diversity of organisms and of any and all relationships among them." Both the interdisciplinary nature and modernization of systematics should be emphasized. Unlike classical systematists who based their conclusions almost entirely on the morphology of museum preserved specimens, modern systematists collect and integrate data from many fields (ecology, genetics, biochemistry, behavior, etc.) in search of biological patterns and explanations. In short, systematists strive to synthesize the knowledge of the separate sciences.

The second problem is somewhat more difficult to contend with and resolve. It involves educating the general public and scientific community on the meanings and contributions of both taxonomy and systematics in such a way that a broad band of communication ensues. Up until now, most articles that emphasize the benefits of taxonomy and systematics to other branches of biology and to mankind as a whole occur in publications frequented principally by taxonomists and systematists. This sort of communication only serves to cycle this valuable information in a confined, closed system. More communication (W. L. Lee's article has the potential to serve well as a model) should appear in places readily accessible to the general public and scientific community. In this way, an appreciation for taxonomy and systematics can be acquired and future funding will be enhanced for both systematics collections and proper training of qualified taxonomists and systematists.

Submitted by Richard M. Hammer, Allan Hancock Foundation, University of Southern California, University Park, Los Angeles, California 90007.

SECTION II.
Museum Collections and Their Uses

The two papers presented in this section are general ones, not specifically related to wildlife management. The first deals with collections of birds and the kinds of information that can be developed from them. The second concerns collections of mammals and their uses and users. With only minor modification, the topics of these papers could be reversed, and their principles easily can be extended to collections of reptiles, amphibians, fishes, invertebrates, or plants. Museum collections are inherently useful, and become even more so if they are both growing and well cared for.

Neither of these papers (nor others in this compilation) discusses the importance of collections in the education of biologists. Perhaps that is because this function is more often carried out with separate, smaller teaching collections than with active research collections. Still, one of the most frequent problems mentioned by teaching ornithologists is the lack of adequate reference collections. With the modern concern of wildlife biologists extending to the well-being of all of the earth's diverse ecosystems, should they not share that concern?

Additional Readings

Cohen, D. M., and R. F. Cressey, eds. 1969. Papers presented at a symposium on natural history collections, past, present, future. *Proc. Biol. Soc. Washington* 82: 559-762.

Cressey, Roger F., and Bruce B. Collette. 1970. Copepods and needlefishes: a study in host-parasite relationships. *Fishery Bulletin* 68: 327-432.

Guides to Collections

Banks, Richard C., Mary H. Clench, and Jon C. Barlow. 1973. Bird collections in the United States and Canada. *Auk* 90: 136-170.

Choate, Jerry R., and Hugh H. Genoways. 1975. Collections of recent mammals in North America. *Jour. Mamm.* 55: 452-502.

Clench, Mary H., Richard C. Banks, and Jon C. Barlow. 1976. Bird collections in the United States and Canada: addenda and corrigenda. *Auk* 93: 126-129.

Collette, Bruce B., and Ernest A. Lachner. 1976. Fish collections in the United States and Canada. *Copeia* 1976: 625-642.

Committee on Resources in Herpetology. 1975. Collections of preserved amphibians and reptiles in the United States. *Herpetological Circular* no. 3, Society for the Study of Amphibians and Reptiles.

Reprinted from *The Living Bird*

THE CONTRIBUTION OF MUSEUM COLLECTIONS TO KNOWLEDGE OF THE LIVING BIRD

KENNETH C. PARKES

"Why do you need so many of each kind?"

Few curators of large museum collections have not heard that question from laymen (ranging from high school students to trustees) on a tour "behind the scenes" in the museum. It is probably asked more frequently of ornithologists than of other curators, as most people seem more sensitive to the fact that museum specimens of birds imply the death of living creatures than to this same fact applied to mice, salamanders, or butterflies.

The question is usually answered with a short lecture, the success of which varies with the ability of the curator to explain his research in laymen's vocabulary, on the study of variation—seasonal, sexual, age, geographic, and individual. The visitor may not understand all of the lecture, but I think he does leave with the realization that just one Robin, one Eastern Bluebird, or one Cardinal is not "enough."

Perhaps even more difficult is the task of communication between the museum man and his colleagues who are engaged in completely different kinds of ornithology, involving the study of birds as living organisms, whether in the field or in the laboratory. Many have somehow picked up the notion that museum ornithology consists primarily of measuring beaks and tail feathers, and giving new scientific names to finer and finer subdivisions. Needless to say, formal descriptive taxonomy is still very much a part of our work, and there are (fortunately) those of us who find it more fun than analyzing sound spectrograms or taking temperatures of nestlings. Taxonomy is not quite so sterile a pursuit as some of our friends seem to believe; we are interested not only in the differences and similarities we note among our trays of dead specimens, but in the significance to the birds themselves of these differences and similarities, and how they came about.

Taxonomy as such, however, is not the subject of the present paper. What I would like to do is to demonstrate some of the other kinds of information that can be obtained from museum skins of birds, and what can be deduced about various biological processes and cycles of the living bird from such information. I am discussing only skins, omitting the whole subject of functional anatomy, with its research material (also housed in museums) consisting of skeletons and alcoholic specimens.

In the following discussion, certain assumptions about the nature of the available material will be made. First, that a reasonably large series of specimens is available of the species being studied, if only by borrowing material from all possible museums. Needless to say, some birds are so rare in collections that the total of all extant specimens cannot be considered "adequate" for many kinds of research. However, there is no shortage of unsolved problems among species for which long series *are* available, as will be quickly seen after thumbing through the first volume of the "Handbook of North American Birds" (Palmer, 1962). And many species from areas of the world other than the *relatively* well-studied Holarctic Region are represented in museum collections by surprisingly large numbers of specimens, even if virtually nothing is known of their habits in life.

The second assumption we will make, at least at first, is that the label of each specimen bears accurate data, but that these data are confined to the bare minimum of sex, date, and locality. Both parts of this assumption are subject, in practice, to qualification. Date and locality are usually accurate, subject to slips of the pen, but the museum worker soon learns to expect a certain proportion of "mis-sexed" specimens. These can usually (but not always) be detected by discrepancies in size or color, and suspected specimens discarded

"THE IMAGE"

from further consideration. On the other hand, data supplementary to the minimum listed above are often to be found on museum labels. Whereas data on condition of gonads and skull ossification, to supplement the simple sex symbol and the notation "ad." or "imm." are most frequently found on modern specimens (though still lamentably missing on many), a gratifying number of Nineteenth Century collectors noted on their labels the colors of soft parts (eyes, bill, feet, naked skin) and the contents of the stomach. Brief mention will be made again, toward the end of the present paper, of the subject of improving label data. The experienced worker with museum collections knows that, much as he would like to have this supplementary information available for all specimens, he will in most instances have to get along without it; we will thus proceed on the premise that only the basic minimum is available for each specimen in the series.

In what non-taxonomic fields of ornithology, then, can the series of museum skins be of use?

Identification

The first point may seem so obvious as to be scarcely worth mentioning, but it is nevertheless of fundamental importance. The collection of museum skins is the chief basis for *identification.* We are now blessed with excellent guides to field identification of the birds of many parts of the world. It should never be forgotten, however, that much of the work of the author of such a guide (especially in areas where field identification has not evolved to the sophisticated level to be found in North America and Europe) and almost all of the work of the illustrator is based upon the use of correctly identified museum specimens. Identification is important in the laboratory as well as in the field. Whereas experimental work on birds was formerly largely confined to domestic species, more and more of the research of ethologists, endocrinologists, physiologists, and other laboratory workers is now being done with suitable wild species. Such birds are often obtained from dealers who neither know nor care about the exact geographic origin of their stock. Most workers are aware that their conclusions apply *with certainty* only to the species being studied, and sometimes only to the *population* of the species from which their experimental birds originally came. There are still, unfortunately, occasional references in the literature to studies of "the weaver finch" or "the quail." Not only is the museum collection the place to turn for verification of the identity of the species or subspecies of bird being studied, but samples of the laboratory stock may well be deposited in the museum collection as additional verification for future workers. If this is done, when later taxonomic work shows the existence of hitherto undescribed geographic variation within a species, examination of the reference specimens may indicate from which population the laboratory or aviary stock originally derived. The Bleeding-heart Pigeon *(Gallicolumba luzonica),* for instance, is a well-known aviary species, most of the captive specimens having come from the Philippine island of Luzon. I recently found (Parkes, 1962) that the population of northern Luzon *(G. l. griseolateralis)* is separable on several characters from that of central and southern Luzon *(G. l. luzonica).* Examination of numerous aviary specimens in museum collections indicates that most, if not all, of the caged stock came from within the range of the subspecies *luzonica.* Generalizations about the behavior of living Bleeding-heart Pigeons based on aviary birds, therefore, apply to the populations of southern and central Luzon, and might not be valid for those of the north.

Among birds commonly kept in captivity are several belonging to genera which include confusingly similar species. Harrison (1957) published an account of the behavior of the Dark Fire Finch *(Lagonosticta rubricata).* Fortunately he kept a reference skin of one of his birds, and this now proved to be Jameson's Fire Finch *(L. rhodopareia)* rather than the species first named. Had this skin not been available for comparison with museum material, this error in the behavior literature might not have been detected (Harrison, 1963).

Plumage Succession and Age Characters

Two major subjects of investigation with museum skins are inextricably bound together, and can hardly be discussed separately. These are the study of the sequence of molts and plumages, and the determination of characters indicative of age. The latter subject, of course, is of great interest to workers in the field of wildlife management as well as to bird-banders, both groups wanting reliable external criteria by which to age the live birds they handle.

The first step in working out age characters, through the study of plumage succession, is to find the *youngest* specimens in the museum series. Museum skins of newly-hatched birds (or birds only a few days old) are not uncommon if the chicks are downy (Anseriformes, Galliformes, Falconiformes, Strigiformes, Charadriiformes, etc.), but relatively rare of passerines and other naked or sparsely-downed hatchlings. Of the latter, however, stub-tailed nestlings are frequently available. The critical point is that the true juvenal plumage, that which succeeds the natal down (if any), be identified as such, to act as the reference point for changes to later plumages. When the characters of the juvenal plumage (which may be rather short-lived) have been determined, the next step is to find, among the series of specimens, individuals in which the plumage now known to be juvenal, is being replaced by the next succeeding, or first basic plumage (variously "first winter," "second juvenal," "immature," etc. of earlier literature; see Humphrey and Parkes, 1959). This transition, as all later molts, is detected by finding incoming, partially-sheathed feathers among those of the previous generation. Individuals in which the molt had progressed further at the time of collection will illustrate progressive stages of this first prebasic molt (always identifiable as such as long as there are recognizable juvenal feathers left), until the characters of the fully-developed first basic plumage become completely evident. The same procedure is followed in determining the characteristics of whatever plumage succeeds the first basic in the species being studied; that is, the search for incoming new feathers among the (often somewhat worn or faded) older feathers of the first basic plumage, and the lining up of individuals showing progressive stages of this molt until no further molt can be detected (in the many species with a partial prealternate molt, only a portion of the first basic plumage is replaced). The same procedure is followed to determine the transition between each identifiable generation, until finally the plumage or plumages show no further significant change from cycle to cycle, and can be conveniently called "definitive" ("adult" in much of the literature, although some species are sexually mature, hence "adult," in pre-definitive plumages).

The series of specimens can now be arranged in order of relative age, from the newly-hatched chick to the definitive plumage or plumages (depending on whether the species has one, two, or rarely three plumages in each cycle). Note that I say *relative* age. Up to this point we have not paid any attention to any of the label data except sex (in order to keep the plumage sequences of sexually dimorphic species clear). We have ignored dates. We can now, with a certain amount of confidence, state that a given specimen represents a plumage stage younger (or older) than another, but we are not prepared to say how much older or younger, nor to give an estimate of the *absolute* age of the bird.

In the world as a whole, the reproductive period of birds varies in spectrum fashion from short, discrete breeding seasons in which the reproductive cycles of all individuals of the population are virtually synchronous (typical, among others, of many arctic birds and of certain colonial species elsewhere) to an apparent total lack of breeding "season," individuals being found in any stage of the reproductive cycle at any calendar date (22 of 60 equatorial cloud forest species; Miller, 1963). For those species with little or no population synchrony in reproduction, no deductions other than *relative* age of the various plumage stages can be made; such species, in museum series, can be identified by the wide scattering of label dates on birds of corresponding plumage stage (especially the very young birds).

On the other hand, population synchrony in breeding is suggested by close correspondence in label dates of young birds of similar degree of plumage development. In a sufficiently large sample, a rough estimate of the length of the breeding season can be determined by the extreme dates of very young or newly-hatched birds. This period will, of course, probably include specimens from later nestings in multiple-brooded species, but (again, with a large enough series) it may be possible to show that dates of young specimens of such species when graphed show two or more peaks. Once it has been shown that a degree of reproductive synchrony exists, and the range of variation of this synchrony determined (for example, hatchlings collected 1 June through 18 July; thus the known range of variation in this population is fifty days), it is possible to give not only relative age but an estimate of absolute age for a given specimen. A young bird at any stage can be aged by dating its hatching arbitrarily at the midpoint of the previously determined hatching season, and counting from that date. The possible error is half the number of days in the hatching season. In the example given above, a young bird collected 25 September will be three months ±25 days of age, 25 days being the extent of the hatching season on either side of 25 June, the midpoint. With this information, the degree of population synchrony of molt can be compared to that of breeding; in some species, for instance, individuals may be found in some stage of the first prebasic molt over a period longer than that corresponding to the range of variation of actual age, whereas other species seem to molt roughly simultaneously, even though individuals from a broad range of hatching dates are involved.

The periodical replacement of the feather coat is, of course, an important part of the biology of birds. The study of the sequence of molts and plumages has traditionally used museum specimens as its research materials. From our large sample of specimens we can deduce the duration of a period of molt for the population of birds being studied, and we can work out the sequence of molt for a hypothetical "typical" individual based on a composite of the individuals in our sample. It would, of course, be desirable to know, in addition, the duration of molt for a single individual, but this cannot be ascertained from museum skins except insofar as it can be no *longer* than the duration of molt in the population as a whole, and is probably substantially shorter. Duration of molt for an individual bird must be determined from the living bird, preferably a wild bird under natural condition. This has been done by several workers by repeated netting or trapping of banded individuals, but this must necessarily involve the assumption that repeated handling of the live bird has no traumatic effect which may alter in any way the normal molt sequence.

Reproduction and Development

Certain information about the reproductive cycle can be obtained from the simple examination of museum skins. We have already seen that hatching dates are readily ascertainable when the museum material contains very young specimens. A certain amount of information on post-hatching development of young birds is also easy to obtain from skins; in addition to plumage development *per se,* this might include loss of egg tooth, approximate date of fledging based on development of remiges, and time of attainment of adult proportions. In the absence of label notations about reproductive stage of adult birds, certain deductions can still be made; the incubation patch is often evident in museum skins, especially in certain groups such as Charadriiformes, Anseriformes, and Galliformes. The timing of "defeathering" and "refeathering" of the incubation patch can be worked out, as well as the relative role of the sexes in incubation (for an example, see Parkes, 1953). Evidence of incubation is also afforded in certain large birds, especially hawks, by excessive wear of feathers of the throat and upper breast where these rub against the edge of the nest while the bird is sitting on eggs (example, Carnegie Mus. No. 114049, *Buteo jamaicensis fuertesi,* 10 miles S. of Marathon, Texas, May 23, 1933).

Migration

Another major field of ornithological study in which the museum collection may play an important part is migration. The most obvious application of museum data to knowledge of migration is in the correlation of dates with presence of the species at a given location, in the absence of data on living birds from reliable observers. Although migration dates within North America are now based largely on field observations of living birds (as published in *Audubon Field Notes* and countless local journals), the status of North American migrants south of our borders is still known chiefly from museum specimens (see, for example, Eisenmann, 1955). In addition to the simple definition of the wintering area and migration routes of a species, museum specimens may be used in other ways in connection with migration studies. Engels (1962) used museum specimens of Bobolinks *(Dolichonyx oryzivorus)* to determine the latitudinal distribution of this transequatorial migrant between August and November, in working out the day-lengths experienced by the birds during this period.

As already mentioned, most of our knowledge about sequences of molts and plumages of birds comes from museum specimens. The chronological relationship between molt and migration is also to be learned from the study of museum collections. Some migratory species molt prior to migration, some along the way, and some not until the wintering grounds have been reached. Information of this sort may be used in ways not immediately apparent. For instance, Phillips and Parkes (1955) showed that the name *Tyrannula richardsonii* Swainson, long applied to the Western Wood Pewee, could not possibly have been based on an individual of that species. Baird, who was the first to use this name for a pewee, realized that the wing formula of Swainson's type specimen (from Saskatchewan) differed from that of any pewee and resembled that of a phoebe, but attributed this discrepancy to "incomplete growth during the moulting season." However, as we showed through examination of museum specimens, the wing molt of adults of the Western Wood Pewee (now *Contopus sordidulus)* takes place on the wintering grounds in South America between December and March, a fact unknown in Baird's time through lack of specimen material. Since, for various reasons, Swainson's type could not have been a *juvenile* wood pewee, the attribution of the non-pewee wing formula to "incomplete growth" is thus shown to be erroneous.

Having already deduced the plumage characteristic of various age classes from our initial study of the museum series, we can make certain secondary deductions, using these primary deductions as evidence. For example, a preponderance of specimens exhibiting the plumage characters of birds of the year from certain areas in winter, and of older birds from other areas, is strong evidence of some sort of differential migration of age classes in such a species. Howell (1953), through censusing museum collections, found that there is differential migration of the sexes in the Yellow-bellied Sapsucker *(Sphyrapicus varius),* as well as differences in extent of migration among the four subspecies. He presents some interesting conjectures as to the possibility of sex-linked multiple genetic factors affecting intensity of pigmentation and extent of migration, since there appears to be a correlation between distance of migration and degree of pigmentation, both as to sex and to population.

Behavior and Other Topics

Although the study of behavior is of necessity restricted to observations of living birds, the specialized structures utilized in displays can be studied in detail in museum specimens. Feathers so modified as to produce sounds in flight can seldom be well observed under field conditions, but I remember as a high school student attaching several of the narrow primaries of a Woodcock *(Philohela minor)* to a cork on the end of a string and whirling it around my head. To the amazement of my fellow-conspirators (the feathers were surreptitiously detached from a mounted bird in the biology lab), a distinct whistling noise was produced. There are, of course, many species of birds well known as museum specimens but virtually unknown as living organisms. The presence of ornamental plumes, modified flight feathers, bright color patches, etc. at least gives us hints as to the possible nature of courtship displays in such species.

As mentioned at the beginning of this paper, the museum collection represents the raw material from which our entire system of naming and classifying birds is derived. Once named and classified, the museum collection becomes the research tool of students of zoogeography and of avian evolution, whose findings may range from straightforward descriptions of things-as-they-are to the wildest conjectures about how-they-may-have-gotten-there.

The reader will, with this impetus, be able to think of many other ways in which the collection of bird skins housed in a museum can lead to additional knowledge about birds as living organisms.

Improving Specimen Labels

The usefulness of museum specimens is vastly increased if the collector will record, at the time of collection, as much information as he can to supplement the minimum of date, locality, and sex. Many of the important points with respect to label data were covered in a well-known paper by the late Josselyn Van Tyne (1952). He stressed the need for unequivocal clarity in the recording of locality and date, the importance of indicating the condition of gonads and degree of skull ossification as evidence to support the sex and age given, and the value of recording weights and perishable colors. "Other matters which should be recorded on the label are the details of physical condition (*i.e.,* degree of fatness, presence of brood patches or other evidences of breeding), readily identifiable stomach contents, and an indication of the habitat." (Van Tyne, 1952: 30.) To this list can be added such additional points as a notation when known families are collected ("mate of No. 3456, parent of No. 3457 and 3458"), notes on parasites, abnormalities or injuries, method of collecting if unusual (such as diving ducks caught in fishing nets at known depths, swifts taken in caves, migrants that hit buildings or TV towers, etc.), vocalizations *(Empidonax),* nest-type *(Collocalia;* see Sims, 1961, and Medway, 1962) and, especially, molt. Although, as mentioned above, most studies of molt are carried out by examining prepared specimens for incoming feathers, such feathers ("pinfeathers") at the growing stage are much easier to detect on the inside of the skin during preparation. If only a few new feathers are appearing, these may be lost during skinning or overlooked in later examination. The preparator's notation on the label such as "trace of molt on flanks" will be of great use in later studies.

Van Tyne mentioned the fact that all of this material *can* be written legibly, using both sides of a "standard 3- by ⅝-inch museum label," using an additional blank label to add data if necessary. However, there should be nothing sacred about either the size or the format of a museum label. Any institution fortunate enough to be receiving specimens from a collector who conscientiously records all of the kinds of information suggested above is ill-advised to force him to conform to a standard small label if a somewhat larger one will offer more flexibility. An institutional tradition which demands, for instance, that one entire side of the label be left blank except for the name of the institution and the scientific name of the bird, is sterile and needlessly confining.

Under the heading "hieroglyphics," Van Tyne stressed the need for legibility, but did not go into detail. I would suggest that *all* label data be *printed* (in so-called "manuscript writing") by the collector, rather than written in his usual handwriting. I have seen labels which conform rigidly to institutional requirements as to format, but written in handwriting usually thought of as typical of physician's prescriptions.

As an indication of the kind of information now being added to museum skins by modern collectors aware of the great importance of such data, I reproduce here (Figure 1) a typical label from one of E. T. Gilliard's recent expeditions for the American Museum of Natural History. I might suggest also that collectors and students who are not used to printing their labels try to emulate the legibility of data recorded by George M. Sutton.

The bulk of this paper suggests some of the information about living birds that can be deduced from museum skins bearing only minimal data. With the addition on the label of the kinds of facts suggested in the last few paragraphs, the museum collection becomes infinitely more useful. For example, among passerine birds whose first basic plumage cannot be distinguished from the definitive basic plumage, a notation about degree of skull ossification will enable young birds to be identified as such well after the completion of the first prebasic molt. In species which may not breed their first year but do return to the breeding grounds along with sexually mature birds, the combination of age as deduced from plumage characters and gonad condition as noted on labels may help to determine the minimum age of breeding. An abnormal internal condition (such as excessive infestation by parasites) noted on the label may explain what appears to be a long-delayed molt. In areas traversed by a migrant population while the resident population is already breeding, a label notation as to reproductive condition is vital if the characters of the resident population are to be correctly assessed (migrant individuals of the northern race of Yellow Warbler, *Dendroica petechia amnicola,* have been collected near Ithaca, New York, when nesting of the local race *D. p. aestiva* was well under way). Soft-part colors are now known to vary with age, sex, and season in many species, and can be correlated with other cyclic manifestations if noted on labels. And so on.

Conclusion

Although new species of birds are still trickling into museums yearly, and the flow of newly-described subspecies may seem (especially to the non-taxonomist) to be endless, systematics of birds has progressed to the point at which it is at least *conceivable* that all valid taxa may some day have been described.

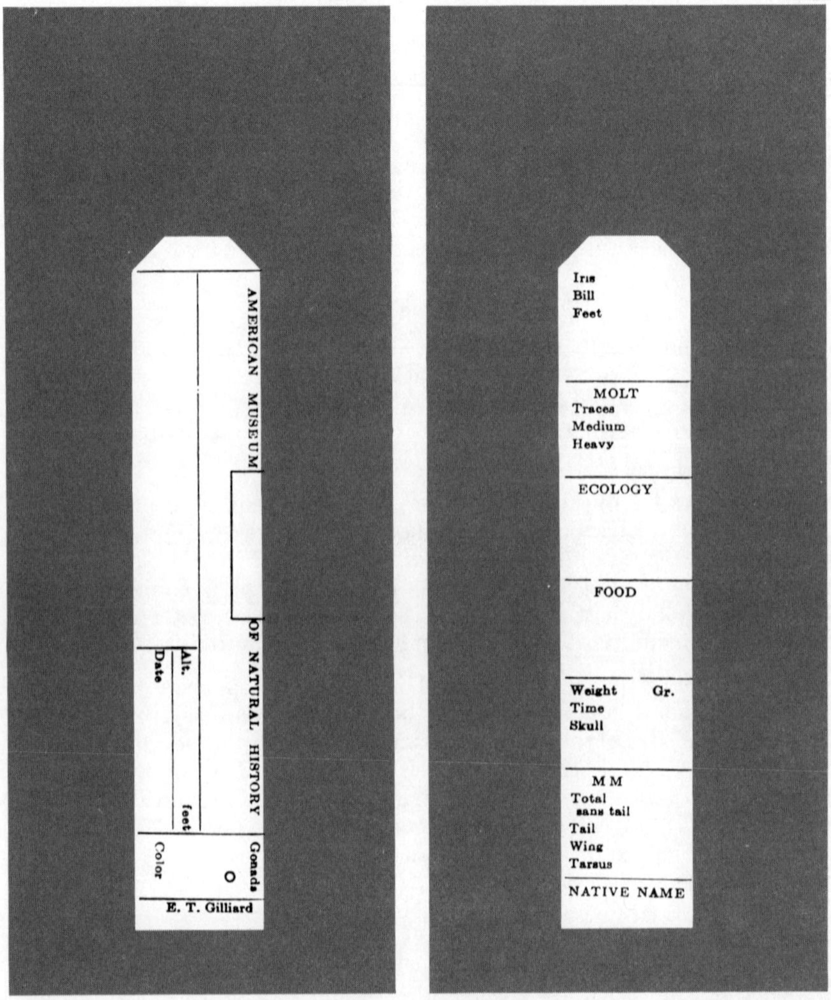

Figure 1. The type of label (both sides shown) used by E. Thomas Gilliard on one of his recent expeditions for the American Museum of Natural History.

That this will by no means call a halt to the usefulness of the museum collections should now be obvious, and, indeed, much additional museum material is needed of many of the best-known (in a taxonomic sense) species in order to understand fully such matters as molt and plumage sequences.

Acknowledgments

I am grateful to E. Thomas Gilliard for permission to reproduce one of his specimen labels, of his own design. This paper is an offshoot of a somewhat different one originally planned and outlined in collaboration with Philip S. Humphrey of the United States National Museum; much of the material on plumage sequence and age characters is derived from what had originally been intended as a joint paper, and I am grateful to Dr. Humphrey for his agreement to incorporate this material in the present paper. The cartoon was drawn by Clifford J. Morrow, Jr., Chief of Special Exhibits, Carnegie Museum.

LITERATURE CITED

Engels, W. L.
1962 Day-length and termination of photorefractoriness in the annual testicular cycle of the transequatorial migrant Dolichonyx (the Bobolink). *Biol. Bull.,* 123:94-104.

Harrison, C. J. O.
1957 Notes on the Dark Fire Finch *(Lagonosticta rubricata). Avicultural Mag.,* 63:128-131.
1963 Jameson's Firefinch and the Dark Firefinch. *Avicultural Mag.,* 69:42.

Howell, T. R.
1953 Racial and sexual differences in migration in *Sphyrapicus varius. Auk,* 70:118-126.

Humphrey, P. S., and K. C. Parkes
1959 An approach to the study of molts and plumages. *Auk,* 76:1-31.

Medway, Lord
1962 The swiftlets *(Collocalia)* of Java and their relationships. *J. Bombay Nat. Hist. Soc.,* 59:146-153.

Miller, A. H.
1963 Seasonal activity and ecology of the avifauna of an American equatorial cloud forest. *Univ. Calif. Publ. in Zool.,* 66:1-78.

Palmer, R. S., Ed.
1962 Handbook of North American birds. Volume 1. Yale University Press, New Haven, Connecticut.

Parkes, K. C.
1953 The incubation patch in males of the suborder Tyranni. *Condor,* 55:218-219.
1962 New subspecies of birds from Luzon, Philippines. *Postilla,* No. 67.

Phillips, A. R., and K. C. Parkes
1955 Taxonomic comments on the Western Wood Pewee. *Condor,* 57:244-246.

Sims, R. W.
1961 The identification of Malaysian species of swiftlets *Collocalia. Ibis,* 103a:205-210.

Van Tyne, J.
1952 Principles and practices in collecting and taxonomic work. *Auk,* 69:27-33.

CARNEGIE MUSEUM, PITTSBURGH, PENNSYLVANIA

Reprinted from *Museology/Texas Tech University*

SYSTEMATISTS, OTHER USERS, AND USES OF NORTH AMERICAN COLLECTIONS OF RECENT MAMMALS

HUGH H. GENOWAYS, JERRY R. CHOATE, EDWARD F. PEMBLETON, IRA F. GREENBAUM, and JOHN W. BICKHAM

In 1972, two committees of the American Society of Mammalogists were established to deal with problems in systematic mammalogy. These committees were comprised as follows: Committee on Systematic Collections—James H. Brown, Jerry R. Choate, Emmet T. Hooper, Murray L. Johnson, Clyde Jones, David Klingener, and James L. Patton; Committee on Information Retrieval—Sydney Anderson, Hugh H. Genoways, Robert S. Hoffmann, Clyde Jones, Donald R. Patten, Henry W. Setzer, and Luis de la Torre. In January 1973, the President of the Society (J. Knox Jones, Jr.) and the two committee chairmen (Choate and Anderson) met to discuss future work of the committees. Three fundamental questions were raised at this time: What and where are the systematic resources of mammalogy? How should they be maintained? How can they be made more accessible and used more effectively?

It was decided that the best way to address these questions was to combine the committees into an Advisory Committee for Systematic Resources in Mammalogy. This is the fourth report produced as the result of the activities of the committee. The first report dealt with standards of maintenance and with accessibility of collections as presented in the *Report and Recommendations* . . . of the Advisory Committee (1974). The second report (Committee on Systematic Collections, 1975) was an initial list of collections of mammals in North America that meet the minimal standards of maintenance. The size and location of systematic collections of Recent mammals in North America were described in the third report by Choate and Genoways (1975). This report discusses another (and extremely important) systematic resource, the systematists themselves. Also included is an evaluation of the other users of systematic collections of Recent mammals and the types of uses that these collections receive.

As part of the survey of systematic collections, a questionnaire was prepared and sent to: (a) curators or owners of all known or suspected collections; (b) all U.S. National Parks, Monuments, Areas, and Sites; (c) all persons who responded to requests for information placed in volumes 54(3) and 55(1) of the *Journal of Mammalogy*; (d) all state fish and wildlife conservation departments. Four of the questions asked by that questionnaire were used in this study (see Choate and Genoways, 1975):

18. Approximately how many loans have been made in the past two years? How many specimens would you estimate were involved in those loans?
21. Who are the professional scientists at your institution who use the collection, and what are their specialties?

22. In the past two years, how many students at your institution (if applicable) used the collection *directly* in their thesis projects? How many for other projects? How many students have used the collection *indirectly* (such as for deposition of voucher specimens) in their thesis projects? How many for other projects?
23. Estimate the number of visitors who have used the collection in the past two years: professional mammalogists? visiting graduate students? other professional scientists? state or federal employees? others?

A total of 824 of these questionnaires was mailed and 565 (69 percent) replies were received.

A second questionnaire was mailed to all members of the American Society of Mammalogists as a supplement to volume 56(1) of the *Journal of Mammalogy*. This questionnaire was divided into questions for systematists and those for non-systematists. Systematists were defined as those scientists who use Recent mammals as the primary object of their systematic research. Systematists were requested to supply the following information: name; address; position; systematic group(s) of specialty; major geographic area(s) of specialty; names of systematic collections used in the past five years; interest in providing systematic services. Non-systematists were requested to provide the following information: name; address; position; professional specialty; names of systematic collections used in past five years; uses made of systematic collections. Four hundred and fifty seven responses were received from systematists and non-systematists to this questionnaire. This represents approximately 15 percent of the membership of the American Society of Mammalogists.

Although the recent survey of systematic collections of mammals (Choate and Genoways, 1975) is the fourth to be conducted by the Society, this is the first attempt to compile a listing of the systematists and other scientists who use collections, and to detail the kinds of uses made of them. We admit numerous shortcomings in this effort because data to answer many questions simply are not available. However, we offer the data that we have gathered as a baseline for future studies. Also, we hope that this report will encourage curators to keep more complete information on the users and uses of their collections; such information is needed to justify future support of systematic collections.

This report was accepted by the American Society of Mammalogists at the fifty-fifth annual meeting in Missoula, Montana, in June 1975, as part of the report of the Committee on Systematics Collections.

This report is divided into three principal sections. The first details the systematists who study Recent mammals and includes their areas of systematic and geographic specialty. The second section lists the disciplines and individuals that use systematic collections of mammals. The last part of this report discusses the types and amounts of use that collections receive.

SYSTEMATISTS

Appendix 1 contains a list of systematists who work on Recent mammals, and relates their systematic and geographic specialty. Appendix 2 is a cross index of systematists by their systematic and geographic specialties. A total of 154 systematists living in North America was identified as working on Recent mammals. This is an extraordinarily low number, particularly when the fact that many of those listed do not consider their primary professional specialty to be systematics and that several others are nearing the end of their professional careers.

We believe that within the next 20 years many of the disciplines that depend heavily on systematic mammalogy (see below), such as ecology (particularly environmental assessment) and wildlife biology, will face a critical shortage of systematists to aid them in their work. We further believe that systematists are a critical element necessary for the future advancement of the science of mammalogy. Systematics provides the theoretical framework for the organization of most of the data generated in the field of mammalogy, and systematists are involved in the synthesis of diverse information as well as in providing the taxonomic and nomenclatural judgments necessary for the functioning of many disciplines.

We believe that the growing shortage of systematic mammalogists is being caused by at least three factors: general reduction of funding in all areas, but particularly in almost total withdrawal of funds in support of classical systematic studies and faunal surveys; general reduction of support for graduate education in most fields; increasing difficulty in doing field studies because of the proliferation of federal and state regulations. Although considerable systematic work has been conducted in the United States and Canada, many groups still await thorough systematic review. Because many geographic areas have never been surveyed for mammals and others need thorough reassessment because of human impact, we support the resolution of the Association of Systematics Collections for re-institution of a national biological survey and increased support at the federal level for surveys conducted by individuals. This is particularly important, at present, because of the increased pressures that the environment will face in the future, because of the energy crisis, and because of the expanding human population. Good systematic and taxonomic work will elevate the cost of environmental impact assessment, but, if these statements are to be of any value, good systematic and taxonomic work is an absolute necessity. Accordingly, steps should be taken to insure the authenticity of systematic work in future environmental impact statements. And, to insure a future supply of systematists, we believe that traineeships for graduate students studying in this area should be re-instituted.

One of the major problems faced by systematists in the near future is the increased restrictions placed upon field studies by federal and state agencies. Presently, at least four federal departments (Interior, Agriculture, Commerce, and Health, Education and Welfare) have regulations governing collection and importation of mammals. Many of these permit regulations are so restrictive as to almost eliminate work outside of the United States. Requiring identification of animals to be collected before field work is done or at time of importation is demanding a degree of systematic sophistication and knowledge that is nearly impossible anywhere in the world. Besides federal regulations, many states are becoming increasingly reluctant to issue scientific collecting permits, especially for faunal survey work. This is particularly unfortunate at the present time, when there is an increasing demand for baseline distributional and taxonomic data for environmental impact studies.

The list of systematic specialties of North American mammalogists reveals several important groups that are receiving little systematic study. Only two people, R. M. Wetzel and T. H. Manning, listed an artiodactyl group (Tayassuidae and *Rangifer*) as a systematic specialty. Surprisingly, only five people listed primates as their specialty, even though this is an extremely important group in biomedical research and has been receiving intensive study by ethologists. Only two persons listed marsupials and edentates. As would be expected because of the size of the groups being studied, most systematists were listed for bats and rodents.

Most geographic areas are covered by the interests of North American mammalogists, although no one listed a special interest in European mammals. All regions are represented within the United States, with the most work (or at least workers) in the southwest.

USERS OF COLLECTIONS OF MAMMALS

The users of systematic collections of Recent mammals are not limited to the systematists or curators whose activities are related directly to the acquisition and study of specimens. To the contrary, most users of collections rely on the facilities or specimens housed therein in indirect ways. Many of these individuals are non-systematists and, in fact, many do not even consider themselves mammalogists.

Appendix 3 lists the names, institutions, and area of specialization of professional scientists who regularly use the collections of mammals at their respective institutions. For convenience, the list has been subdivided into the broadly overlapping categories of "Mammalogists" (including systematists as well as others), "Ecologists," "Wildlife Biologists," "Paleontologists," and "Miscellaneous" (including persons who describe their areas of specialty in a variety of ways).

Appendix 4 lists the names, institutions, and areas of specialization of individuals who responded to the non-systematist section of the questionnaire inserted with the *Journal of Mammalogy* and mailed to all members of the American Society of Mammalogists, and who are not listed in Appendices 1 or 3.

Based on responses to the two questionnaires, it can be seen that systematic collections of Recent mammals are used by persons representing diverse pure and applied disciplines. Appendix 4 alone includes representatives of more than 30 different disciplines. If this list of disciplines were expanded to include all those in the "Miscellaneous" category of Appendix 3, more than 50 scientific disciplines could be counted among those which depend upon systematic collections of Recent mammals for various kinds of services.

The disciplines outside systematic mammalogy which probably use collections of mammals most extensively fall into four groups: 1) ecology and wildlife biology; 2) anatomy, paleontology, and archaeology; 3) ethology and physiology; 4) parasitology and biomedical sciences. A few of the many ways in which collections are used by these disciplines are described below.

1) *Ecologists* and *wildlife biologists* constitute the largest category of nonsystematic users of collections of Recent mammals. Together, they outnumber the systematists who manage or work directly with collections, and the demands that they place on collections and their curators are rapidly increasing. Ecological data of several kinds are available in museum collections, but these data often are not employed to the extent that they should be. Specimens in collections often enable estimation of population age structure, annual and population cycles, population turnover rates, time and age at which reproduction occurs, reproductive potential and/or success, preferred habitats, seasonal changes in preferred habitats, periods of activity or inactivity, actual or potential competitors or predators, and hibernal proclivities, to mention a few. Collectors of specimens necessarily are natural historians, and their familiarity (as described in field notes) with the mammals they work with often provides considerable insight into complex ecological problems. However, perhaps the most important way in which collections of mammals are of use to ecologists and wildlife biologists is as a source of historical and contemporary data to be used in the preparation of environmental impact statements. Unfortunately, collections often are both inadequately and improperly used in this regard, and we contend that few impact statements should ever be filed without prior review by a qualified systematic mammalogist.

2) *Anatomists*, *paleontologists*, and *archaeologists* use collections of Recent mammals as a source of both research and reference specimens. Systematic collections include representatives of all known species of mammals prepared in diverse ways. Specimens preserved in alcohol or formalin are available for dissection by gross anatomists interested in various aspects of the skeletal, muscular, digestive, reproductive, nervous, circulatory, endocrine, excretory, or integumentary systems. Similar preservation of specimens of different taxa enables comparative studies, and thereby facilitates interpretation of the specializations and adaptations associated with the different ways in which diverse species of mammals cope with their environment. Specimens preserved in other fluids (Bouin's solution, buffered formalin, and others) are available for use by microanatomists who wish to examine tissues microscopically (including electron microscopes and scanning electron microscopes) and thereby shed light on the form and function of different mammals at the level of cells or tissues. Osteological preparations are especially useful to paleontologists and archaeologists, but also to anatomists and others, who need comparative material in order to identify and better understand hard parts of mammals (including man) preserved as fossils or subfossils. In this regard, paleontologists are increasingly assessing variability in Recent taxa as an aid to interpreting variability in related but now extinct taxa. Likewise, archaeologists routinely compare the skeletal remains of mammals found in association with prehistoric man with osteological preparations of Recent mammals in order to determine what the activities of man were and

how extensively man relied on other mammals for food, clothing, shelter, and other necessities of life.

3) *Ethologists* and *physiologists* often depend on the curators of collections of Recent mammals for assistance in identification and for basic natural history information. Reciprocally, they add to collections of mammals by depositing voucher specimens.

4) Perhaps the most effective way to demonstrate the close relationship between *parasitologists* and *biomedical researchers* and collections of Recent mammals is to quote from a letter received from the Chairman of the Department of Entomology at the Bernice P. Bishop Museum in Honolulu:

"The collection of mammals of the Papuan Subregion, principally the island of New Guinea, at Bishop Museum is one of the foremost in the world, and may be unsurpassed as a mammalogical research base for that area. The collection was assembled initially and primarily as documentation for studies of ectoparasites of the Papuan Subregion. The ectoparasite collections, with vector and potential vector implications as well as major use in studies of parasite systematics, zoogeography, and parallel evolution with hosts, are maintained in the Department of Entomology; they are by far the richest such collections for the area and are being studied by a number of collaborating specialists as well as Museum staff. The mammal collections are maintained in the Division of Vertebrate Zoology, and we recognize them as a major part of the Museum collections per se, as well as documentation as host specimens."

"The interesting aspect relative to your current survey is that a major collection of mammals (and other vertebrates) was assembled first as a user resource, and secondarily acquired importance as a research collection mammalogically."

In short, parasitologists and biomedical researchers are very much dependent on collections of hosts and, conversely, their activities contribute much to the growth and development of host collections. This is especially true for the relationship between ectoparasitologists and mammalogists.

USES

This section presents data on the intensity and types of uses that systematic collections of Recent mammals receive. The emphasis in this section is on service and non-systematic uses of these collections. However, we must point out that these are secondary uses of the collections, with systematic studies performed by systematists being the primary function. We believe that systematic collections need to be supported for their primary function quite apart from their service activities.

A curator is required to divide his time between research, curation of the collection, teaching, administrative duties, and providing systematic services. Therefore, any increase in the latter four types of duties impairs the total systematic output of the curator. This becomes particularly critical when curators are called upon to consult and provide other services in environmental studies, but are not funded adequately to provide replacements for their other responsibilities.

Visitors and Loans

Appendix 5 presents data on the loans, student usage, and visitation at collections of over 1000 specimens in North America. These collections have made just over 1000 loans per year in the two years covered by the collection survey questionnaire and involved a total of 20,718 specimens per year. This is an average of 6.6 loans per collection per year and 20.5 specimens per loan. The National Museum of Natural History made the most loans during this two year period (271) and University of Kansas collections loaned the most specimens (6650).

An average of 856 students per year used these collections directly in their research and an average of 537 students used them indirectly, such as for deposition of voucher specimens.

Visitations to these 152 collections averaged 1695 per year or just over 11 visitations per collection. Visitations per year for the categories for which data were obtained were as follows: professional mammalogists, 432; graduate students, 550; other professional scientists, 227; state and federal employees, 251; others, 235. Collections receiving the largest number of visitors during the two years were the National Museum of Natural History (400), Field Museum of Natural History (239), American Museum of Natural History (200), and The University of Kansas, Museum of Natural History (180).

Collections used by systematists and non-systematists are given in Appendix 6. These results are from the questionnaire sent as a supplement to the *Journal of Mammalogy*. A total of 145 collections were listed on the returned questionnaires. One hundred and fifteen collections were listed by systematists and 113 were listed by non-systematists. This would indicate that systematic collections are receiving about equal usage by these two groups. Collections listed most often by systematists were the National Museum of Natural History (74), American Museum of Natural History (60), Field Museum of Natural History (39), University of California-Berkeley, Museum of Vertebrate Zoology (39), The University of Kansas, Museum of Natural History (35), University of Michigan, Museum of Zoology (27), and The Museum of Texas Tech University (23). Among the collections receiving high usage by non-systematists were the National Museum of Natural History (77), American Museum of Natural History (52), University of California-Berkeley, Museum of Vertebrate Zoology (37), Field Museum of Natural History (34), University of Michigan, Museum of Zoology (24), The University of Kansas, Museum of Natural History (24), and Harvard University, Museum of Comparative Zoology (19).

Types of Uses

Systematic collections of Recent mammals are established and built around the research interests of their curators and other professional staff. Most often these research interests are in the systematic relationships of the animals being collected. Systematics research is, therefore, the primary product of these collections.

Appendix 7 gives those uses made of systematic collections by non-systematists. Identification of specimens is by far the most common use made by this group of scientists. This is not unreasonable because any scientific work must contain proper and confirmed identifications of the animals cited. This is true whether the research is in physiology, anatomy, or environmental studies. Other common uses made of these collections are for teaching, deposition of voucher specimens, non-systematic research, and consulting services. Training of students in mammalogy, vertebrate natural history, wildlife biology, and ecology should, and generally does, involve the use of scientific specimens. Good scientists will deposit specimens from their research in collections so that future scientists can verify their findings. Non-systematic research uses of collections are nearly as varied as the number of non-systematists. One of the increasing demands being made on collections is consulting with curators for systematic, distributional, and other data.

Recent Literature Survey

The Recent Literature sections accompanying the *Journal of Mammalogy* for the years 1970-1974, inclusive, were surveyed for papers written by North American scientists using specimens of North American mammals housed in North American collections. The following information was obtained from this survey: total number of papers produced by this group of scientists; what disciplines were using North American collections of North American mammals and at what intensity; number of North American scientists using North American collections of North American mammals. It should be pointed out that this is a very conservative estimate of the intensity of use of North American collections of Recent mammals because papers written by North American scientists using specimens of non-North American mammals housed in North American collections were not included nor were papers written by non-North American scientists using North American collections.

A total of 24,934 citations was listed in the Recent Literature section for the five years surveyed. Of this total, 1306 papers were judged to be written by North American scientists using North American collections of North American mammals. Thus, this group of scientists contributed 5.24 percent of the total scientific literature concerning mammals produced in the World for this five-year period. This does not include papers that made use of systematic information published by systematists but only those that made direct use of specimens.

Categories used to classify authors and papers in Appendix 8 are those of the indexers of the Recent Literature section of the *Journal of Mammalogy*. The percentage contributed by each category (Appendix 8) to the total 1306 papers was as follows: anatomy, 12.2; behavior, 3.1; conservation and management, 1.1; disease and parasites, 13.3; distribution and fauna, 27.1; ecology, 11.1; evolution and genetics, 13.4; physiology, 2.3; taxonomy and paleontology, 16.4. Clearly, specimens of mammals are being used most heavily in papers dealing with distribution and fauna and taxonomy and paleontology, and most of these papers are written by systematists. However, specimens are being used in a significant number of papers dealing with anatomy, disease and parasites, ecology, and evolution and genetics, and most of the authors of these papers are non-systematists.

The other interesting facts learned from this survey concern the number of authors involved in producing the papers using mammal specimens. The number of different authors producing the papers in each year were as follows: 1970, 319; 1971, 222; 1972, 280; 1973, 332; 1974, 265. This means that an average of 284 different authors were producing papers using collections of mammals per year for this five-year period. For the entire five-year period 970 different authors produced papers.

Specific Uses of Collections

In order to document specific uses made of collections and the tasks that a curator is called upon to perform as a result of his association with a collection of mammals, a number of curators were asked to supply an anecdotal list of these services that had been performed at their institutions. Requests were made to selected institutions to give a diversity of size, types of support, and geographic location.

The 11 collections included here are only a small portion of the nearly 400 collections of Recent mammals in North America. Most of the 11 have reported the collection-related services that they have provided during a single year. We do not believe that these lists represent all of the kinds of services provided by systematic collection of mammals, but they should give an idea of the range of services provided.

Most of these 11 collections have omitted from these lists the systematic research conducted by their professional staff and visiting scientists. However, we would like to reiterate that we believe that these are the basic functions of systematic collections and deserve support quite apart from the service operations conducted by the collections.

American Museum of Natural History

The American Museum of Natural History possesses the largest collection of Recent mammals held by a private institution in North America. The following list of uses, which was supplied by Dr. Sydney Anderson, are those dealing with specimens only and does not include calls made upon the people and library resources associated with the collection. These items are drawn from the experiences of the Department of Mammalogy, and do not include the basic systematic studies done by the professional staff or similar studies by visiting systematists.

1) Inspectors of imported goods, in commerce and in the possession of tourists, for the U.S. Department of Interior brings items to the Museum so that the specific identity of these items may be determined by comparison with specimens. Some of these are of endangered species. Adequate and impartial enforcement depends on these contacts.

2) Identified specimens taken in an ecological survey needed to prepare an environmental impact statement for a proposed power plant site in New York State.

3) A feasibility study of the use of a scanning electron microscope to identify individual hairs of endangered species was begun.

4) Selected specimens were used in graduate courses in mammalogy.

5) Specimens were used by a teacher in our education department for popular lectures.

6) A paleontologist interested in the origin of mammals studied skeletons.

7) Consulted by the New York City Board of Health on species that should be prohibited in private possession.

8) Specimens were used in preparing and testing an illustrated key to the mammals of Uruguay, for use by entomologists, parasitologists, agriculturists, and others, as well as mammalogists.

9) A voucher specimen of a hamster used in studies of behavior was accepted, prepared, and verified as to identification.

10) Mustelid skeletons were loaned to a colleague in Leningrad.

11) Four trophy heads were given for display in a children's ward of the Hospital for Special Surgery.

12) Accepted and prepared dozens of skulls of dolphins obtained in an ecological and management study by colleagues on the West Coast of accidentally killed dolphins in the Pacific tuna fishery.

13) Accepted and prepared hundreds of skulls of mongooses obtained by a colleague in the Virgin Islands who is studying mongoose ecology there.

14) Accepted and prepared several dozen lynx from Alaska that were being used in an ecological and management study by an Alaskan biologist.

15) Accepted and catalogued several dozen trophy specimens, some of potential scientific value, others having other values in education or exhibition.

16) Specimens were used in checking the accuracy of figures and texts of accounts being edited for the *Journal of Mammalogy*, *Mammalian Species*, and other publications.

17) An archeologist borrowed skeletons for use in the analysis of material from excavations in Panama.

18) A graduate student from Yale University came to get hair samples of a variety of species of Insectivora for comparative study.

19) An author working on an autobiography came to read original field notes and other records in our file.

20) A physiologist studying the high frequency hearing of bats came to check on the identity of a rare species he had obtained.

21) Sent information on a type specimen of a rare South American marmoset collected more than a hundred years ago to the reviser working on the group.

22) A functional anatomist in a Florida medical school studied skulls and ear regions of cetaceans in our collection. We loaned him some skulls and took photographs of others for his use.

23) Checked features of young tapirs of different species for information that might be used to identify the species by persons inspecting imported skins.

24) A graduate student from Florida who is studying tropical bat communities visited to see certain species.

25) The collection was checked for data on old specimens of bison for the Ohio Historical Center.

26) A piece of scrimshaw work for a whaling exhibit at the Southport Seaside Museum was compared with specimens.

27) Carved chessmen were compared with bone and ivory for Macy's Department store.

28) The collections at a small private college were reviewed and a report submitted with recommendations on the better use of these neglected materials.

29) Identified specimens of the geographically restricted Chihuahuan mouse recently discovered more than 100 miles from the nearest known locality.

30) Identified bones found in the country by a small boy (potential biologist or legislator?).

31) Specimens were used in checking paintings for calendar of wildlife being prepared for an industrial concern that has recently contributed a tract of swamp to a public park.

32) Identified a substance thought by the finder to be ambergris, by comparison with specimens of ambergris. It was not. It usually is not.

33) A graduate student in vertebrate paleontology from the University of Kansas visited to study skeletons and to dissect preserved specimens of manatees and dugongs.

34) A professor from a private school upstate and her student came to use our large collections of artiodactyls in a study of sexual dimorphism and how this may relate to behavior.

35) A professor and his student from the Department of Anatomy of a nearby state university visited several times to make comparative studies of teeth and lower jaws.

36) A technician-preparator from a large museum in another state visited to study the techniques used in our osteology laboratory.

37) Students in the Undergraduate Research Participation Program in various departments here were shown around the Department of Mammalogy as part of their training.

38) Showed parts of the collections and talked on techniques to a group from the Rochester Museum of Science as part of their museology training program.

39) Showed parts of the collections to visiting corporation executives invited by the Museum to familiarize them with our institution and its relevance to business and other areas of society.

40) Commented on illustrations of locomotion in apes and man being prepared for an encyclopedia.

National Fish and Wildlife Laboratory

The collections of the National Fish and Wildlife Laboratory are housed together with those of the National Museum of Natural History. These collections form the largest collection of Recent mammals in North America. The following information supplied by Dr. Clyde Jones, Director of the Laboratory, concerns conservation issues involving his organization.

Advice to Endangered Species Program for Fiscal 1974

1) Upon request, review and make recommendations on applications for permits to import endangered species. Applications reviewed—26; Environmental Statements reviewed—11.

2) Assist the Division of Law Enforcement in the identification of endangered vertebrates and their parts or products. Items identified—birds, 53; mammals, 13,200; herps, 70.

3) Prepare criteria for "endangered" and "threatened with becoming endangered" species and prepare lists of currently "endangered" and "threatened with becoming endangered" animals both native and foreign for the 1973 Endangered Species Conservation Act and for the Convention on International Trade in Wild Fauna and Flora, including locations or areas where the animals are "endangered" or "threatened," probable cause of their status, and needed protection or management.

4) Complete recovery plans on the Indiana bat. Due December 1, 1973. Recovery plan completed and accepted—July 22, 1973. Additional research being conducted under contract. Completion date, December 31, 1975.

Services Routinely Performed in Connection with
the Bureau's Endangered Species Program

1) Provides information on the biology and status of species included on the United States' List on Foreign Fish and Wildlife.

2) Reviews data sheets and provides information on the species listed in the Bureau's "Redbook."

3) Reviews applications for permits to import endangered species and makes recommendations based on assembled biological data.

4) Identifies specimens and provides information for the identification of materials for field personnel.

5) Loans specimens for educational purposes to the Office of Conservation Education, and provides information on endangered and unique species.

6) Prepared specimens of endangered species for research purposes and for preservation.

7) Participates in the development of recovery programs for endangered species.

8) Conducts field work in order to determine the status of endangered species and species of uncertain status.

9) Identifies specimens and parts of specimens, such as skins, feathers, hair, pieces of ivory, leather, and the like, sent in by Game Management Agents.

10) Provides information directly to Game Management Agents with regard to the distribution and identification of specimens and other materials.

11) Serves as a temporary depository for confiscated specimens held as legal exhibits, and ships such materials to designated places for use in legal actions.

12) Signs affidavits with regard to the positive identification of specimens with regard to legal actions against illegal importations.

13) Provides testimony in court with regard to prosecution of various cases.

14) Reviews proposed rules and regulations.

15) Conducts research for information in order to provide data on wildlife imported into the United States, and makes recommendations with regard to the advisability of allowing additional importations in view of the dangers of accidental establishment.

16) Arranges for the deposition of specimens of endangered species that were imported into the country under permit or were confiscated at ports of entry.

17) Provides biological information directly to The Division of Law Enforcement for use in legal interpretations of Bureau responsibilities.

18) Provides specimens, information, and help in teaching wildlife identification at the Bureau's school for new special agents.

Major Conservation Issues in which Laboratory has Participated

Marine Mammals

1970 Evaluated data on the status of whales, provided general information, and prepared status reports.

1970 Reviewed proposed Federal legislation for the protection of marine mammals.

1971 Served on the Whale Policy Committee of the Council of Environmental Quality.

1971 Served on the Executive Committee responsible for the administrative organization of the International Conference on the Biology of Whales.

1971 Served as an observer for the Department at the International Conference on the Biology of Whales.

1971 Provided general information on biology and status of the polar bear.

1971 Served as an observer for the Department on the NOAA Marine Mammal Committee charged with the development of a national program for research on marine mammals.

1972 Participated in the census of the pacific walrus by aerial surveys of marine mammals of the Bering Sea.

1972 Assigned to the Interior-Commerce Marine Mammal Task Force that has the responsibility of recommending policies for the agencies with regard to the Marine Mammal Protection Act of 1972.

1972 Participated in research planning for those marine mammals that are the responsibility of the Bureau and the Department.

1973 Served as Chairman, Marine Mammal Program Development Committee.

1973 Prepared status reports on marine mammals in accordance with the Marine Mammal Protection Act of 1972.

1974 Prepared status report on the dugong.

1974 Prepared status reports on manatees.

1974 Participated in testimony at the Oversight Hearings for the Marine Mammal Protection Act.

1974 Prepared the annual status reports on marine mammals in accordance with the Marine Mammal Protection Act of 1972.

Antarctica

1970 Appointed as the Antarctic observer for the Department.
1971 Participated in the Antarctic inspection.
1972 Reviewed proposed international convention with regard to conservation of Antarctic seals.
1972 Reviewed proposed Federal legislation for the protection of Antarctic flora and fauna.
1973 Served on the Bureau's Antarctic Task Force.

Primates

1970 Provided general information on biology and status of gorillas.
1971 Status report on gorillas, especially the lowland or western gorilla.
1973 Initiated a program to develop a data base for preparation of status reports on primates of the world.
1974 Provided a detailed report on the status of the primates of Costa Rica.

Carnivores

1970 Participated in development of a program for management of the red wolf.
1971 Conducted taxonomic studies to develop methods of positive identification of specimens of red wolf.
1971 Status report on cats of the world.
1972 Participated in development of recovery plans for the red wolf.
1974 Prepared a report on the status of the gray wolf in North America.
1974 Prepared a report on the status of the cougar in North America.
1974 Participated in preparation of an environmental impact statement on proposed management of the grizzly bear.

Bats

1971 Status report on populations of bats in the United States.
1972 Developed the Bureau policy with regard to bat conservation.
1973 Prepared a recovery plan for the Indiana bat.
1973 Review of the status of the Hawaiian hoary bat.
1974 Review of the status of endangered bats on Guam.

Imported Wildlife

1970-1973 Prepared lists of imported birds, mammals, reptiles, and amphibians from declaration of importations.
1971 Served on a committee to evaluate the interpretation of the Lacey Act.
1972 Prepared lists of wildlife and corresponding justifications for inclusion on the list of low risk wildlife.
1973 Reviewed the proposed importation regulations for injurious wildlife.
1973 Prepared lists of birds, mammals, reptiles, and amphibians in accordance with the Endangered Species Act of 1973.

1973 Prepared appendices of lists of birds, mammals, reptiles, and amphibians in connection with the Convention on International Trade on Wild Flora and Fauna.

1974 Participated in discussions and meetings between representatives of the pet industry, aquarium fish trade, scientists, Bureau, and Department in connection with the proposed importation regulations for injurious wildlife.

Museum of Zoology, University of Michigan

The Museum of Zoology contains one of the largest collections of Recent mammals associated with an institution of higher learning. The following information was compiled by Dr. Emmet T. Hooper. Each year, the Mammal Division (Museum of Zoology, University of Michigan) accommodates many requests. Some of these are requests for loan of specimens for taxonomic revisionary studies of faunal reports. In addition, much personnel time is spent in supplying other information from the collections. To fulfill such requests typically involves collating data from specimen tags, abstracting records from our museum catalogs and species-card file, or condensing information from field notebooks; the time spent per request often is as great as that necessary to process a loan. Dispensing public information either in the form of telephone conversations or to "walk-in" visitors to the museum also requires much time. The following anecdotal list of specific requests processed in 1974 conveys the multifaceted nature of a collection's purpose and utility. While a given example may be unique and atypical, the lot taken as a whole is commonplace.

1) Loan of vampire and other bats for demonstration purposes in parasitology course.

2) Loan of fruit bats for examination of hair microstructure in study of bats as potential transport vectors for certain pollens and seeds.

3) Loan of skulls of various Felidae for use in making endocranial casts in study of brain evolution in Carnivora.

4) Loan of spiny anteater mounted skeleton for preparation of hologram.

5) Loan of specimens, from teaching collection, for a mammal course in forestry camp.

6) Examination of pongid skulls by anthropologist in comparison with subrecent fossils from Kenya, Africa.

7) Use of specimens for teaching (diversity of mammals) to primary school children.

8) Information on mole control to disgruntled home owner.

9) Numerous queries on bats: why are they in *my* house; how do we get rid of them; will we get rabies?

10) Information on probable inhabitant of den found near citizen's home.

11) Tuft of fur recovered from wire fence enclosing chicken coop sent for identification of the miscreant (raccoon).

12) Identification of bones found near home by local citizen.

13) Information on current classification of mammals to local high school teacher.

14) Instruction on preparation and mounting of skeletons to undergraduate zoology students.

15) Information on type of traps available for live trapping mammals.

16) Detroit Free Press, Action Line: Do Dolphins Talk?

17) Examination of armadillo skeletons by industrial design student for models in designing comfort chairs.

18) Citizen, Ann Arbor: Do camels spit?

19) Representative of insurance company inquired to find out if the squirrel is considered a rodent; claimant had filed for house damage by rodent.

20) Obtained list of *Sciurus carolinensis* in our museum collected prior to 1900 for student at Duke University.

21) List of species occurring in Ottawa National Forest (Ironwood, Michigan) prepared for state official from specimens contained in our collection.

22) Identification of fur on confiscated winter parkas (wolf or coyote?) submitted by a special agent.

23) Examination of confiscated lynx pelt submitted by a special agent to determine if it is summer or winter; that is, was it taken out of season.

24) Identification of tooth (horse) found in backyard by youngster.

25) Identification of *Peromyscus* collected in Jalisco, Mexico, for colleague at Oklahoma State University.

26) Guide and instructional tour of mammal division facilities given for boy scouts working for mammalogy merit badge.

27) Use of skeletons by art students for sketching exercises.

28) Guide through mammal collection for students taking paleontology course.

Museum of Natural History, University of Kansas

The Museum of Natural History also possesses a large collection of Recent mammals and is associated with an institution of higher learning. The information supplied by Dr. Robert S. Hoffmann, Dr. Norman Slade, and Mr. James Bee contains requests for information or services directed to the curators and assistant curators other than those coming from systematic biologists who regularly use the collection.

1) Inspector of Customs, Kansas City International Airport, requested identification of species of wild cat fur in coat, and of carved tooth, both confiscated from incoming overseas travelers.

2) Persons both working in local office of environmental assessment firm (private) requested list of species of mammals likely to be found in area of proposed reservoir site.

3) Several calls from persons experiencing trouble with striped skunks who have taken up residence in or around their property.

4) Graduate students in archaeology at the University of Kansas frequently request assistance in identifying bone fragments found in archaeological sites that they are studying.

5) Specimens for demonstration only are used in graduate and undergraduate courses in population biology, mammalogy, and other courses at the University of Kansas.

6) Specimens were used in setting up a special exhibit in connection with The Museum of Art and entitled "Hands in Nature" to accompany a parallel exhibit on the human hand in The Museum of Art.

7) Special teaching specimens are made available about monthly to one of several persons in the Museum's Public Education Division for use in special lectures and demonstrations to grade and high school students.

8) Locality records were checked for a parasitologist working on specimens of lice combed from the hair of pocket gophers in the Museum.

9) A private documentary film company requested assistance in determining the presence or absence of certain endangered species in a little-known region of the Soviet Far East.

10) Assistance was provided a colleague in the University's Medical School who was trying to obtain information on certain species of Near Eastern hamsters in connection with his research on toxoplasmosis.

11) Information was provided an official from the Department of Interior concerning the nearest museum which might have skeletal material of sperm whales for use in identifying a questioned ivory object.

12) "Behind the Scenes" tours of the Division of Mammals were provided to in-service training course for executives from the Hallmark Corporation and Southwestern Bell Company.

13) Responded to a request for information on identity of certain pinniped specimens collected during the last century, some of which were catalogued in the collection. The inquiry came from scientists concerned with the ecology and commercial significance of pinnipeds in a certain section of the Pacific coast.

14) Provided certain specimens for a temporary exhibit on Greenland Eskimos being constructed for display by the Department of Anthropology.

15) Identified specimens from a retired paleontologist who no longer had access to comparative material.

16) Loaned some skeletal material from the teaching collection to the Art Department of the University for use in drawing class.

17) Prepared a number of skulls and skeletons of cotton rats being utilized by a colleague in the Department of Systematics and Ecology in population genetics study.

18) Checked identity of several specimens in the collection for a colleague preparing volume on mammals of a particular state.

19) Conducted brief workshop on recent advances in mammalian systematics for advanced students at small college nearby.

20) Identification of kinds of animals that qualify for rodent damage clause for insurance company. For example, can reparation in money be claimed for damage from skunk odor.

21) Identification of hair as human or non-human for local police.

22) General questions concerning kinds of mammals, their identification, behaviorism, food habits, care, etc., from citizens of community. Averages about one inquiry per week.

23) Examination of cochlear areas of skulls to determine most suitable species for experimental purposes, for medical researcher.

24) Identification of endoparasites in game species of mammals and birds, the most frequent request being concerned with nematodes in rabbit muscle.

25) Information for feature and publicity articles (newspapers, radio, television).

26) Assistance in compiling a synoptic skeletal collection of mammals for the Department of Anthropology.

27) Identification of tissue composition (bone or integument) of Indian masks of South America for the Nelson Art Museum of Kansas City, Missouri.

28) Centron Films of Lawrence has employed specimen subjects in many of their documentary films.

29) Response to requests from scholars seeking illustrative material for publications.

30) Obtained information on reproduction from specimens of female bats in collection for graduate student at another institution.

Carnegie Museum of Natural History

The Carnegie Museum of Natural History is a private institution housing approximately 60,000 Recent mammals. The following list, which was furnished by Dr. Duane A. Schlitter, includes the activities of the Section of Mammals that were directly related to use of actual specimens during 1974. Not included are inquiries that required the expertise or knowledge of the curators or use of the mammal library.

1) Identified North African jerboas taken by paleontologist at another museum, misidentified as gerbils.

2) Sent *Phenacomys* loan to colleague at a museum on west coast.

3) Explained mammal collection and research to President of Institute and some members of the Board of Trustees.

4) Helped local high school teacher in charge of environmental curriculum in county schools with identification of local fauna and relevant problems that should be included in courses.

5) Consulted with exhibit staff on plans for elk exhibit.

6) Went to Algeria to consult with persons in Algerian government and local company on effects of irrigation scheme on rodent populations and survey potential species involved.

7) Research associate from a state college visited several times to study skunks and the occurrence of a particular parasite.

8) Gave tour of collection and lectured on mammals to biology class from branch campus of local university.

9) Returned bats used in comparative study by paleontologist of Recent and Pleistocene species to collection.

10) Furnished information on holdings of *Blarina* to colleague from northeastern United States.

11) Furnished information on holdings of *Scalopus* to colleague from south-central United States.

12) Packed two skulls of specimens of Marco Polo sheep for shipment to another museum so skulls could be duplicated for exhibit.

13) Assisted graduate student from local university in searching out specimens of primates for research project on comparative study of hand bones.

14) Assisted graduate student from local university in establishing a comparative osteology collection for identification of mammal remains from archeological digs.

15) Gave lecture on mammals and tour collection to high school biology class interested in mammals.

16) Assisted colleague from Alaska who visited to study *Phoca* collection.

17) Sent loan of some northeastern U.S. *Blarina* to colleague.

18) Installed returned loan of Antillean bats to collection.

19) Identified specimen of coyote sent by state game commission as part of their program to monitor populations of canids in state.

20) Checked and identified 22 specimens of monkeys in collection for research project of graduate student from local university.

21) Checked collection and sent information on museum holdings for black-footed ferrets to colleague.

22) Identified specimen of dog sent by state game commission as part of their program to monitor populations of canids in the state.

23) Identified two dog skulls for two visitors to Museum.

24) Assisted colleague who visited museum to study the distribution of *Pitymys* in Pennsylvania.

25) Colleague visited museum to study squirrel skeletons as part of a comparative study of postcranial anatomy of squirrels.

26) Identified cow skull brought in by visitor.

27) Identified red bat captured on construction site for worker who was concerned about the presence of the bat.

28) Surveyed and identified mammal fauna in a county park study, done to assist county officials in their park management program.

29) Identified small collection of mammals from Angola taken by staff member ancillary to other work.

30) Curated and identified collection of mammals from Nepal sent for identification and deposit by medical ecologist on a long term project there.

31) Curated and identified collection of rodents from South West Africa sent on exchange basis for my research.

32) Gave lecture on mammals and tour of mammal collection to group of students from a state college.

33) Sent loan of *Monodelphis* from various countries in South America to colleague at another museum to be included in revisionary study of that genus.

34) Sent a starter colony of dermestid beetles to colleagues at a state university.

35) Listed specimens in collection originating in British Honduras for colleague working on checklist of that country.

36) Talked with taxidermist from another museum about anatomy, behavior, and distribution of Old World big-horn sheep.

37) Returned loans of Old World gerbils used in my research to various museums in the United States.

38) Received and returned to collection loan of South American chiroptera sent to colleague and another of North American chiroptera used by paleontologists for comparative study of Recent and fossil species.

39) Assembled reference specimens from collection and gave a lecture to Junior High School students on mammalogy as part of biology program sponsored by Western Pennsylvania branch of Pennsylvania Junior Academy of Sciences.

40) Worked with a predoctoral student from Argentina visiting the collection for six months to work on taxonomy of northwestern Argentine mammals and to learn about curatorial methods in a medium-sized U.S. museum and how they could be used in a small mammal collection in Argentina.

41) Worked with a sophomore student from a North Carolina college who spent one month working in the mammal collection as part of college interim program to learn about mammals, particularly their natural history, and natural history collections.

42) Consulted with colleague who visited to examine the Recent and fossil peccaries in the Museum's collection as part of his study of South American peccaries.

43) Identified coyote skull for state game commission sent to them by a trapper for identification.

44) Answered questionnaire sent by colleague surveying curatorial policies and procedures.

45) Prepared report on mammals of the local engineering district for Corp of Engineer's environmental inventory.

46) Assisted mammalogist for private company doing environmental impact statements with identification and distributional problems of small mammals in various states in the U.S.

47) Identified bats analyzed for rabies by local county and state public health units.

48) Identified collections of small rodents from Ethiopia taken by colleagues at a state university medical school as part of long-term research of murine typhus and plague in that country.

49) Monitored scientific accuracy of exhibits being prepared by exhibit staff, identified specimens of mammals being included in various exhibits, wrote appropriate label copy for various exhibits, and furnished scientifically accurate source material for graphics to be included in various exhibits.

50) Assembled appropriate specimens from lecture collection of mammals and gave natural history lecture on African mammalogy to local natural history club sponsored by county government.

51) Identified human bones brought in by South American visitor who thought they were non-human.

52) Worked with local microscopist who wanted to examine hairs with scanning electron microscope in study of adaptation by mammals to winter climate.

53) Identified raccoon skull brought in by a group of young girls from the inner city who found the skull in a park.

54) Research Associate from a state college visited to examine shrew specimens as part of research on taxonomy and distribution of selected species of shrews.

The Museum, Texas Tech University

The collection of Recent mammals at Texas Tech University consists of 25,000 specimens and is used primarily in research, which is conducted by the professional staff and graduate students. Information on uses of this collection during 1974 were provided by Dr. Hugh H. Genoways and Mr. Stephen L. Williams.

1) Received on two occasions mammals from Baja California and mammals from New Mexico that were deposited by Wayland Baptist College in Plainview, Texas, as voucher specimens.

2) Provided material for a medical parasitologist at Texas Tech University Medical School for the study of internal and external parasites.

3) Twelve students from Texas Tech, Department of Biological Sciences, deposited specimens resulting from independent research.

4) Four students from Texas Tech, Department of Biological Sciences, used specimens already in the collection for degree requirements.

5) Seven students from Texas Tech, departments of Biological Sciences and Museum Science, used specimens already in the collection for independent research.

6) Gave three tours to high school classes that were interested in the collection and its function.

7) Received and assisted a visitor from Central College in Pella, Iowa, who used the collection for a study of bats from Yucatan, Mexico.

8) Received and assisted a visitor from the American Museum of Natural History who used the collection for a study of Neotropical bats.

9) Received and assisted a visitor from the Royal Ontario Museum who used the collection for a study of *Molossus*.

10) Received and assisted a visitor from the University of Minnesota who used the collection for a study of pocket gopher parasites.

11) Provided lecture services on several occasions as guest speaker in the Medical School and in the Department of Biology.

12) Identified species of *Rattus* from central Texas for U.S. Wildlife Service in San Antonio, Texas.

13) Made special efforts to obtain large mammal skeletons for comparative material for the Division of Anthropology and the Division of Vertebrate Paleontology at The Museum.

14) Received and processed about 300 geomyid rodent skeletons that were deposited by the University of Texas at Austin, as voucher specimens.

15) Trained ten students from Texas Tech University in collecting, preparing, and properly documenting mammalian specimens.

16) Trained five students in various aspects of technical work in mammal collections.

17) Responded to a request from the University of California at Davis for frozen mammalian specimens and specific species for exchange.

18) Responded to a request from the University of Nevada for a specimen of *Dasypus.*

19) Received 75 Neotropical bats that were deposited by Hofstra University at Hempstead, New York, as voucher specimens.

20) Received 100 Neotropical bats that were deposited by Central State College in Pella, Iowa, as voucher specimens.

21) Received 600 Indonesian mammals that were deposited by World Health Organization as voucher specimens resulting from a plague survey.

22) Sent a small colony of dermestids to Arkansas State University to help start the "bug" facilities at that institution.

23) Responded to an inquiry from University of Arkansas at Monticello concerning collection supplies and management.

24) Responded to an inquiry from Texas A & I at Corpus Christi, Texas, concerning collection supplies and management.

25) Responded to request from Central State University in Edmond, Oklahoma, for list of Oklahoma mammal specimens in collections.

26) Responded to request from University of Texas at El Paso for a list of bats from El Paso County, Texas.

27) Received 135 Chilean mammals that were deposited as voucher specimens resulting from outside individual research.

28) Responded to a request from National Park Service personnel concerning skin preservation techniques and sources of various supplies.

29) Responded to a request from Shippensburg State College in Shippensburg, Pennsylvania, concerning the specimens of *Mephitis* in the collection.

30) Responded to a request from the California State University at Fullerton concerning techniques for capturing pocket gophers alive.

31) Responded to a request from the University of California at Berkeley concerning the specimens of *Bassariscus* in the collection.

32) Loaned 97 pocket gophers to Texas A & M University for comparative material in graduate research on *Geomys bursarius* of Texas.

33) Loaned 35 New World bats to the University of Vermont for graduate research on the jaw, hyoid, and laryngeal musculature of bats.

34) Loaned 14 *Tylomys* to the Department of Biology at Hood College in Frederick, Maryland, for systematic studies of the genus.

35) Loaned six *Micronycteris* to Texas A & M University for systematic studies of the genus.

36) Loaned eight *Sylvilagus* to the Department of Biology at Sam Houston State University at Huntsville, Texas, for studies of the distribution of the genus in eastern Texas and Louisiana.

37) Loaned 31 *Gerbillus* to Carnegie Museum of Natural History for systematic studies of the group.

38) Loaned 20 skeletons of *Dipodomys* to the Department of Zoology at the University of Oklahoma in Norman for systematic studies of the genus.

39) Made intracampus loan of 34 New World bats to the Department of Biological Sciences at Texas Tech for photographing.

40) Made intracampus loan of 58 representative skeletons of local mammals to the Division of Paleontology at The Museum for use as comparative material with items recovered from cave excavations in the Guadalupe Mountains National Park.

41) Made intracampus loan of 7 mammal skeletons to the Division of Anthropology at The Museum for use as comparative material with items recovered from excavations from the Lubbock Lake site.

42) Performed special collecting at the Lubbock Lake site for the Division of Anthropology in order to acquire comparative material.

43) Made intracampus loan of representative mammals to the Department of Biological Sciences at Texas Tech for teaching purposes in Vertebrate Natural History.

44) Made two intracampus loans of representative mammals to the Department of Biological Sciences at Texas Tech for teaching purposes in Mammalogy.

45) Made intracampus loan of representative mammals to the Division of Education of The Museum for teaching purposes in special summer programs for elementary school children.

46) Contributed many specimens to The Museum and to the Department of Biological Sciences at Texas Tech for respective teaching collections.

47) Made intracampus loan of 5 mammal specimens to the Exhibition Staff of The Museum for public exhibition in the museum.

48) Provided consultation services for two graduate students from Fort Hays State College concerning computer utilization and research data.

49) Provided consultation services for United States Air Force concerning wildlife management on Matagordo Island, Texas.

50) Provided consultation services for National Park Service concerning the development of a master plan for management of Guadalupe Mountains National Park.

51) Received and assisted a visitor from Milwaukee who used the collection for a study of the dentition of *Artibeus lituratus.*

52) Received and assisted a visitor from Sul Ross University at Alpine, Texas, who used the collection for a study of North Texas mammals.

53) Received and assisted a visitor from Texas A & I in Corpus Christi who wished to show his accompanying students the mammal collection.

54) Received and assisted a visitor from Texas A & M at College Station, who used the collection for a study of Neotropical bats.

55) Received and assisted two visitors from Texas A & M who used the collection for individual studies concerning *Geomys* and rare and endangered mammals of Texas.

56) Received and assisted a visitor from Pan American University at Edinburgh, Texas, who wished to show his accompanying students the mammal collection.

57) Received and assisted a visitor from Purdue University at Lafayette, Indiana, who used the collection for studies of Indiana mammals and Neotropical bats.

58) Received and assisted a visitor from the University of New Mexico at Albuquerque who used the collection for a study of Sonoran mammals.

California Academy of Sciences

The California Academy of Sciences is a private institution that houses a collection of 21,000 specimens of Recent mammals. Dr. Robert T. Orr supplied the following list of non-systematic uses of their collection during 1974.

1) Numerous tours given for prospective donors, especially corporation executives.

2) Tours given for special groups including certain biology classes and foreign scientists.

3) Specimens made available for use by artists.

4) Materials selected from the collection for public lectures.

5) Several days of preparation involved in the development of research displays for the Members' Annual Open House.

6) Selection of specimens for weekly lectures in connection with the Academy's docent training program.

7) Selection of specimens used for symposia preceding members' trips abroad.

8) Numerous identifications of bones for the public, such bones usually consisting of parts of cattle, dogs, horses, deer, pinnipeds, and cetaceans.

9) Assistance to archeological students working on shell mounds locally.

10) Assistance in the identification of various materials used in archeological artifacts in conjunction with the Department of Exhibits.

11) Identification of substances suspected of being ambergris brought in by the public.

12) Identification of both raw products and artifacts for the U.S. Fish and Wildlife Service and the U.S. Customs when such material is suspected of being sold in violation of federal legislation.

13) Identification of collection of Mexican specimens for a student at Indiana State University.

14) Identification of rats found on an Army depot in central California where it was suspected foreign species had been introduced.

15) Identification of mammals involved in a study of Chagas disease and hemorrhagic fever in Oaxaca, Mexico, by the University of California Hooper Foundation for Medical Research.

16) Identified but dataless specimens provided for the San Francisco Zoo docents as well as Academy docents.

17) Identification of sheep skulls that were possible evidence of theft of mouflon sheep at the San Francisco Zoo.

18) Blood and tissue samples of certain specimens saved for research workers in local universities.

19) Countless telephone inquiries answered on various subjects ranging from the migration of gray whales to the feeding of ringtailed cats.

20) Many graduate students and visiting scientists assisted in the use of the collections.

Museum of the High Plains,
Fort Hays Kansas State College

The collection of Recent mammals in the Museum of the High Plains is a relatively small collection that is dedicated primarily to research. The associated institution is a small state-supported school that offers master's degrees. Dr. Jerry R. Choate provided the following anecdoctal list which includes uses of the collection and demands on the curator except research and graduate teaching.

1) A total of 15 classes (high school classes from throughout western Kansas) was given tours through the Museum and lectures on the research activities and capabilities associated with the collection of mammals.

2) A group of conservationists visited the Museum to express their concern regarding the destruction of wildlife, and were counseled regarding population dynamics of mammals.

3) The curator was called upon numerous times to remove bats from buildings, and used the collection as a reference in identifying those bats.

4) The curator removed a live squirrel from a house near campus, and on two occasions was called upon to examine pet squirrels and assess the extent of their maladies.

5) The collection was used at least three times as a source of information requested by the Kansas Forestry, Fish and Game Commission.

6) The collection was used as a reference in identifying a weasel that was brought in by a State Game Protector.

7) The collection was used as a reference on two occasions to identify mammals in photographs brought in by the Wildlife Extension Agent for the state of Kansas.

8) The library of the curator was used to identify a cast of a footprint brought in by the Kansas Wildlife Extension Agent.

9) At the request of the Bureau of Land Management, the collection was used as a reference by the curator in preparing a list of mammals known to inhabit a valley scheduled for a possible impoundment.

10) Using the collection as a reference and basing his comments on knowledge derived from activities associated with the collection, the curator served as a consultant to the Kansas Academy of Science in preparing a list of Rare and Endangered Mammals in Kansas.

11) The curator removed gophers from a cemetery.

12) Using collection materials, the curator supervised the preparation of an exhibit for Junior Science Day.

13) The curator prepared an exhibit of collection materials for visiting counselors from various junior colleges in Kansas.

14) Using the collections as a reference, the curator provided information (regarding distribution and abundance of skunks) to employees of the Kansas Department of Health.

15) The curator loaned specimens from the collection to be exhibited by the Kansas Department of Health as potential carriers of rabies.

16) The curator loaned specimens of vampire bats from the collection to be used in a campus exhibit commemorating the 75th anniversary of the publication of the novel "Dracula."

17) Using skeletal materials in the collection for reference, the curator identified bones brought in by an archaeologist from Kansas State University.

18) The curator helped determine the race, sex, and age of a human skull found by a soil conservation agent working on a terracing project at the Kansas Agricultural Experiment Station.

19) The curator advised a local farmer as to how he could reduce losses resulting from bobcats raiding his pig pens and chicken house.

20) The curator took part in a coyote census conducted by the U.S. Fish and Wildlife Service.

21) The curator responded to a claim by a local rancher that he had seen a black-footed ferret in a prairie dog town located on his range. He had actually seen a long-tailed weasel.

22) The following scientists visited the Museum of the High Plains during 1974: Hugh H. Genoways, Robert J. Baker, Stephen L. Williams, and Edward F. Pembleton (Texas Tech University); Clyde Jones and Michael Bogan (U.S. Fish and Wildlife Service); Mike Kennedy (University of Oklahoma); John P. Farney (Kearney State College); James D. Smith (California State University, Fullerton).

23) Loans of specimens were made to the following institutions during the last three years: University of Texas at El Paso, 20; University of Connecticut, 93; Texas A&M University, 9; Pennsylvania State University, 11; Bird and Mammal Laboratories, 6; Texas A&M University, 8; University of Arizona, 48; Kansas Department of Health, 3; various in-house loans, 61.

24) Loans of specimens were received from the following institutions during the past three years: University of Kansas, 558; Bird and Mammal Laboratories, 2; University of Kansas, 3; University of Kansas, 4; University of Kansas, 14; University of Connecticut, 29; American Museum of Natural History, 80; American Museum of Natural History, 1; University of Connecticut, 8; University of Michigan, 11; University of Texas at El Paso, 16.

The Museum, Oklahoma State University

This institution houses one of the largest collections of Recent mammals (10, 000) in North America that is used primarily in teaching. Dr. Bryan P. Glass supplied the following list of uses made of this collection during 1974.

1) Loaned specimens to art classes for use as models in free-hand drawing class.

2) Specimens were used in teaching a course in mammalogy to senior level and graduate students.

3) Conducted tours through the collections for Brownies and Cub Scouts.

4) Identified bones for County Sheriff (carnivore).

5) Identified bones for bulldozer operation (bison).

6) Identified fossilized mammal teeth for several citizens (horse, bison, and mammoth).

7) Loaned trophy heads (one at a time) to grade school for a rotating display.

8) Loaned a special portable exhibit on Oklahoma mammals to three grade schools.

9) Accepted and catalogued two collections of trophy heads from Africa, Asia, and North America.

10) Identified live wild rodents used in physiology experiments by physiological psychologist.

11) Accessioned specimens collected by mammalogy students.

12) Loaned our collection of *Zapus* to Indiana State University at Terre Haute.

13) Loaned our collections of *Scalopus* to Texas A&M University.

14) Accepted and processed 15 species of exotic mammals donated from the Oklahoma City Zoo.

15) Exchanged U.S. specimens for 30 species of small mammals from Russia.

16) Exchanged U.S. specimens for 25 species of small mammals from Poland.

17) Exchanged U.S. specimens for 22 species of mammals from Australia.

18) Used specimens in preparing manuscript for identifying mammal skulls.

Department of Biology,
New Mexico State University

The collection of Recent mammals at New Mexico State University is small (5000) but is used about equally in research and teaching. The research activities are carried on by the curator and graduate students in master's and Ph.D. programs. Dr. Charles S. Thaeler, Jr., stated that the following activities were conducted at this collection in 1974.

1) A number of people working on mammal studies in connection with the Desert Biome of the IBP used the collection to familiarize themselves with mammals from the IBP study sites.

2) Specimens from the teaching part of the collection were used to help students in Biology 464 learn the local mammal fauna.

3) Specimens from the collection were used as illustrations of several principles of evolution for a beginning course, Biology 110.

4) Specimens from the teaching portion of the collection were loaned to a teacher for use in a public school.

5) Provided identification of local rodents collected by a U.S.D.A. representative as part of a study involving the agricultural use of pesticides.

6) Provided identification of rodents caught by vector control personnel at Holloman A.F.B.

7) Received, cataloged, and processed specimens collected as part of the course requirements by students in a mammalogy class.

8) Received, cataloged, and processed voucher specimens from a thesis research problem dealing with pocket gopher distribution and physiology.

9) Reviewed specimens of black-tailed prairie dogs in the collection to provide information about these rodents in connection with the endangered species program of the U.S. Fish and Wildlife Service.

10) Loaned skulls of kangaroo rats to a psychologist studying problems of hearing.

11) Three university students used the collection to familiarize themselves with local mammals.

School of Forest Resources, Pennsylvania State University

A small collection of Recent mammals (1500) at Pennsylvania State University is used primarily in teaching. The uses of this collection that are listed below were provided by Dr. Robert G. Anthony.

1) Selected specimens used in one undergraduate and three graduate courses.

2) Used specimens to present a short television show on seasonal color changes in mammals.

3) Graduate students referred to collection for their field research projects.

4) Specimens loaned to high school teachers for natural history talks.

5) Graduate students borrowed specimens to present seminars and popular presentations.

6) Wildlife extensionists referred to the mammal collection for their news releases on wildlife.

7) Used collection in connection with research on *Microtus pinetorum*.

LITERATURE CITED

ADVISORY COMMITTEE FOR SYSTEMATIC RESOURCES IN MAMMALOGY. 1974. Report and Recommendations. . . . Processed by The Graduate School, Texas Tech Univ., 30 pp. (Available from Jerry R. Choate, Museum of the High Plains, Fort Hays Kansas State College, Hays, Kansas 67601).

CHOATE, J. R., and H. H. GENOWAYS. 1975. Collections of Recent mammals in North America. J. Mamm., 56:452-502.

COMMITTEE ON SYSTEMATIC COLLECTIONS. 1975. Collections that meet minimal standards. J. Mamm., 56:293-295.

Addresses of authors: HUGH H. GENOWAYS, *The Museum, Texas Tech University, Lubbock, 79409*; JERRY R. CHOATE, *Museum of the High Plains, Fort Hays Kansas State College, Hays, 67601*; EDWARD F. PEMBLETON, IRA F. GREENBAUM, and JOHN W. BICKHAM, *Department of Biological Sciences, Texas Tech University, Lubbock, 79409*. Present addresses of: Bickham, *Department of Wildlife and Fisheries Sciences, Texas A&M University, College Station, 77843*; Genoways, *Section of Mammals, Carnegie Museum of Natural History, 4400 Forbes Avenue, Pittsburgh, Pennsylvania 152213*; Pembleton: *Museum of Science and History, MacArthur Park, Little Rock, Arkansas 72202*.

APPENDIX 7.—*Types of uses of systematic collections of Recent mammals made by non-systematists. Figures indicate the number of non-systematists making this type of use. Total number of non-systematists using collections that responded to this questionnaire was 272. Data were obtained from the questionnaire sent as the supplement to the* Journal of Mammalogy.

Type of use	Number using
Identification of specimens	138
Teaching	87
Deposition of voucher specimens	71
Non-systematic research	37
Consulting services	29
Distributional records	25
Anatomical comparisons	23
Identification of fossils	17
Identification of archaeological remains	10
Reproductive data	7
Identification of hair	5
Functional anatomy	4
Preparation of keys	4
Habitat records	3
Loans	3
Parasites	2
Exhibits and displays	2
Preparation of skeletons	2
Food habit studies	2
Identification of endangered species pelts	1
Pelage variation analysis	1
Photography	1

SECTION III.
Identification

Proper identification of wildlife—or the fragmentary remains of wildlife—is important for many reasons. In legal cases involving the taking or possession of a protected species, verified identification of the evidence is essential in proving that a violation did in fact occur. Identifications of the remains of birds that collide with aircraft lead to knowledge of weight and other characteristics of the most likely candidates for air strikes, useful in setting specifications for coverings on cockpits. Although identifications are important in particular instances, they are generally routine service functions of museums and seldom lead to publication unless some other potentially useful information is derived as a by-product. Manville's paper is an example of such information. The second paper in this section—an extract from an Annual Report of the Division of Wildlife Research of the U.S. Fish and Wildlife Service—indicates the number and kinds of routine identifications made annually by one federal laboratory.

Although routine identifications may be based entirely on comparison with museum specimens, consultation with literature unrelated to the identification process may be necessary or useful. Thorough, illustrated analyses of skeletal anatomy may be helpful in the identification of bones, for example. Similarly, clues leading to the discovery of characters that will permit identification of age and sex classes within a species may come from studies of molt or from examination of skulls, scales, or other parts.

Information derived from museum specimens often serves as the basis for guides or manuals to be used in making identifications in the field. The paper by King and Brazaitis is an outstanding example of such a guide, one that has become indispensable in wildlife law enforcement. (Unfortunately, space limitations do not permit the printing of the 40 additional pages of illustrations in that report.)

It has long been traditional for distributional records of wildlife to be based on properly identified and documented museum specimens. In recent years, especially where birds are concerned, it has been increasingly accepted that such records may be based on identifiable photographs. Several records of this sort are included in Lawson's paper. It is still usual for the identity of the subjects of these photographs to be verified by comparison with museum specimens, and often the pictures are stored in a museum for future reference. Unfortunately, because of delusive qualities of light or an odd posture of the subject, it is not always possible to make unassailable identifications of photographs. Subspecific identification is seldom possible, and much valuable information available from a specimen is not available from a photo.

Properly identified museum specimens remain the standard against which critical comparisons are made. John Farrand's paper illustrates one way in which the information contained in these specimens can be made available to workers in the field.

Another way in which museum collections assist field identification is by providing specimens to serve as models for artist's illustrations. Nearly every illustrated field guide to birds, mammals, reptiles, or fishes includes an acknowledgment to one or more museums whose specimens the artist has consulted.

The identification of animal remains found in stomach contents, pellets, feces, and in and around nests and dens has been the basis of food-habits studies for decades. The paper by Fitch and Brownell is but one of many possible examples. The need in such studies of ample comparative material as is found in museum collections, albeit sometimes only very specialized ones, is immediately apparent, even though the three plates accompanying this paper are omitted.

Identification of material from kitchen middens yields information on former ranges and abundance and on use of wildlife resources by native peoples, of interest to both biologists and anthropologists.

Additional Readings

Devillers, Pierre. 1970. Identification and distribution in California of the *Sphyrapicus varius* group of Sapsuckers. *California Birds* 1: 47-76.

Harrison, Bill. 1976. Plain-capped Starthroat in Arizona. *Birding* 8: 74-75.

Hunn, Eugene. 1973. First record for the swamp sparrow in Washington state. *Western Birds* 4; 31-32.

Laybourne, Roxie C. 1974. Collision between a vulture and an aircraft at an altitude of 37,000 feet. *Wilson Bull.* 86: 461-462.

Nelson, Colleen Helgeson. 1976. A key to downy cygnets with analysis of plumage characters. *Wilson Bull.* 88: 4-15.

Phillips, Allan R. 1974. The need for education and collecting. *Bird-Banding* 45: 24-28.

Williams, Cecil S. 1938. Aids to the identification of mole and shrew hairs with general comments on hair structure and hair determination. *Jour. Wildl. Mgt.* 2: 239-250.

Reprinted from *The Wilson Bulletin*

ALTITUDE RECORD FOR MALLARD

Richard H. Manville

Recently the Operations Division of the Air Transport Association submitted for identification one feather salvaged after an aircraft-bird strike. It was a right primary, in good condition, and was determined by Mrs. Roxie C. Laybourne, of our Bird and Mammal Laboratories, as coming from a Mallard (*Anas platyrhynchos*).

The strike occurred at 4:15 PM on 9 July 1962, between Battle Mountain and Elko, Nevada. The plane, a Western Airlines L–188 Electra, was cruising at an air speed of 345 knots and an altitude of 21,000 feet. The bird was not seen by any of the crew, all of whom were looking outside at the time. The pilot, Captain Markle Sparks of Los Angeles, believed the bird was at a higher altitude than the plane, that it attacked the plane as it approached, and lost control just before impact. He suspected an eagle as the only bird in that area that could climb so high.

The pilot reported that he "felt a light thud and about a minute later the stewardess came forward and said that the passengers in the rear of the airplane felt a small explosion. Upon landing, there was a dent in the leading edge of the right horizontal stabilizer about the size of a football. It hit just at the side of a rib and there was a tear approximately 9 inches long beside the rib. There was down around the tear, and the feather inside. There was no blood nor any other indication that it was a bird that we had struck." The strike did not cause any immediate change in the flight procedure, but did necessitate replacing a 5-foot section of skin on the tail assembly.—RICHARD H. MANVILLE, *U.S. Fish and Wildlife Service, Washington 25, D.C., 17 October 1962.*

Reprinted from *Wildlife Research 1971*

IDENTIFICATION SERVICES

The staff of the Bird and Mammal Laboratories identify specimens, portions of animals, and fragmentary animal remains for other units of the Bureau, other Federal and State agencies, professional colleagues, and private individuals. An increasing proportion of this work concerns the identification of endangered wildlife species received by Federal Game Management Agents at ports of entry. Much of the avian material submitted by the agents consists of dyed or otherwise altered feathers intended for use as decorative material. Mammals were identified from photographs, pelts, and bundles of hair. In 1970, 125 identifications related to wildlife or their products entering the United States.

Barn owl pellets submitted by the Welder Wildlife Foundation contained remains of 338 birds of 19 species, ranging in variety from rails to swallows. These identifications were made in support of a study of nesting and feeding habits of an owl population in an unusual and fluctuating habitat in Texas. Fragmentary remains from 58 birds, ranging in size from warbler to goose, and two mammals (bats) struck by aircraft were identified for the Federal Aviation Agency and the Air Force. Several lots of material, both birds and mammals, were identified for the Federal Bureau of Investigation. Specimens identified for professional colleagues at other institutions totaled 589.

Ornithologists of the Bird and Mammal Laboratories identify the remains of many birds in collision with aircraft. In this case the bird was a red-tailed hawk. (U.S. Army Photo)

Reprinted from *Zoologica*

2

Species Identification of Commercial Crocodilian Skins

F. WAYNE KING[1] AND PETER BRAZAITIS[2]

(Figures 1-41)

Gross similarities in the morphology of crocodilian skins has made specific identification of individual commercial hides extremely difficult. Qualitative and quantitative differences between the hides of various species are defined and form the basis of a key to commercial crocodilian hides. The distribution, common, and commercial names, distinguishing characteristics of the hides, and the status of the wild populations of each of the 27 species and subspecies is given.

INTRODUCTION

TO THE LAYMAN most crocodilians look similar – all have relatively long toothy snouts, scaly backs, flattened oar-like tails, large webbed hindfeet, and most have crossbanded color patterns. As a consequence, most Americans mentally lump all crocodilians under the collective heading "alligator." Profound differences exist between the species of crocodilians, but the untrained eye notices the gross similarities rather than the less obvious differences. When it comes to identifying a species of crocodilian from a commercial hide, however, even a trained herpetologist faces serious difficulty. All commercial skins are grossly alike. All crocodilian leather is retailed throughout the United States as "alligator," while in Europe, Africa, and Asia the same hides are sold as "crocodile."

In this paper, we attempt to provide means to identify commercial crocodilian hides. Since the paper will be read by layman, trained herpetologist, government inspector, and commercial dealer alike, we have endeavored to use terminology comprehensible to all. Where there is a chance of confusion, we have provided photographs and line drawings for clarification.

MATERIALS AND METHODS

Comparisons were made between the skins of live specimens in zoos and private collections, preserved specimens and dried skins in museum collections, and tanned and finished commercial skins supplied by the Reptile Products Association of the United States. A total of over 350 specimens were examined. Museum specimens of every species and subspecies were studied. Living specimens of every form, except *Caiman crocodilus apaporiensis* and *Crocodylus siamensis*, were seen. Commercial hides of most species were examined. The notable exceptions were *Gavialis gangeticus, Alligator sinensis, Paleosuchus palpebrosus, Paleosuchus trigonatus, Crocodylus palustris*, and *Crocodylus rhombifer*. The raw data are on deposit at the New York Zoological Park. The characters and terms used in this study are defined below.

COMMERCIAL HIDES. Hides used in the crocodilian hide trade for the manufacture of leather goods are termed commercial hides, whether they are raw skins or are in the process of being tanned and finished.

HORNBACK HIDES. Rough dorsal (back) skins obtained by skinning the animals beginning from an incision made along the midventral (belly) line. Large bony dorsal scales, usually with raised keels, occupy the center of the hide. Smooth squarish scales from the ventral surface are located along the lateral edges of the hide. Hornback hides usually are skinned from relatively small specimens since the heavily ossified dorsal scales of adults make their hides stiff and limits its use for leather. Skin from the tail and proximal portion of the legs is attached to the hide (figure 1).

BELLY HIDES. Smooth ventral (belly) skins obtained by skinning the animal beginning at an incision just below the large bony dorsal scales high on one side and continuing down the side, under the body, and up the other side to the

[1] Curator of Herpetology, New York Zoological Park, Bronx, New York 10460.

[2] Assistant Animal Manager, Department of Herpetology, New York Zoological Park, Bronx, New York 10460.

edge of the large dorsal scales on that side of the back. The scales of belly skins are squarish or rectangular in shape over most of the center of the skin. The round or oval scales from the side of the body are located along the lateral edges of the hide. The vent is represented by a hole along the midline of the skin. Skin from the tail and proximal portion of the legs is attached (figure 1).

BUTTON-BELLY AND SOFT-BELLY HIDES. Belly skins that possess osteoderms, or buttons (the commercial name), are called button hides. Belly skins that lack osteoderms are called soft-belly hides. Button-belly hides will not flex through the middle of a scale because of the osteoderm button. Soft-belly hides not only flex between scales but also bend to a lesser degree through the middle of scales (figure 2). As a result the most sought after hides are soft-bellies. Button-bellies with large osteoderms are most frequently used for making the flat, non-flexing sides of purses or attaché cases. The stiffness of these skins limits their use where flexibility is demanded, as in shoes, belts, and watch-bands.

SIDES. A narrow strip of soft skin taken from a point under the lower jaw and extending back over the front leg, along the side of the trunk, under the rear leg and ending near the vent (figure 15). Such strips are skinned from large caimans (*Caiman, Melanosuchus, Paleosuchus*) which have such heavy osteoderm buttons in the belly skin, as to make this skin almost worthless as leather. The hide-hunters avoid both the bony dorsal scales and the bony belly scales by removing only the soft side skin. The scales on sides consist of large, round to diamond-shaped scales separated by soft skin or small scales. The largest scales usually possess osteoderm buttons and may have a low keel.

Sides could, of course, be cut from soft-bellied alligators and crocodiles, as well as button-bellied caimans. The reason they are not is that soft-belly hides are worth more with the sides attached than they are with the strips removed. The best tanned soft-belly hides sell for as much as $12.00 per square foot. Tanned sides sell for only $5.00 to $15.00 apiece.

THROATS, SIDES, AND GIRDLES. Throats are V-shaped pieces skinned from under the chin and the sides of the neck. Girdles are taken from the thighs and belly immediately anterior to the vent. Sides, which accompany throats and girdles, are stripped from the sides of the trunk between the legs (figure 20). These cuts of skin are from exceptionally large caimans (*Melanosuchus*, and possibly *Caiman*) which have heavy bony belly osteoderms. The large size of the scales on these pieces attests to the size of the animals they come from.

FLIPPERS. The small, irregular-shaped pieces of skin from the legs. The scales are usually uniformly small, smooth, and squarish.

RAW HIDES OR SALTED HIDES. Untanned commercial hides. They may be dry or moist from the salt. They usually are rolled up for shipment, and retain the color pattern of the live specimen – most frequently dark spots or dark crossbands on the side of the trunk and tail.

CRUSTS. Hides which have been tanned, but not dyed or polished. Crusts are usually ash gray or tan, and have a dull, unpolished finish (figures 9, 10, and 15). The next step in the finishing process is to bleach or dye the hide its final color. If it is a button-belly, the osteoderms are shaved from the inside of the hide to eliminate as much of them as possible. It is not possible to remove every osteoderm button in its entirety, but it is possible to remove enough of them to make a stiff hide much more flexible (figures 14 and 19). Unskilled tanners may shave the hide too closely and leave thin, weak sections between the rows of scales.

POLISHED AND FINISHED HIDES. After dyeing, skins customarily are given a high-gloss finish by burnishing the scales under the pressure of a polishing wheel. The glossy finish is characteristic of most crocodilian products. After the hide has been finished, it is ready for cutting into the pieces that are to be made into the manufactured product.

FLAT FINISH AND BOMBE FINISH. Finished skins with flat, level scales are flat finished skins. Those in which the individual scales are slightly curved, with the center of each one arching up from the crease where it meets adjacent scales, are given the French name *bombe*.

SAUVAGE FINISH. Not all finished hides have a highly polished surface. Some are finished with a process that retains much of the texture of the original crust. The result is a textured, non-glossy, oiled-leather appearance called *sauvage* (figure 9).

VENTRAL SCALES AND LATERAL SCALES. Scales from the underside of the throat, body, and tail are ventral scales (figure 1). They are large and squarish in all crocodilians. The individual square scales are in contact with adjacent scales and are arranged in rows across the belly of the animal. Scales from the side of the body are lateral scales (figure 1).

They frequently are arrayed in two distinct size classes, the larger composed of oval and diamond-shaped scales. These individual scales usually are not in contact with adjacent scales.

Ventral and lateral scales possess a number of characteristics which may be used to identify a particular species or group of species. Most commercial skins used in the United States are either belly skins or sides, and since even hornback hides normally retain ventral scales along their edges, only the ventral and lateral scales are considered in this paper.

BUTTONS. The bony osteoderms present in the scales of many crocodilians. The presence of buttons in the ventral scales of belly skins gives them their commercial name, button-belly hides. Buttons can be seen if the hide is turned over and the backside (inside) examined. Even though they are decalcified during tanning, the osteoderm buttons are a different color. In crusts and light-colored finished skins, buttons are usually slightly darker than the rest of the skin (figures 13 and 14). In dark finished skins, they may be lighter than the surrounding tissue, although they are usually darker (figures 19, 34, 35, 36, and 37).

In many cases buttons also can be detected from the front side (outside) of a finished skin as ill-defined light-colored blotches in the center of the ventral scales (figures 18 and 34). They may also be evident as raised or slightly sunken areas on the surface of the ventral scales. They can best be seen by holding the polished surface of the scale in a position where light reflected off its surface reaches your eye. In this position any surface irregularity becomes apparent (figures 35 and 41).

SURFACE PITTING. During the tanning process, the bony osteoderms do not shrink as much as the non-calcified tissue. As a result, button skins may exhibit some of the pockmarked or wrinkled texture of the underlying osteoderms. This condition is called surface pitting. Surface pitting is best seen where the osteoderms are most heavy, as in the dorsal scales of hornback hides, and the ventral scales of the skins of caiman (*Caiman, Melanosuchus, Paleosuchus*), slender-snouted crocodile (*Crocodylus cataphractus*), dwarf crocodile (*Osteolaemus*), or button-belly Nile crocodile (*Crocodylus niloticus*) (figures 10, 11, 12, 18, 36, and 37).

The skins of some species are more strongly pitted than others; adult specimens tend to be more pitted than juveniles; and not all individuals exhibit the same degree of pitting throughout all their scales.

SINGLE BUTTONS AND DOUBLE BUTTONS. Those crocodile and alligator specimens which have osteoderms are characterized by having single buttons, one osteoderm button per scale (figures 8, 34, 36, and 37), while all caimans have double buttons, two osteoderms per scale (figures 13, 14, and 19). In caimans, the posterior of the two buttons occupies almost the entire area of the scale. The anterior button occupies the anterior one-quarter of the scale and curves upward and inward (mesially) to protectively overlap the posterior margin of the next anterior scale and the intervening soft skin. The overlapping of osteoderms is an evolutionary adaptation that affords more protection to the weak hinge point between the scales, although it causes the loss of some freedom of movement. The dwarf caimans (*Paleosuchus*), with proportionally the largest double buttons of any species of crocodilian, are well on the way to evolving the stiff analog of a turtle plastron. The two parts of the caiman double button are more evident before the buttons are shaved during finishing (figure 14). During shaving, the anterior, inward-curving button is nearly completely removed. In most cases, however, part of the anterior button remains even after the shaving is complete (figure 19).

FOLLICLE GLAND. Belly scales of gavials and all crocodiles have a pit-like structure, the follicle gland, near the posterior margin (figures 22 and 24). The glands are clearly visible in live specimens, raw skins, and crusts. In finished skins dyed dark, the follicle glands may be lighter in color (figure 25). The glands may be partly obscured during the polishing process. If this happens, the gland usually can be seen as a short deep wrinkle running from the gland to the posterior edge of the scale (figure 35). In finished skins, follicle glands are detected most easily on the throat and in the area just anterior to the vent as well as on the underside of the tail (figures 28 and 30).

SPIDER-WEB UMBILICUS. Alligator hides have neither surface pitting nor follicle glands. The ventral scales of finished alligator skins are glossy smooth without ornamentation. It is also possible to distinguish the belly skin of the American alligator (*Alligator mississippiensis*) from all other species on the basis of the shape of the umbilicus scar. Crocodilians with buttons in the ventral scales tend to lose all evidence of the umbilicus once it is healed and the osteoderms are formed. Other species may retain the posterior one-third of the scar as a zigzag arrangement of small scales scattered along the midline just anterior to the vent. In the American alligator, and no other species, this same posterior portion of the scar remains as an area of soft skin lacking scales, and because of the profusion of creases and lines it has a distinctly spider-web appearance (figure 7).

VENTRAL COLLAR. Most crocodilians have a prominent row of enlarged scales, called a ventral collar, across the throat just anterior to the front legs (figure 1). A few species lack an enlarged row of scales, so the collar is not conspicuous. One species has a double collar, two enlarged rows of scales.

TRANSVERSE VENTRAL SCALE ROWS. Ventral scales are arranged in transverse rows in all crocodilians, but the number of rows found between the neck and vent differs between species. The transverse rows of ventral scales are counted from the first row posterior to the ventral collar to, but not including, the row of scales encircling the vent (figure 1). The row of scales around the vent may be missing from a commercial hide because the hide-hunter was careless when skinning the animal. In that event the position of the missing rows must be estimated. Only rows which cross the midline are counted. Incomplete or missing rows will add confusing variation to the count, so to eliminate doubt, the count first should be made only to the right of the midline and then repeated on the left side, and the two counts compared.

LARGE-SCALE AND SMALL-SCALE HIDES. Soft-belly crocodilians which have 26 to 35 transverse ventral scale rows are called small-scale hides by the hide trade. Soft-belly species with 20 to 25 transverse ventral scale rows are called large-scale hides (see the key that follows and figure 26).

TAIL WHORLS. The transverse rows of scales under the tail are the ventral portions of the whorls of scales that completely encircle the tail. The ventral portion of these whorls, like the rows of ventral scales on the body, are usually complete and evenly arranged (figure 28). Morelet's crocodile (*Crocodylus moreletii*), however, possesses irregular or incomplete whorls 66 percent of the time (figure 30). No other species shows as high an incidence of irregularity in this character.

Identification of Crocodilian Hides

The following keys can be used to identify the species, or species groups, of commercial belly skins, hornbacks, and sides. The keys are of limited use in identifying throats and girdles, and are useless for flippers. They may be useless in identifying skins already manufactured into finished products.

Keys are identification tools which employ a series of alternative choices. To use the keys, first decide whether or not the hide you wish to identify is a side, belly, or hornback. Once this determination has been made, proceed to the appropriate key. Each set of alternative choices, or couplets, is numbered. Starting with couplet 1, decide which of the two choices, "a" or "b," best describes the hide to be identified. The number that follows the correct choice indicates the next couplet. By moving from couplet to couplet following the numbers shown after each correct choice, you will arrive at a final choice which indicates the species, or species group, of crocodilian from which the hide was taken. Once the identification has been made, you should turn to the text that follows the keys for information on the distribution of the species, the commercial names under which it is sold, additional distinguishing characteristics, and status of the wild populations.

Species identifications supplied by manufacturers are not to be relied on until verified by means of the keys. In the past two years, the authors have seen live African slender-snouted crocodiles and South American caimans shipped into the United States from Bangkok, Thailand, as Siamese crocodiles; finished African dwarf crocodile hides enter from a tanner in France who labelled them gavial; and wallets made from South American caimans arrive from an Italian manufacturer who declared they were Nile crocodile.

A Key to Commercial Caiman Sides

The key to sides is based on the assumption that the sides being identified are from caimans (*Caiman*, *Melanosuchus*, or *Paleosuchus*), and not from other species. At the present time, caimans are the only crocodilians being skinned in this manner. This may not be the case at some future date. In addition, small finished products such as belts may be pieced together from scrap left over from the manufacture of large belly hide products. These small pieces may come from any species, therefore, the key is of little use in identifying pieced items.

1. a) Rows of large oval scales alternating with rows of small scales (figure 21) . . . *Melanosuchus niger*.
   b) Rows of large scales alternating with network of creases and small irregular scales (figure 16) . . . 2

2. a) Large oval scales, usually smooth, and arranged in distinct rows . . . *Caiman crocodilus* (four subspecies), *Caiman latirostris*.

b) Large oval scales usually keeled, and usually not arranged in distinct rows . . . *Paleosuchus palpebrosus, Paleosuchus trigonatus.*

A KEY TO CROCODILIAN BELLY SKINS AND HORNBACK SKINS

This key is for use with belly skins. Hornback hides can be identified if you limit your attention to the belly scales found along the lateral edges of the hide. Surface pitting is not evident in untanned hides.

1. a) Ventral (belly) scales with follicle glands (figures 22 and 24) . . . 2
   b) Ventral (belly) scales without follicle glands (figures 4, 11, and 18) . . . 4
2. a) Osteoderm buttons present (figures 34 through 37) . . . *Crocodylus cataphractus, Crocodylus niloticus, Osteolaemus tetraspis* (two subspecies).
   b) Osteoderm buttons not present (figure 27) . . . 3
3. a) Transverse rows of ventral scales 20 to 25 . . . *Crocodylus acutus* (south of Panama), *Crocodylus intermedius, Crocodylus johnsoni, Crocodylus novaeguineae* (two subspecies), *Tomistoma schlegelii.*
   b) Transverse rows of ventral scales 26 to 35 . . . *Crocodylus acutus* (north of Panama), *Crocodylus moreletii, Crocodylus niloticus, Crocodylus palustris* (two subspecies), *Crocodylus porosus, Crocodylus rhombifer, Crocodylus siamensis, Gavialis gangeticus.*
4. a) No osteoderm buttons present in midbelly (figure 6), or single buttons present (figure 8) . . . 5
   b) Double osteoderm buttons present in midbelly (figures 14 and 19) . . . 6
5. a) Umbilicus scar has spider-web appearance (figure 7); transverse rows of ventral scales 29 or more . . . *Alligator mississippiensis.*
   b) Umbilicus scar not evident or lacks spider-web appearance; transverse rows of ventral scales 28 or fewer . . . *Alligator sinensis.*
6. a) Large osteoderm buttons present medially only, not over pelvic girdle (figure 19); surface pitting slight; transverse rows of ventral scales 25 or more . . . *Melanosuchus niger.*
   b) Large osteoderm buttons in all large ventral scales, throat to pelvis (figure 13); surface pitting slight to pronounced; transverse rows of ventral scales 18 to 30 . . . 7
7. a) Surface pitting pronounced; transverse rows of ventral scales 20 to 30 . . . 8
   b) Surface pitting slight or absent; transverse rows of ventral scales 18 to 22 . . . *Paleosuchus palpebrosus, Paleosuchus trigonatus.*
8. a) Transverse rows of ventral scales 26 to 30; double ventral collar . . . *Caiman latirostris.*
   b) Transverse rows of ventral scales 20 to 27; single ventral collar . . . *Caiman crocodilus* (four subspecies).

In the text that follows, the species and subspecies are listed alphabetically by scientific name within each family. The systematic arrangement follows Wermuth and Mertens (1961).

Family ALLIGATORIDAE

AMERICAN ALLIGATOR

Alligator mississippiensis (Daudin)

DISTRIBUTION. Southeastern United States – the states of North Carolina, South Carolina, Georgia, Florida, Alabama, Mississippi, Louisiana, Arkansas, and Texas. This species does not occur outside the United States (Schmidt, 1953; U.S. Department of Interior, 1968; Wermuth, 1953; Wermuth and Mertens, 1961; Werner, 1933).

OTHER COMMON NAMES. It is also called the Florida and the Mississippi alligator, or gator. Hides are marketed as American, Florida, or Louisiana alligator or soft-belly.

DISTINGUISHING CHARACTERISTICS OF COMMERCIAL HIDES. *Belly skins:* No follicle glands. No osteoderm buttons (large specimens from Florida have single osteoderm buttons on the throat). Transverse ventral scale rows 29 or more. Umbilicus scar prominent and with spider-web appearance. Maximum length of live specimen is 18 feet.

STATUS OF WILD POPULATIONS. Endangered (Honegger, 1968; Pan American Union, 1967; U.S. Department of Interior, 1968). Now protected by state law in every state in which it occurs; by federal prohibition on interstate traffic in illegal hides; and by local and state prohibitions on sales of live specimens, hides, and hide products.

CHINESE ALLIGATOR

Alligator sinensis Fauvel

DISTRIBUTION. The lower Yangtze River drainage of China (Pope, 1935; Wermuth, 1953; Wermuth and Mertens, 1961; Werner, 1933).

OTHER COMMON NAMES. In China it is called *T'o, Tou Lung, Yow Lung.*

DISTINGUISHING CHARACTERISTICS OF COMMERCIAL HIDES. *Belly skins:* No follicle glands. Usually no osteoderm buttons, but occasionally single buttons may be present in the midbelly and collar areas. Transverse ventral scale rows 28 or fewer. Umbilicus scar not evident. Maximum length of live specimen is 6½ feet.

STATUS OF WILD POPULATIONS. Probably endangered (Honegger, 1968). This species is known from a strip of territory only a few hundred miles long. A. F. Oeming of the Alberta Game Farm, Canada, recently returned from a trip to China and reported (in *litt.*) that the species is totally protected by law and the law is rigidly enforced. Dr. Cheng, of the Institute of Zoology, Academia Sinica, Peking, is studying the species.

SOUTH AMERICAN CAIMAN

Caiman crocodilus crocodilus (Linnaeus)

DISTRIBUTION. Northern South America – Colombia east of the Andes, Peru, Ecuador, Venezuela, Guyana, Surinam, French Guiana, Trinidad, and, with the exception of a few southern tributaries, the Amazon drainage of Brazil [the exceptions are listed under *Caiman crocodilus yacare*] (Carvalho, 1955; Medem, 1968; Schmidt, 1928b; Wermuth, 1953; Wermuth and Mertens, 1961; Werner, 1933).

OTHER COMMON NAMES. In the United States it is frequently called spectacled caiman (and equally frequently given the old synonymous scientific name *Caiman sclerops*). In Central and South America it is called "alligator," *baba, babilla, cachirré, caimán, caimán blanco, caimán del Paraguay, cascarudo, cocodrillo, jacaré, jacaré de Lunetas, jacaretinga, lagarto, lagarto blanco, lagarto negro, ocoroche, tinga,* and *yacaré blanco.* Hides are frequently marketed under these names.

DISTINGUISHING CHARACTERISTICS OF COMMERCIAL HIDES. *Belly skins:* No follicle glands. Full double osteoderm buttons present. Surface pitting evident. Transverse ventral scale rows 20 to 24. Single prominent ventral collar. Highest point of ventral scale slightly anterior of center (figure 39). *Sides:* Rows of large oval scales alternating with network of small irregular scales and creases. (The network is actually soft skin folds and creases without scales.) Maximum length of live specimen is 8½ feet.

STATUS OF WILD POPULATIONS. Most wild populations are declining and some have all but disappeared due to slaughter by hide hunters and capture by live animal collectors (Pan American Union, 1967). The subspecies is considered endangered by some experts (Honegger, 1968). South American countries require that hides be tanned before export. Colombia protects specimens less than 1.2 meters in length (Medem, 1970, in *litt.*) and Peru protects those less than one meter long (Crowe, 1965).

RIO APAPORIS CAIMAN

Caiman crocodilus apaporiensis Medem

DISTRIBUTION. Colombia – known only from the Apaporis River and its tributaries between the Falls of Jirijirimo and Puerto Yaviya (Medem, 1955, 1968; Wermuth and Mertens, 1961).

OTHER COMMON NAMES. In Colombia it is called *babilla, cachirré, cocodrillo, jacaretinga, lagarto negro* and *ocoroche.* If marketed, hides would be sold under these names, or as *tinga.*

DISTINGUISHING CHARACTERISTICS OF COMMERCIAL HIDES. *Belly skins:* No follicle glands. Full double osteoderm buttons present. Surface pitting evident. Transverse ventral scale rows 20 to 24. Single prominent ventral collar. Highest point of ventral scale is slightly anterior of center (figure 39). *Sides:* Rows of large oval scales alternating with network of small irregular scales and creases. Maximum length of live specimen is 7 feet.

STATUS OF WILD POPULATIONS. Critically endangered. This subspecies has the most restricted range of any crocodilian. It is known only from an area 125 miles long in one river. Hide hunters can completely decimate this form in one or two years unless hunting is prohibited immediately. Colombia prohibits the export of untanned hides and protects specimens less than 1.2 meters in length (Medem, 1970, in *litt.*).

BROWN CAIMAN

Caiman crocodilus fuscus (Cope)

DISTRIBUTION. Central America – southern Mexico to Colombia, west of the Andes (Medem, 1968; Schmidt, 1928b; Smith and Taylor, 1950; Wermuth, 1953; Wermuth and Mertens, 1961; Werner, 1933).

OTHER COMMON NAMES. In the United States it is also called spectacled caiman, Central

American caiman, dusky caiman, and Magdalena caiman. In Central America it is called "alligator," *caimán, cocodrillo,* and *cuajipal.* In Colombia it is known as *babilla, lagarto negro,* and *jacaretinga.* Hides are marketed under these names, and Central American *tinga.*

DISTINGUISHING CHARACTERISTICS OF COMMERCIAL HIDES. *Belly skins:* No follicle glands. Full double osteoderm buttons present. Surface pitting evident. Transverse ventral scale rows 20 to 24. Single prominent ventral collar. Highest point of ventral scale slightly anterior of center (figure 39). *Sides:* Rows of large oval scales alternating with network of small irregular scales and creases. Maximum length of live specimen is 7 feet.

STATUS OF WILD POPULATIONS. Many wild populations are disappearing due to hide-hunting (Pan American Union, 1967). The subspecies is considered endangered by some experts (Honegger, 1968). South American countries prohibit the export of untanned hides. Colombia and Panama protect specimens less than 1.2 meters in length (Medem, 1970, in *litt;* D. Tovar, 1970, in *litt.*). Peru protects specimens less than 1.5 meters in length (Honegger, 1968). Mexico's laws regulate hunting of this species.

YACARE

Caiman crocodilus yacare (Daudin)

DISTRIBUTION. Southern South America – specifically the Paraguay and Parana river drainage systems of Paraguay, Uruguay, Argentina, and Brazil, and the southern tributaries of the Amazon in Bolivia [the Mamore, Itenez, and Beni drainages] and Brazil [the Guapore drainage, and the Araguaia River above its confluence with the Tapirape] (Carvalho, 1955; Medem, 1968; Schmidt, 1928b; Wermuth, 1953; Wermuth and Mertens, 1961; Werner, 1933).

OTHER COMMON NAMES. It is also called the Paraguay caiman and red caiman in the United States. In South America it is called *caimán del Paraguay, cascarudo, jacaré, jacaré de Lunetas, jacaretinga, lagarto, tinga, yacaré,* and *yacaré de hocico angosto.* Hides are marketed under these names.

DISTINGUISHING CHARACTERISTICS OF COMMERCIAL HIDES. *Belly skins:* No follicle glands. Full double osteoderm buttons present. Surface pitting evident. Transverse ventral scale rows 20 to 25. Single prominent ventral collar. Highest point of ventral scale is slightly anterior of center (figure 39). *Sides:* Rows of large oval scales alternating with network of irregular shaped small scales and creases. Maximum length of live specimen is 8 feet.

STATUS OF WILD POPULATIONS. Endangered (Pan American Union, 1967; U.S. Department of Interior, 1970). Most wild populations declining in numbers (Jose Cei, 1970, in *litt.*). South American countries prohibit the export of untanned hides. Its import is prohibited under provisions of the Endangered Species Conservation Act (U.S. Department of Interior 1970).

BROAD-SNOUTED CAIMAN

Caiman latirostris (Daudin)

DISTRIBUTION. Southern South America – the drainages of the Paraguay, Parana, and Uruguay rivers in Argentina, Uruguay, Paraguay, and Brazil, and the rivers emptying into the southeast coast of Brazil south of Recife (Carvalho, 1955; Medem, 1968; Schmidt, 1928b; Wermuth, 1953; Wermuth and Mertens, 1961; Werner, 1933).

OTHER COMMON NAMES. In South America it is called *jacaré de Papo Amarelo, overo, ururau,* and *yacaré de hocico ancho.* Hides are marketed under these names.

DISTINGUISHING CHARACTERISTICS OF COMMERCIAL HIDES. *Belly skins:* No follicle glands. Full double osteoderm buttons present. Surface pitting slight or absent. Transverse ventral scale rows 26 to 30. Double ventral collar. Highest point of ventral scale located at center of scale. *Sides:* Rows of large oval scales alternating with network of irregular shaped small scales and creases. Maximum length of live specimen is 9 feet.

STATUS OF WILD POPULATIONS. Endangered (Pan American Union, 1967). This species is nearly extinct from excessive hide hunting (Jose Cei, 1970, in *litt.*). South American countries prohibit export of untanned hides.

BLACK CAIMAN

Melanosuchus niger (Spix)

DISTRIBUTION. Northern and central South America – Amazon basin drainages of Brazil, Colombia, Venezuela, Guyana, Peru, and Bolivia (Carvalho, 1955; Medem, 1963, 1968; Schmidt, 1928b; Wermuth, 1953; Wermuth and Mertens, 1961; Werner, 1933).

OTHER COMMON NAMES. In South America it is called *asu, caimán, caimán negro, cocodrillo, jacaré açú, jacaré assú, jacaré asú, jacaré uassú, jacaré una,* and *yacaré assú.* Hides are marketed under these names.

DISTINGUISHING CHARACTERISTICS OF COMMERCIAL HIDES. *Belly skins:* No follicle glands. Full double osteoderm buttons, at least medially. Lateral scales may lack osteoderms or possess small osteoderms in center of scales. Surface pitting slight. Transverse ventral scale rows 25 to 28. *Sides:* Parallel rows of large oval scales alternating with rows of small oval scales. Maximum length of live specimen is 16 feet.

STATUS OF WILD POPULATIONS. Endangered (Honegger, 1968; Pan American Union, 1967). Rapidly declining everywhere, and exterminated in many areas. South American countries prohibit the export of untanned hides. Peru prohibits the killing of specimens less than 2 meters in length (Honegger, 1968).

DWARF CAIMAN

Paleosuchus palpebrosus (Cuvier)

DISTRIBUTION. Northern and central South America – Amazon and Orinoco river drainages of Colombia, Venezuela, Guyana, Brazil, Peru, Ecuador, and Bolivia (Carvalho, 1955; Medem, 1967, 1968; Schmidt, 1928b; Wermuth, 1953; Wermuth and Mertens, 1961; Werner, 1933).

OTHER COMMON NAMES. It is also called musky caiman and Cuvier's smooth-fronted caiman. In South America it is called *cachirré, jacaré coroá,* and *yacaré coroá.* Hides are marketed under these names.

DISTINGUISHING CHARACTERISTICS OF COMMERCIAL HIDES. *Belly skins:* No follicle glands. Full double osteoderm buttons on all large ventral scales. Surface pitting slight or absent. Transverse ventral scale rows 18 to 22. Single prominent ventral collar. *Sides:* Large scales scattered, not in well-defined rows, and separated by wide areas of soft skin. Maximum length of live specimen is 5½ feet.

STATUS OF WILD POPULATIONS. Declining in numbers (Pan American Union, 1967). *Paleosuchus* is possibly the least persecuted of the crocodilians at the present time. Its small size and heavy osteoderm buttons make the skins less desirable than skins from the larger caimans and crocodiles of South America. South American countries prohibit the export of untanned hides.

SMOOTH-FRONTED CAIMAN

Paleosuchus trigonatus (Schneider)

DISTRIBUTION. Northern and central South America – the Amazon and Orinoco river drainages of Colombia, Venezuela, Guyana, Brazil, Ecuador, Peru, and Bolivia (Carvalho, 1955; Medem, 1967, 1968; Schmidt, 1928b; Wermuth, 1953; Wermuth and Mertens, 1961; Werner, 1933).

OTHER COMMON NAMES. It is also called Schneider's smooth-fronted caiman. In South America it is called *cachirré, jacaré coroá, jacaré curuá,* and *yacaré coroá.* Hides are marketed under these names.

DISTINGUISHING CHARACTERISTICS OF COMMERCIAL HIDES. *Belly skins:* No follicle glands. Full double osteoderm buttons on all large ventral scales. Surface pitting slight or absent. Transverse ventral scale rows 18 to 22. Single prominent ventral collar. *Sides:* Scattered large keeled oval scales, not in well-defined rows, and separated by wide areas of soft skin. Maximum length of live specimen is 7 feet.

STATUS OF WILD POPULATIONS. Declining in numbers (Pan American Union, 1967). *Paleosuchus* is possibly the least persecuted of the crocodilians. Its small size and heavy ossification of the osteoderms makes the skins less desirable than skins from the larger caimans and crocodiles of South America. South American countries prohibit the export of untanned hides.

Family CROCODYLIDAE

AMERICAN CROCODILE

Crocodylus acutus Cuvier

DISTRIBUTION. Florida, West Indies, Central and northern South America – southern Florida, Cuba, Hispaniola (Haiti and Dominican Republic), Jamaica, Mexico south to Colombia and Venezuela, exclusive of the Orinoco river drainage system (Cochran, 1941; Medem, 1968; Smith and Taylor, 1950; Wermuth, 1953; Wermuth and Mertens, 1961; Werner, 1933).

OTHER COMMON NAMES. In Central America and Cuba it is called *caimán,* and in South America it is known as *caimán* and *caimán de aguja.* Hides may be marketed under these names, or simply as Central or South American "alligator," crocodile, soft-belly, small scale (north of Panama) or large scale (south of Panama).

DISTINGUISHING CHARACTERISTICS OF COMMERCIAL HIDES. *Belly skins:* Follicle glands present. No osteoderm buttons. Transverse ventral scale rows 25 to 35. Tail whorls regular ventrally. Maximum length of live specimen is 23 feet.

STATUS OF WILD POPULATIONS. Declining everywhere due to excessive hide-hunting (Pan American Union, 1967). The species is considered endangered by some experts (Honegger, 1968). Many populations in Central and South America have been totally exterminated. The species is protected by state law in Florida, and South American countries prohibit the export of untanned hides. Mexico regulates the hunting of the species, as does Nicaragua. Jamaica prohibits the export of crocodiles, their eggs, or skins (K.C. Hall, 1970, in *litt.*). The species is protected in Cuba and Colombia, although the law is not enforced in the latter (Honegger, 1968).

AFRICAN SLENDER-SNOUTED CROCODILE

Crocodylus cataphractus Cuvier

DISTRIBUTION. Western and central Africa – the Congo, Niger, and Volta river drainages, and the coastal rivers from Senegal south to northern Angola. Only once recorded from East Africa at Ujiji, Tanzania, on Lake Tanganyika (Schmidt, 1919; Wermuth, 1953; Wermuth and Mertens, 1961; Werner, 1933).

OTHER COMMON NAMES. It is sometimes called the West African crocodile, African long-nosed crocodile, African gavial, or sub-water crocodile. Hides are sold under these names or as Nigerian, Congo, or Cabinde "alligator," crocodile, or button hides.

DISTINGUISHING CHARACTERISTICS OF COMMERCIAL HIDES. *Belly skins:* Follicle glands present. Round or elliptical single osteoderm buttons present. Surface pitting may or may not be present. Transverse ventral scale rows 25 to 29. Hides from Nigeria usually are missing the tip of the tail, due to local hunting practices. Skins from other parts of Africa usually have complete tails. Maximum length of live specimen is 13 feet.

STATUS OF WILD POPULATIONS. Critically endangered (A. C. Pooley, 1971, personal communication). This species is limited to large rivers, and is rarely abundant anywhere. Populations are declining everywhere due to hide hunting and the spread of human population (Lowes, 1970).

ORINOCO CROCODILE

Crocodylus intermedius Graves

DISTRIBUTION. Northern South America – the Orinoco river drainage of Colombia (east of the Andes), Venezuela, and possibly Guyana (Medem, 1968; Wermuth, 1953; Wermuth and Mertens, 1961; Werner, 1933).

OTHER COMMON NAMES. It is called *caimán* in South America. It is marketed under this name, or as Colombian, Venezuelan, or Venezuelan delta "alligator," crocodile, large scale, or soft-belly.

DISTINGUISHING CHARACTERISTICS OF COMMERCIAL HIDES. *Belly skins:* Follicle glands present. No osteoderm buttons. Transverse ventral scale rows 20 to 25. Tail whorls usually regular. Maximum length of live specimen is 23 feet.

STATUS OF WILD POPULATIONS. Endangered (Pan American Union, 1967; U.S. Department of Interior, 1970). Because of excessive hide hunting the species is now rare in Venezuela, and apparently exterminated in Colombia (Honegger, 1968). South American countries prohibit the export of untanned hides. Colombia has legislation prohibiting the hunting of crocodiles, but it is not enforced (Honegger, 1968). Import is prohibited under provision of the Endangered Species Conservation Act (U.S. Department of Interior, 1970).

JOHNSON'S CROCODILE

Crocodylus johnsoni Krefft

DISTRIBUTION. Northern Australia—from the Fitzroy River in northern Western Australia to Mackay in eastern Queensland (Wermuth, 1953; Wermuth and Mertens, 1961; Werner, 1933; Worrell, 1963).

OTHER COMMON NAMES. In Australia it is called the freshwater crocodile, Johnson's river crocodile, Johnstone's crocodile, and fish crocodile. It may be marketed under these names, or as Australian or Singapore "alligator," gator, crocodile, soft-belly, or large scale.

DISTINGUISHING CHARACTERISTICS OF COMMERCIAL HIDES. *Belly skins:* Follicle glands present. No osteoderm buttons. Transverse ventral scale rows 22 to 24. Tail whorls usually regular. Maximum length of live specimen is 9½ feet.

STATUS OF WILD POPULATIONS. Rare (Honegger, 1968). The species is completely protected by law in Western Australia and Northern Territories, but skins are still shipped from Queensland (Fauna Preservation Society, 1970b; Green, 1969; Honegger, 1968).

MORELET'S CROCODILE

Crocodylus moreletii Duméril, Bibron and Duméril

DISTRIBUTION. Northern Central America – Atlantic and Pacific coasts of Mexico, British Honduras, and Guatemala (Smith and Taylor,

1950; Wermuth, 1953; Wermuth and Mertens, 1961; Werner, 1933).

OTHER COMMON NAMES. It is sometimes called Belize crocodile or Central American crocodile. In Central America it is called "alligator," *caimán*, and *lagarto de El Petén*. Hides are marketed under these names, or as Mexican "alligator," crocodile, small scale, or soft-belly.

DISTINGUISHING CHARACTERISTICS OF COMMERCIAL HIDES. *Belly skins:* Follicle glands present. No osteoderm buttons. Transverse ventral scale rows 27 to 32. Tail whorls irregular (66 percent of the time). Maximum length is 8 feet.

STATUS OF WILD POPULATIONS. Endangered (Honegger, 1968; Pan American Union, 1967; U.S. Department of Interior, 1970). This species has all but been eliminated from British Honduras and parts of Guatemala (Charnock-Wilson, 1970). It is still locally abundant in parts of Mexico (Fauna Preservation Society, 1969b). Mexico has protective laws but they are unenforced (Honegger, 1968). Guatemala began enforcing its protective legislation in 1970. Importation is prohibited under provision of the Endangered Species Conservation Act (U.S. Department of Interior, 1970).

NILE CROCODILE

Crocodylus niloticus Laurenti

DISTRIBUTION. Africa (all of Africa except the northwest corner and central Sahara); east along the Mediterranean coast to Syria; Malagasy Republic (Madagascar); and Seychelles, Comoros, and Mauritius (Schmidt, 1919; Wermuth, 1953; Wermuth and Mertens, 1961; Werner, 1933).

OTHER COMMON NAMES. It is also called the Nilotic crocodile. Hides are marketed as African, Ethiopian, Kenya, Madagascan, or Nile "alligator," "caiman," crocodile, small scale, button-belly, or soft-belly.

DISTINGUISHING CHARACTERISTICS OF COMMERCIAL HIDES. *Belly skins:* Follicle glands present. Usually no buttons, but occasionally single buttons may be present in the midbelly and collar area. Transverse ventral scale rows 26 to 32. Tail whorls usually regular. Hides from Nigeria have the tip of the tail missing due to local hunting practices. The tails are complete on hides from elsewhere. Maximum length of live specimen is probably 18 feet.

STATUS OF WILD POPULATIONS. Endangered (Cott, 1961; Honegger, 1968; Pooley, 1969; U.S. Department of Interior, 1970). This species has been exterminated over large areas of Africa by hide hunters (Fauna Preservation Society, 1969c, 1969d, 1970a; Lowes, 1970; Pooley, 1970, in *litt.*). It can be found in numbers only in small local populations. It is extinct in the Seychelles and Mauritius. It is protected by law in most East African countries and in national parks and game preserves (Cott, 1969). Hunting of this species is to be regulated throughout all of Africa by the African Convention for the Conservation of Nature and Natural Resources (Burhenne, 1970; Honegger, 1968). South Africa has set up a research program in hopes of saving the species and restocking it in areas where it has been exterminated (Pooley, 1970). Importation is prohibited under provision of the Endangered Species Conservation Act (U.S. Department of Interior, 1970).

NEW GUINEA CROCODILE

Crocodylus novaeguineae novaeguineae Schmidt

DISTRIBUTION. New Guinea (Schmidt, 1928a, 1932; Wermuth, 1953; Wermuth and Mertens, 1961; Werner, 1933).

OTHER COMMON NAMES. It is also called the New Guinea freshwater crocodile. Hides may be marketed as Australia, New Guinea, or Singapore "alligator," crocodile, soft-belly, or large scale.

DISTINGUISHING CHARACTERISTICS OF COMMERCIAL HIDES. *Belly skins:* Follicle glands present. No osteoderm buttons. Transverse ventral scale rows 24 to 25. Tail whorls usually regular. Maximum length of live specimen is 9½ feet.

STATUS OF WILD POPULATIONS. Rare (Honegger, 1968). Populations are declining rapidly due to hide hunting. Specimens over 20 inches in belly width are protected by laws in most of Papua and Northeast New Guinea (Bustard, 1970; Fauna Preservation Society, 1969a; Honegger, 1968).

PHILIPPINE CROCODILE

Crocodylus novaeguineae mindorensis Schmidt

DISTRIBUTION. Philippine Islands – Luzon, Mindoro, and Mindanao Islands (Schmidt, 1935; Wermuth, 1953; Wermuth and Mertens, 1961).

OTHER COMMON NAMES. Also called the Mindoro crocodile and Philippine freshwater crocodile. Hides may be marketed under the name Philippine or Singapore "alligator," crocodile, soft-belly, or large scale.

DISTINGUISHING CHARACTERISTICS OF COMMERCIAL HIDES. *Belly skins:* Follicle glands present. No osteoderm buttons. Transverse ventral scale rows 24 to 26. Tail whorls usually regular. Maximum length of live specimen is 8 feet.

STATUS OF WILD POPULATIONS. Rare, possibly endangered. Hide hunting is eliminating the species from parts of its former range.

MUGGER CROCODILE

Crocodylus palustris palustris Lesson

DISTRIBUTION. India and Pakistan – from the Dasht River in West Pakistan through all the river systems of India to the Brahmaputra River drainage in the east (De Rooij, 1915; Schmidt, 1935; Smith, 1931; Wermuth, 1953; Wermuth and Mertens, 1961; Werner, 1933).

OTHER COMMON NAMES. Also called the marsh crocodile, broad-snouted crocodile, swamp crocodile, and Indian freshwater crocodile. Hides may be marketed as Indian "alligator," crocodile, soft-belly, or small scale.

DISTINGUISHING CHARACTERISTICS OF COMMERCIAL HIDES. *Belly skins:* Follicle glands present. No osteoderm buttons. Transverse ventral scale rows 26 to 32. Ventral collar not distinct (no enlarged scales). Tail whorls usually regular. Maximum length of live specimen is 13 feet.

STATUS OF WILD POPULATIONS. Endangered. The species is protected in India by a ban on the export of crocodile hides, and in Pakistan by a ban on the export of all wild animal hides (Fauna Preservation Society, 1967, 1970c; Mountfort, 1969).

CEYLON MUGGER CROCODILE

Crocodylus palustris kimbula Deraniyagala

DISTRIBUTION. Ceylon (Deraniyagala, 1936, 1939, 1953; Wermuth, 1953; Wermuth and Mertens, 1961).

OTHER COMMON NAMES. It is also called the Ceylon swamp crocodile, Ceylon marsh crocodile, and lake crocodile. In Ceylon it is known as *hale kimbula, ala kimbula,* and *kulathi muthele.* It may be marketed as Ceylon "alligator," crocodile, soft-belly, or small scale.

DISTINGUISHING CHARACTERISTICS OF COMMERCIAL HIDES. *Belly skins:* Follicle glands present. No osteoderm buttons. Transverse ventral scale rows 26 to 32. Ventral collar present and distinct. Tail whorls regular. Maximum length of live specimen is 18 feet.

STATUS OF WILD POPULATIONS. Declining in numbers. Hunting is regulated by the Ceylon government (Fauna Preservation Society, 1970e).

SALTWATER CROCODILE

Crocodylus porosus Schneider

DISTRIBUTION. India and Ceylon east to Australia and New Guinea – the coastal rivers, lagoons, and marshes from Cochin in extreme southwestern India east to Ceylon, Burma, Malaysia, Thailand, Cambodia, Vietnam, Indonesia, the Philippines, Palau Islands, northern Australia, New Guinea, Solomon Islands, New Hebrides, and Fiji (Deraniyagala, 1939, 1953; De Rooij, 1915; Schmidt, 1932; Taylor, 1970; Wermuth, 1953; Wermuth and Mertens, 1961; Werner, 1933; Worrell, 1963).

OTHER COMMON NAMES. It is also called the estuarine crocodile, gator (in Australia), and sea-going crocodile. In Ceylon it is known as *pita gatteya, gatte kimbula, gorekeya,* and *semmukhan;* in Indonesia, *buaja;* in Malaysia, *buaja, buaya, baya,* and *rawing.* Hides may be marketed under these names, or as Indian, Javan, Philippine, Singapore, Sumatran, or Thailand "alligator," crocodile, soft-belly, or small scale.

DISTINGUISHING CHARACTERISTICS OF COMMERCIAL HIDES. *Belly skins:* Follicle glands present. No osteoderm buttons. Transverse ventral scale rows 30 to 35. Tail whorls regular. Maximum length of live specimen is probably 25 feet.

STATUS OF WILD POPULATIONS. Most populations are declining rapidly due to hide hunting, and the species is non-existent in some parts of its former range where it was once abundant (Fauna Preservation Society, 1970d; Honegger, 1968). It is partially protected in most of Papua and North East New Guinea, where specimens over 20 inches belly width may not be killed (Fauna Preservation Society, 1969a; Bustard, 1970). The species is completely protected in Western Australia until 1980 (Fauna Preservation Society, 1970b; Honegger, 1968). Indonesia has imposed size limits. Ceylon, India, and Pakistan protect the species completely by banning the export of all crocodile skins or the skins of all wild animals (Fauna Preservation Society, 1967; Honegger, 1968; Mountfort, 1969). Singapore requires export licenses. Deraniyagala (1939, 1953) mistakenly listed this species as occurring on the Seychelles and Mauritius where *Crocodylus niloticus* was known to occur in the past.

CUBAN CROCODILE

Crocodylus rhombifer Cuvier

DISTRIBUTION. Cuba and the Isle of Pines (Barbour and Ramsden, 1919; Varona, 1966; Wermuth and Mertens, 1961; Werner, 1933).

OTHER COMMON NAMES. In Cuba it is called *cocodrilo, cocodrilo perla, cocodrilo criollo, cocodrilo legitimo, caimán,* and occasionally *zaquendo.*

DISTINGUISHING CHARACTERISTICS OF COMMERCIAL HIDES. *Belly skins:* Follicle glands present. No osteoderm buttons. Transverse ventral scale rows 32 to 33. Tail whorls regular. Maximum length of live specimen is 16 feet.

STATUS OF WILD POPULATIONS. Endangered (Honegger, 1968; U.S. Department of Interior, 1970). The species once occurred on the Isle of Pines from which it has been exterminated. Today it only occurs in remnants of the Zapata Swamp on the south coast of Cuba, but hide hunting and land drainage has made it very nearly extinct even there. The Cuban government protects this species rigidly and has established a captive breeding facility in the Zapata Peninsula National Park in an attempt to save it from extinction (Honegger, 1968). Importation is prohibited under provision of the Endangered Species Conservation Act (U.S. Department of Interior, 1970).

SIAMESE CROCODILE

Crocodylus siamensis Schneider

DISTRIBUTION. Southeast Asia—Thailand, Cambodia, Vietnam, and Java (De Rooij, 1915; Smith, 1931; Taylor, 1970; Wermuth, 1953; Wermuth and Mertens, 1961; Werner, 1933).

OTHER COMMON NAMES. It may also be called the Siamese freshwater crocodile. In Indonesia it is called *buaja.* Hides may be sold as Java, Singapore, or Thailand "alligator," crocodile, soft-belly or small scale.

DISTINGUISHING CHARACTERISTICS OF COMMERCIAL HIDES. *Belly skins:* Follicle glands present. No osteoderm buttons. Transverse ventral scale rows 30 to 34. Tail whorls regular. Maximum length of live specimen is 13 feet.

STATUS OF WILD POPULATIONS. Endangered. It has always been a rare animal in Indonesia, and became scarce in Thailand 30 years ago due to hide hunting. Today fewer than 200 remain in the wild in Thailand, but approximately 9,000 specimens are protected in the Sumatprakan Crocodile Farm in Bangkok (U. Youngparpakorn, 1971, personal communication).

WEST AFRICAN DWARF CROCODILE

Osteolaemus tetraspis tetraspis Cope

DISTRIBUTION. West Africa – the Niger and Senegal river drainages and other rivers south of the Sahara and north of the Congo River drainage (Schmidt, 1919; Wermuth, 1953; Wermuth and Mertens, 1961; Werner, 1933).

OTHER COMMON NAMES. It is also called the broad-snouted crocodile. Hides may be marketed as African "caiman," button-belly, bony crocodile, black crocodile, or rough-back crocodile.

DISTINGUISHING CHARACTERISTICS OF COMMERCIAL HIDES. *Belly skins:* Follicle glands present. Large single osteoderm buttons present. Surface pitting usually evident. Transverse ventral scale rows 21 to 27. Maximum length of live specimen is 6½ feet.

STATUS OF WILD POPULATIONS. Endangered (A. C. Pooley, 1971, personal communication). Populations declining due to hide hunting, destruction of habitat, and live animal collecting (Lowes, 1970). This species has never been as abundant as the other African species.

CONGO DWARF CROCODILE

Osteolaemus tetraspis osborni (Schmidt)

DISTRIBUTION. Central Africa—the Congo River drainage (Schmidt, 1919; Wermuth, 1953; Wermuth and Mertens, 1961; Werner, 1933).

OTHER COMMON NAMES. Also called the Central African dwarf crocodile, Osborn's dwarf crocodile, and African broad-snouted crocodile. Hides may be marketed as African "caiman," button-belly, bony crocodile, black crocodile, or rough-back crocodile.

DISTINGUISHING CHARACTERISTICS OF COMMERCIAL HIDES. *Belly skins:* Follicle glands present. Large single osteoderm buttons present. Surface pitting usually evident. Transverse ventral scale rows 21 to 27. Maximum length of live specimen is 5 feet.

STATUS OF WILD POPULATIONS. Endangered (A. C. Pooley, 1971, personal communication). This species does not occur in large populations. Its numbers are declining due to hide hunting.

FALSE GAVIAL

Tomistoma schlegelii (Muller)

DISTRIBUTION. Southeast Asia – Indonesia (Kalimantan and Sumatra) and Malaysia

(De Rooij, 1915; Taylor, 1970; Wermuth, 1953; Wermuth and Mertens, 1961; Werner, 1933).

OTHER COMMON NAMES. It is also called the Malay gavial, Malayan gharial, and Malayan fish crocodile. In Indonesia it is called *bediai sampit* and *buaja sapit;* in Malaya, *buaya senjulong;* in Sarawak, *baya kanulong*. Hides may be sold under these names, or as Singapore "alligator," crocodile, soft-belly, or large scale.

DISTINGUISHING CHARACTERISTICS OF COMMERCIAL HIDES. *Belly skins:* Follicle glands present. No osteoderm buttons. Transverse ventral scale rows 22 to 24. Tail whorls usually regular. Maximum length of live specimen is 16 feet.

STATUS OF WILD POPULATIONS. Declining in numbers, soon to be endangered. Hide hunters have so decimated the populations of this animal in Malaysia that protective legislation is being considered (Lucas Chin, 1970, personal communication).

Family GAVIALIDAE

GAVIAL

Gavialis gangeticus (Gmelin)

DISTRIBUTION. India, Pakistan, and Burma – specifically the Indus, Mahandi, Ganges, Brahmaputra, and Kaladan river drainage systems, and possibly parts of the Irawaddy system in northwestern Burma (Smith, 1931; Wermuth, 1953; Wermuth and Mertens, 1961; Werner, 1933).

OTHER COMMON NAMES. In India it is called *gharial*. Hides may be sold as Indian soft-belly, small scale, "alligator," "crocodile," or gavial.

DISTINGUISHING CHARACTERISTICS OF COMMERCIAL HIDES. *Belly skins:* Follicle glands present. No osteoderm buttons. Transverse ventral scale rows 30 to 31. Ventral collar not prominent. Maximum length of live specimen is 21½ feet.

STATUS OF WILD POPULATIONS. Endangered (U.S. Department of Interior, 1970). Protected in India by a ban on the export of all crocodilian hides, and in Pakistan by a ban on the export of all wild animal hides (Fauna Preservation Society, 1967, 1970c; Mountfort, 1969). Importation is prohibited under provisions of the Endangered Species Conservation Act (U.S. Department of Interior, 1970).

ACKNOWLEDGMENTS

We wish to thank David Klapisch (Southern Trading Corporation, Newark, New Jersey), Hyman Marx (Field Museum of Natural History, Chicago), Henry Molt (Philadelphia Reptile Exchange), Charles Myers (American Museum of Natural History, New York), Ray Pawley (Chicago Zoological Park), Barry Wakeman (Cincinnati Zoological Society), George Zug (U.S. National Museum, Washington, D.C.) for the opportunity of examining specimens and skins in their care. We also wish to thank Edward Baker and Warren Difendall (United States Department of Interior, Department of Sport Fish and Wildlife, New York), Claire Hagen (Hagen and Company, New York), and Glenn Kindler (Connecticut Import Export Corporation, New York) for their assistance. Rene Honegger kindly lent us an advance copy of the International Union for the Conservation of Nature and Natural Resources Red Data Book on amphibians and reptiles so that we might include some of the information here. Jose Cei, Lucas Chin, Federico Medem, A. C. Pooley, James Powell, and others provided personal observations on wild populations of crocodilians. William Meng (New York Zoological Society) provided photos.

LITERATURE CITED

BARBOUR, T., AND C. T. RAMSDEN
1919. The herpetology of Cuba. Mem. Mus. Comp. Zool., 47(2): 73-213.

BURHENNE, W. E.
1970. The African Convention for the Conservation of Nature and Natural Resources. Biol. Conservation, 2(2): 105-114.

BUSTARD, H. R.
1970. A future for crocodiles. Oryx, 10(4): 249-255.

CARVALHO, A. L. DE
1955. Os jacarés do Brasil. Arquivos Mus. Nac., 42(1): 127-139.

CHARNOCK-WILSON, J.
1970. Manatees and crocodiles. Oryx, 10(4): 236-238.

COCHRAN, D. M.
1941. The herpetology of Hispaniola. U.S. Nat. Mus. Bull. 177: 1-398.

COTT, H. B.
1961. Scientific results of an inquiry into the ecology and economic status of the Nile crocodile (*Crocodilus niloticus*) in Uganda and Northern Rhodesia. Trans. Zool. Soc. London, 29(4): 211-356.

1969. Tourists and crocodiles in Uganda. Oryx, 10(3): 153-160.

CROWE, P. K.
1965. What is happening to the wildlife of South America. Oryx, 8(1): 28-37.

DERANIYAGALA, P. E. P.
1936. A new crocodile from Ceylon. Ceylon J. Sci. (B) 19: 279-286.
1939. The tetrapod reptiles of Ceylon. Vol. 1, Testudinates and Crocodilians. Colombo Mus. Nat. Hist. Ser. xxxii + 412 pp.
1953. A colored atlas of some vertebrates from Ceylon. Vol. 2, Tetrapod Reptilia. Ceylon Nat. Mus. Publ. vii + 101 pp.

DE ROOIJ, N.
1915. The reptiles of the Indo-Australian archipelago. Vol. 1, Lacertilia, Chelonia, Emydosauria. E. J. Brill Ltd., Leiden. xiv + 384 pp.

FAUNA PRESERVATION SOCIETY
1967. No crocodiles from India. Oryx, 9(3): 184.
1969a. Protection for Papua's crocodiles. Oryx, 10(2): 81.
1969b. Morelet's crocodile near extinction. Oryx, 10(2): 88.
1969c. Crocodiles decrease in Natal. Oryx, 10(3): 144.
1969d. Crocodiles in Blue Nile gorge. Oryx, 10(3): 144.
1970a. Crocodiles going in Uganda. Oryx, 10(4): 208-209.
1970b. Crocodiles in Australia. Oryx, 10(4): 213.
1970c. Two crocodiles in trouble. Oryx, 10(5): 287.
1970d. Sarawak crocodiles. Oryx, 10(5): 295.
1970e. Poaching in Ceylon. Oryx, 10(5): 295.

GREEN, K.
1969. Rare fauna. Walkabout, 35(4): 6.

HONEGGER, R.
1968. Red Data Book. Volume 3 – Amphibia and Reptilia. International Union for the Conservation of Nature and Natural Resources, Survival Service Commission, Morges, Switzerland. Loose leaf n.p.

LOWES, R. H. G.
1970. Destruction in Sierra Leone. Oryx, 10(5): 309-310.

MEDEM, F.
1955. A new subspecies of *Caiman sclerops* from Colombia. Fieldiana: Zool., 37: 339-344.
1963. Osteologia craneal, distribucion geografica y ecologia de *Melanosuchus niger* (Spix) (Crocodylia, Alligatoridae). Revista Acad. Colombiana Cien. Exactas, Fis. y Nat., 12(45): 5-19.
1967. El genero *Paleosuchus* en Amazonia. Simposio Biota Amazonica, 3: 141-162.
1968. El desarrollo de la herpetologia en Colombia. Revista Acad. Colombiana Cien. Exactas, Fis. y Nat., 13(50): 149-199.

MOUNTFORT, G.
1969. Pakistan's progress. Oryx, 10(1): 39-43.

PAN AMERICAN UNION
1967. Listas de especies de fauna y flora en vias de extincion en los estados miembros. La Convencion para la Proteccion de la Flora, de las Fauna y de las Bellezas Escienicas Naturales de los Estados Americanos. Organization of American States. ii + 48 pp.

POOLEY, A. C.
1969. Rearing crocodiles in Zululand. African Wildlife, 23(4): 314-320.
1970. Crocodile rearing in Zululand. Animals, 13(2): 76-79.

POPE, C.
1935. The reptiles of China. Nat. Hist. Central Asia. Vol. 10. Amer. Mus. Nat. Hist. lii + 604 pp.

SCHMIDT, K. P.
1919. Contributions to the herpetology of the Belgian Congo based on the collection of the American Congo Expedition, 1909-1915. Bull. Amer. Mus. Nat. Hist., 39(2): 385-624.
1928a. A new crocodile from New Guinea. Field Mus. Nat. Hist. Zool. Ser., Publ. 247, 12(14): 177-181.
1928b. Notes on South American caimans. Field Mus. Nat. Hist. Zool. Ser., Publ. 252, 12(17): 205-231.
1932. Notes on New Guinea crocodiles. Field Mus. Nat. Hist. Zool. Ser., Publ. 310, 18(8): 167-172.
1935. A new crocodile from the Philippine Islands. Field Mus. Nat. Hist. Zool. Ser., 20(8): 67-70.
1953. A checklist of North American amphibians and reptiles. 6th ed. Univ. Chicago Press. vii + 280 pp.

SMITH, M.
1931. The fauna of British India. Reptilia and Amphibia. Vol. 1, Loricata, Testudines. Taylor and Francis, London. xxviii + 185 pp.

SMITH, H. M., AND E. H. TAYLOR
1950. An annotated checklist and key to the reptiles of Mexico exclusive of the snakes. U.S. Nat. Mus. Bull. 199: 1-253.

TAYLOR, E. H.
1970. The turtles and crocodiles of Thailand and adjacent waters. Univ. Kansas Sci. Bull., 49(3): 87-179.

U.S. DEPARTMENT OF INTERIOR
1968. American alligator. Rare and endangered fish and wildlife of the United States. U.S. Dept. Interior, Bur. Sport Fish. Wildlife, Washington, D.C., Sheet RA-2.

1970. List of endangered foreign fish and wildlife. Title 50—Wildlife and Fisheries. Federal Register, 35(233): 18319-19322.

VARONA, L. S.
1966. Notas sobre los crocodilidos de Cuba y descripcion de una nueva especie del Pleistocene. Poeyana Inst. Biol., Ser. A, No. 16: 1-34.

WERMUTH, H.
1953. Systematik der Rezenten Krokodile. Mitteil. Zool. Mus. Berlin, 29(2): 375-514.

WERMUTH, H., AND R. MERTENS
1961. Schildkroten, Krokodile, Bruckenechsen. Veb Gustav Fischer Verlag, Jena. xxvi + 422 pp.

WERNER, F.
1933. Reptilia, Loricata. Das Tierreich. Gruyter and Co., Berlin. xiii + 40 pp.

WORRELL, E.
1963. Reptiles of Australia. Angus and Robertson, Sydney. xv + 207 pp.

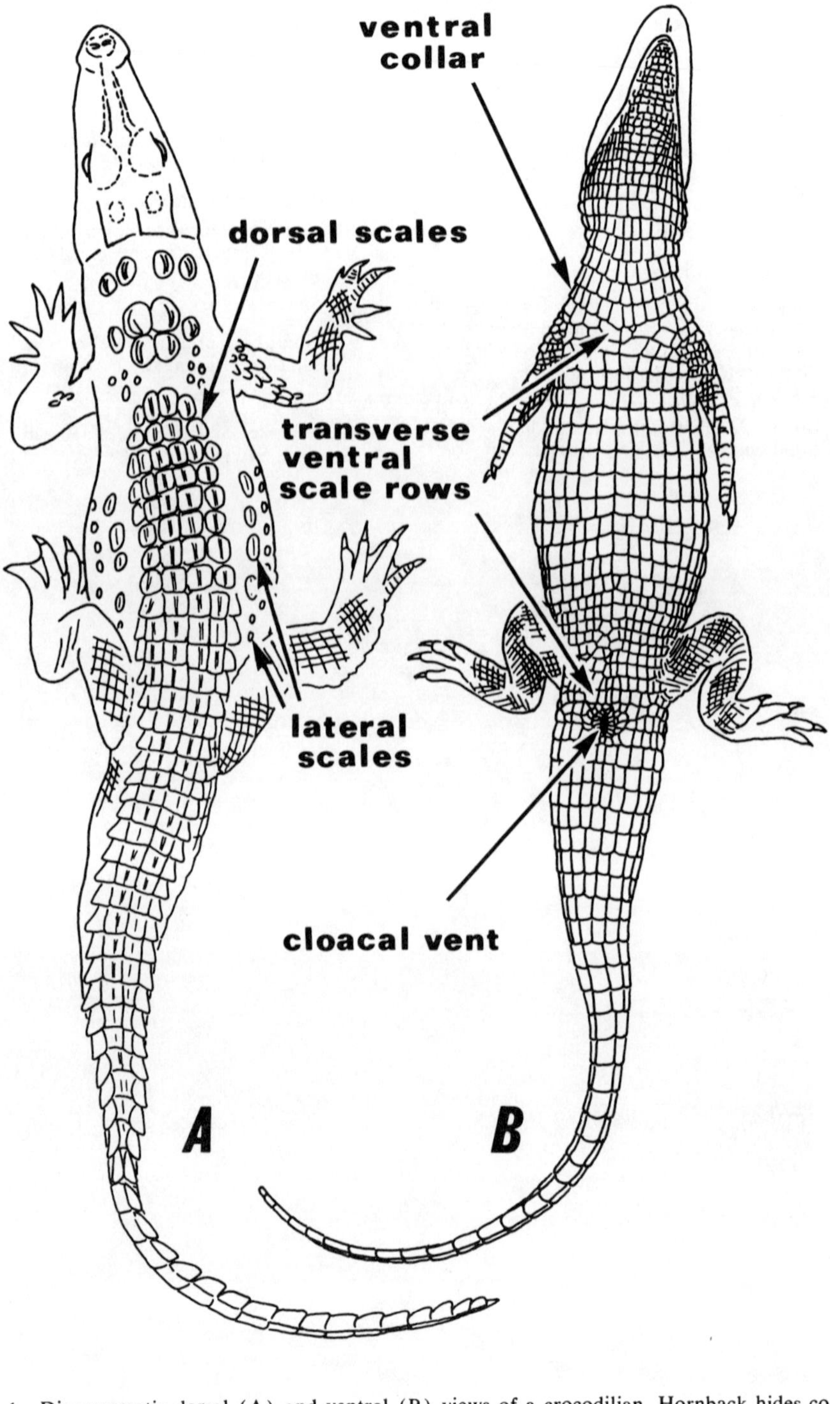

FIGURE 1. Diagrammatic dorsal (A) and ventral (B) views of a crocodilian. Hornback hides consist of most of the skin seen in A (skull and feet are absent). Belly hides consist of most of the skin seen in B (skull and feet are absent and the lateral [side] skin is attached). Transverse scale rows are counted by beginning and ending with the rows indicated by the arrows.

Reprinted from *Western Birds*

CHARADRIIFORMES NEW TO NEVADA

C. S. Lawson

The following records add further knowledge to the status of Charadriiformes in Nevada. Specimens and the original color transparencies of all photographs are deposited in the University of Nevada Museum of Biology (UNMB), University of Nevada, Reno. Species identification of all photographic records has been verified by Joseph R. Jehl Jr. and Guy McCaskie of the San Diego Natural History Museum (SDNHM). Unless otherwise noted, all records are from Clark County, Nevada. Photographs were taken by the author.

RED PHALAROPE

On 26 July 1972 I collected a Red Phalarope (*Phalaropus fulicarius*) from a flock of about 250 phalaropes at the tailing ponds in Las Vegas Wash 10 miles southeast of Las Vegas, Nevada. The flock consisted of 4 or 5 Red Phalaropes in winter plumage, 225 Wilson's Phalaropes (*Steganopus tricolor*), and 25 Northern Phalaropes (*Lobipes lobatus*). The specimen (UNMB 1748) was an adult male in breeding plumage (both testes 2.5 x 4.0 mm), moderately fat, with two feathers of the back in molt. The weight was not obtained.

This first specimen record for Nevada supports a sight and photographic record from the same location 5 June 1971 (Snider, Am. Birds, 25:781, 1971).

GLAUCOUS GULL

On 30 November 1972 Gertrude Lawson and I discovered a juvenal Glaucous Gull (*Larus hyperboreus*) resting ashore at Boulder Beach, Lake Mead, Nevada. The bird was in view about 3 minutes. I took 4 photographs before the bird was flushed by an approaching car. The accompanying photograph (Figure 1) was taken at that time. The following description was obtained:

> The bird was very large, very light grayish-white with some light brownish flecking on the body. The bill was very large, whitish with a black tip. The primaries were white. The tail was white with some light brownish flecking.

The bird was last seen on 16 December. This is the first Nevada record.

Figure 1. Juvenal Glaucous Gull (*Larus hyperboreus*) at Boulder Beach, Lake Mead, Clark County, Nevada 30 November 1972.

Figure 2. Juvenal Thayer's Gull *(Larus thayeri)* at Lake Mead, Clark County, Nevada 30 October 1971. Note the small headed and small billed appearance and the checkered pattern of the back.

Figure 3. Juvenal Thayer's Gull. Note the white wing linings, pattern of the wing coverts and the white tipped primaries.

GLAUCOUS-WINGED GULL

R. W. Campbell (pers. comm.) reports he banded a Glaucous-winged Gull (*Larus glaucescens*) as a local (band 967-19765) at a colony 8 miles northwest of Tofino on the west coast of Vancouver Island, British Columbia 11 July 1971. The disarticulated remains of this bird were found one mile from the east shore of Lake Tahoe near Glenbrook, Douglas County, Nevada in late July 1972 by Tom Gorman of San Mateo, California. The band was recovered but the skeletal remains were left at the discovery site.

The presence of this species in Nevada was first noted on 30 October 1971 when M. Vincent Mowbray, John Koch and Glade Koch discovered a juvenal at Las Vegas Wash Marina, Lake Mead. On 5 November 1971 Nora Poyser, Pauline Long and I observed a juvenal, presumably the same bird, at the same location. Efforts to secure photographs failed. The following description was obtained:

> The bird was larger than a nearby Herring Gull (*Larus argentatus*) and similar in color but larger than a nearby Thayer's Gull (*Larus thayeri*). The bill was large and black. The eye was dark. The body, primaries, and tail were a uniform gray-brown with some slightly darker brownish streaking on the back. The legs were pinkish-brown.

Subsequently, from one to three individuals were observed until 27 February 1972. During November and December of 1972 another juvenal appeared at the same site. These are the first Nevada records.

THAYER'S GULL

On 11 February 1972 I collected a juvenal Thayer's Gull (*Larus thayeri*) at the Silver State Disposal Company garbage dump about 10 miles east of Las Vegas, Nevada. The bird (UNMB 1746, female, weight 737.1 g) was prepared as a study skin by Fred Ryser, UNMB.

The presence of this species in Nevada was first noted 30 October 1971 when I photographed a gull I could not identify at Las Vegas Wash Marina, Lake Mead. The accompanying photographs (Figures 2, 3, and 4) were taken on that date. Duplicate sets of the original color transparencies are on deposit at SDNHM and the National Museum of Canada, Ottawa, Quebec. During December 1972 another juvenal appeared at the same site. These are the first Nevada records.

MEW GULL

On 9 January 1972 M. Vincent Mowbray and I discovered an immature Mew Gull (*Larus canus*) resting on the beach at Las Vegas Wash Marina, Lake Mead, Nevada. The accompanying photograph (Figure 5) was taken at that time. The following description was obtained:

> The bird was smaller than a nearby Ring-billed Gull (*Larus delawarensis*). The bird had a small black bill. The eye was dark. The head, neck, back, chest, sides and stomach were a uniform gray-brown. The tail was gray-brown with the terminal one-third banded by a dark gray-brown to black very narrowly tipped with buffy-white.

The Mew Gull in juvenal plumage most nearly appears like an immature Thayer's Gull, has a small, black bill and the tips of the rectrices are so narrowly tipped with buffy-white that these markings are easily missed.

The bird was last seen 27 February 1972. In November 1972 another juvenal of this species was sighted at Boulder Beach, Lake Mead by Mowbray. These are the first Nevada records.

Figure 4. Juvenal Thayer's Gull, flight. Note the light primaries and the banded tail.

Figure 5. Juvenal Mew Gull (*Larus canus*) at Lake Mead, Clark County, Nevada 9 January 1972. Note the small black bill. Compare size to Ring-billed Gull (*Larus delawarensis*) on the left.

COMMON TERN

On 15 August 1972 I collected an adult Common Tern (*Sterna hirundo*) at the tailing ponds in Las Vegas Wash about 10 miles southeast of Las Vegas, Nevada. The bird is now UNMB 1749 (female, weight not taken, heavily covered with unidentified ectoparasites, moderately fat). The tail was white with dark gray outer rectrices. The tarsi measured 20.5 mm. This is the first Nevada specimen.

Common Terns are probably overlooked among Forster's Terns (*Sterna forsteri*) by most observers in the state. In northern Nevada, McCaskie (pers. comm.) has observed the species in the Reno-Carson City-Lake Tahoe area. Ed Scovill of Boulder City, Nevada photographed the species in Iceberg Canyon of Lake Mead in May 1968 and at the 19-20 mile marker in Lake Mohave in May 1969. In southern Nevada, M. Vincent Mowbray and I have observed the species during spring and fall migrations for several years. In addition to the fall, 1971 and spring, 1972 observations at Lake Mead (Monson, Am. Birds 26:102, 793, 1972), I have observed Common Terns at the following Nevada locations: Stillwater National Wildlife Refuge, Churchill Co.—one 18 May 1971; Las Vegas Wash—seven 3 May 1964, six 14 September 1965, two 16 May 1967, two 23 August 1968, thirteen 24 August 1972; Pahranagat Lake near Alamo, Lincoln Co.—one 6 September 1973; Walker Lake, Mineral Co.—one 22 May 1970; and Cave Rock, Lake Tahoe, Douglas Co.—nine 20 May 1964.

ACKNOWLEDGEMENTS

I wish to thank R. W. Campbell, M. Vincent Mowbray and Ed Scovill for use of their previously unpublished records; John Isola for permission to collect at the Silver State Disposal Company dump; Joseph Jehl and Guy McCaskie for species identification and verification of photographic records; and Fred Ryser for preparation of the specimen of Thayer's Gull. *C. S. Lawson, 509 Altamira Road, Las Vegas, Nevada 89128.*

What to look for: Eskimo and Little Curlews compared

by John Farrand, Jr.

Reprinted from *American Birds*

[Ed. note — Although the many specimens of Eskimo Curlew now lying in museum trays in North America are apparently correctly labeled, and all recent sightings at locations normal to its known migration routes, the Little Curlew of Asia, considered by some to be conspecific with the Eskimo, might conceivably appear sometime, somewhere in North America. For this reason we have asked John Farrand, Jr., to compare the two. His thesis is that the two species (?) are separable in the field under ideal conditions, and most certainly in the hand. Unfortunately, the specimens selected for the accompanying illustrations seem to indicate that the characteristic differences are not always obvious, and that there is considerable overlap in size, bill length, etc.]

The Little Curlew (*Numenius minutus*) of Siberia is very similar to the Eskimo Curlew (*Numenius borealis*), and were one to occur as an accidental in North America, great care would be required to distinguish it from the nearly-extinct Eskimo. However, an examination of specimens of both species in the American Museum of Natural History, New York, reveals several characters by which such a distinction would be possible under favorable conditions. The Eskimo Curlew is a more boldly and coarsely marked bird, with heavier streaking on the sides of the face and neck, and dark chevrons on the breast and flanks; the Little Curlew has a relatively more finely streaked face and neck, and the breast is streaked rather than marked with chevrons, the chevrons being few in number and confined to the flanks. In both species the under wing coverts and axillaries are barred with dark brown, but in the Eskimo Curlew these feathers are a rich cinnamon, while in the Little Curlew they are a much paler buff or sandy color. Finally, the pale pinkish area at the base of the

Fig. 1. Ventral view comparing two Little Curlews, *Numenius minutus*, left, with two Eskimo Curlews, *N. borealis*, right. Note similarity of size and the difference in ventral markings (especially two center specimens). The apparent bill length differential is not a consistent characteristic — lengths vary by sex and age and overlap is normal. Specimens shown, l to r: *minutus*, ♂ Shanghai, April 1886, AMNH[1] 737920; ♀ Shanghai, April 1911, AMNH 114475; *borealis* ♀? Mar del Plata, Argentina, Sept. 19, 1914, AMNH 752067; unsexed, Montauk, N.Y., undated, AMNH 335779.

[1]American Museum of Natural History, New York.

lower mandible is more extensive in the Little Curlew, reaching to or beyond the middle of the bill, whereas in the Eskimo Curlew this light area occupies less than half the length of the bill.

In the hand, the two species may readily be distinguished not only by these characters but also by the fact that in the Eskimo Curlew the rear surface of the tarsus is covered with very small, more or less round scales, while in the Little Curlew the rear surface is covered by transverse scutes like those seen on the front of the tarsus in both species.

The Little Curlew breeds in northeastern Siberia and winters in the Moluccas, New Guinea, and Australia, where it is called the 'Little Whimbrel.' It has occurred casually in the Philippines, Borneo, Celebes, and New Zealand. While there are no records of the species in North America, its appearance on this continent is conceivable, and because of the great rarity of the Eskimo Curlew, and small curlew suspected of being the latter species should be distinguished carefully not only from the Whimbrel (*Numenius phaeopus*) but also from the Little Curlew.

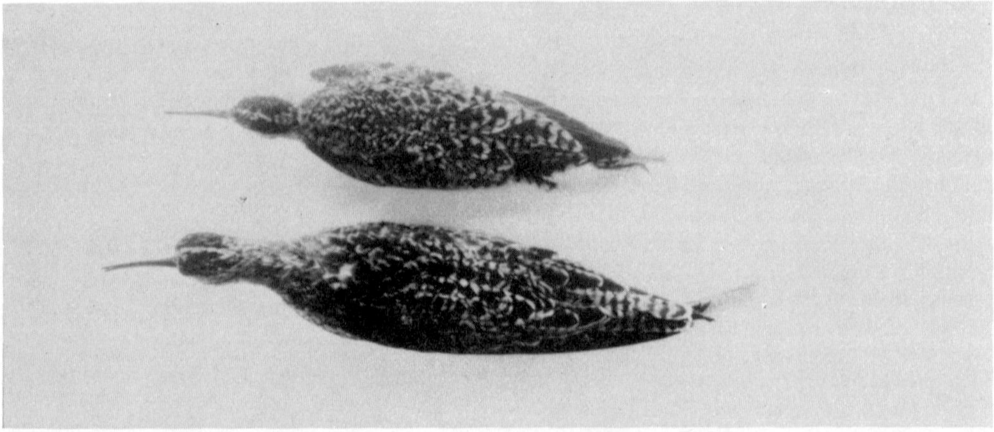

Fig. 2. Dorsal view, showing light median crown stripe in both curlews, a variable characteristic. Note similar feather edging on back and upper wing surfaces. Above, *minutus,* ♀ Western Australia, Oct. 24, 1964, AMNH 803832; below, *borealis,* unsexed, Labrador, Aug. 20, 1918 AMNH 752069. *Photos by Doug Rod*

Fig. 3. Ventral view, showing comparative bill length and body size similarity. In these two specimens *borealis* soft parts (bill and legs) seem darker, but this is not always a reliable characteristic. Above, *minutus,* ♀ Shanghai, May 5, 1911, AMNH 114475; below *borealis,* ♂ Magdalene Islands, undated, AMNH 149312.

Reprinted from *Journal of the Fisheries Research Board of Canada*

Fish Otoliths in Cetacean Stomachs and Their Importance in Interpreting Feeding Habits[1]

By John E. Fitch

California State Fisheries Laboratory
Terminal Island, Calif., USA

and

Robert L. Brownell, Jr.

Los Angeles County Museum of Natural History
Los Angeles, California, USA

ABSTRACT

The stomachs of 17 cetaceans of seven species (3 *Kogia simus*, 5 *Stenella longirostris*, 3 *S. graffmani*, 2 *Lagenorhynchus obliquidens*, 2 *Delphinus delphis*, 1 *Lissodelphis borealis*, and 1 *Phocoena sinus*) yielded 18,164 fish otoliths representing over 51 species, 40 genera, and 22 families. Lanternfish (family Myctophidae) otoliths accounted for more than 89% of the total, and they had come from at least 19 species belonging to nine genera. Only two fish species could have been identified if we had not been able to utilize otoliths. To be useful in food studies involving otoliths, cetacean stomachs cannot be placed in formalin solutions, even for short periods.

The kinds and diversity of species preyed upon by several of these cetaceans indicate that they routinely forage in depths of 650–800 ft (200–250 m), and a few such as *Kogia simus* may descend to 1000 ft (300 m) or deeper.

INTRODUCTION

Although many references are available concerning the few entire fishes encountered in cetacean stomachs, or observed being fed upon, no publication presents a thorough or extensive discussion of the piscivorous activities of whales, porpoises, and dolphins. This paucity of information apparently reflects the failure of cetacean researchers to recognize fish otoliths per se, and to realize the usefulness of these objects in food habit studies. Scales, vertebrae, eye lenses, teeth, bits of smelly flesh, and other fish fragmentia making up the soupy, often worm-ridden mixture typically encountered in a cetacean stomach seldom can be identified to species, but teleost otoliths are slow to digest and can furnish accurate qualitative as well as quantitative data. Their ultimate identification, however, requires a fairly extensive comparative collection from the species most likely to be fed upon.

During the past several years, we have routinely examined stomachs of cetaceans that have been stranded, harpooned, or captured incidental to purse-seine and gillnet fisheries, and have saved otoliths, cephalopod beaks,

[1]Received for publication May 31, 1968.

and other identifiable remains that we have encountered. The stomachs of seven species (pygmy sperm whale, *Kogia simus*, Pacific spotted dolphin, *Stenella graffmani*, spinner dolphin, *Stenella longirostris*, white-sided dolphin, *Lagenorhynchus obliquidens*, common dolphin, *Delphinus delphis*, northern right whale dolphin, *Lissodelphis borealis*, and Gulf of California harbor porpoise, *Phocoena sinus*) have yielded otoliths, and all but *Phocoena sinus* have also contained cephalopod beaks. Although each individual bony fish has three pairs of otoliths (sagittae, asterisci, and lapilli), the asterisci and lapilli usually are extremely small, and even if found, they rarely are used in food studies involving fish ear stones. All otoliths referred to in this report are sagittae unless otherwise noted.

Because formalin will dissolve fish otoliths and render them unrecognizable in a matter of days, stomachs should be examined as soon as possible, or they should be frozen until they can be worked on. Preservation of stomachs in alcohol is satisfactory for a short period, but subsequent separation of otoliths and other identifiable remains is much more difficult and time consuming than when working with fresh material. Otoliths can be preserved dry or in alcohol, but first they should be rinsed in water to remove extraneous amorphic material. When making comparisons, a binocular microscope at 6–10 magnifications is almost a necessity.

Although otoliths are the most important tools available for identifying teleost fishes eaten by whales, porpoises, and dolphins, only 12 reports have noted their occurrence in cetacean stomachs since the first was published in 1903.

Thomas Scott (1903) opened the door more than 65 years ago for researchers to use teleost otoliths in determining food habits of fish-eating cetaceans. He identified and illustrated 280 gadid (Gadidae), sand lance (*Ammodytes* sp.), and other earstones found in the stomach of a common porpoise, *Phocoena phocoena*, captured off the coast of Scotland. In spite of Scott's pioneer work, we know of only 12 additional published reports of otolith occurrence in cetacean digestive tracts, and 20 years were to pass before the first of these appeared (Schmidt, 1923). In that year, Johannes Schmidt, in a letter to the editor of *Nature*, reported having sent to the British Museum of Natural History 15,191 sagittae that he had removed from the stomach of a "common long-nosed porpoise," *Delphinus delphis*, harpooned off southern Spain in the Mediterranean during a research cruise on the *Thor* in June 1910. G. A. Frost (1924), after seeing Schmidt's note, examined 4338 of these otoliths and assigned them to six species in three families (4324 Myctophidae, 10 Macrouridae, and 4 Scomberesocidae). His outline drawings are crude, but there is no question that he correctly identified to family the four kinds of lanternfishes and possibly the scomberesocid, but his "*Macrurus*" very likely is a barracudina, family Paralepididae.

Vladykov (1946) was the next to note fish otoliths in cetacean stomachs, reporting more than 1000 cod (*Gadus* and *Microgadus*), smelt (*Mallotus*), and other sagittae from the digestive tracts of white whales, *Delphinapterus leucas*, taken in the Saint Lawrence River, Canada. In 1948, Scheffer and Slipp noted "otoliths and bones of fish" in the stomach of a pygmy sperm whale, *Kogia breviceps*, from the coast of Washington. They questionably identified a partially digested fish as *Trichodon*, but were unable to assign names to the 21 otoliths, none of which appears to have been saved (V. B. Scheffer, personal communication).

The remaining eight reports we were able to find that mentioned otoliths in cetacean stomachs have been published during the last 12 years. Three of these (Cadenat, 1959a, b; and Cadenat and Doutre, 1959) concern *Tursiops truncatus*, *Delphinus delphis*, and *Stenella longirostris* from off the west coast of Africa, whereas the other five publications (Brown and Norris, 1956; Norris and Prescott, 1961; Best, 1963; Daugherty, 1965; and Fiscus and Niggol, 1965) pertain to dolphins from the eastern north Pacific, primarily off California.

The African *Tursiops* stomach (Cadenat, 1959a) contained several partially digested snake mackerel, *Gempylus serpens*, and numerous otoliths from at least four kinds of deep-living macrourids and morids (Moridae), but the earstones noted in the stomachs of 10 of 25 *D. delphis* (Cadenat, 1959b) and 4 of 4 *S. longirostris* (Cadenat and Doutre, 1959) were neither enumerated nor identified, except for the notation that a large percentage of the otoliths in the *Stenella* stomachs appeared to be "*a la famille des Scopelidae*" [= Myctophidae].

The large, pointed, first stomach chamber of a *Lagenorhynchus obliquidens* from off California was "filled with partially digested anchovies, fish bones, otoliths, fish eye lenses and one blanched but otherwise undigested squid..." (Brown and Norris, 1956). Neither these otoliths nor the "unidentified fish otoliths" noted in the stomach of a *Delphinus delphis* (Fiscus and Niggol, 1965) are available for examination. A *Tursiops gilli* from southern California had eaten mostly shallow-water bottom-living forms as evidenced by 102 otoliths from 8 species (Norris and Prescott, 1961). On the other hand, the stomach of a *Phocoenoides dalli* reported by these same authors yielded 30 otoliths from two species (hake, *Merluccius productus*, and anchovies, *Engraulis mordax*) that it very likely had consumed more than 400 ft (120 m) beneath the surface. Finally, Best (1963) in discussing *Merluccius productus*, stated that "hake otoliths have been identified among the stomach contents of... *Lagenorhynchus obliquidens* and *Phocoenoides dalli*," and Daugherty (1965) reported that otoliths often are found in stomachs of eastern Pacific cetaceans mentioning particularly *Stenella*, *Tursiops*, *Lagenorhynchus*, and *Phocoenoides*.

FISHES EATEN BY *KOGIA SIMUS*

Brownell examined the stomachs of three pygmy sperm whales harpooned offshore from Taiji, Wakayama, Japan in July 1967, and found 153 otoliths that had come from a minimum of 92 fish representing 18 species in seven families (Table I, Fig. 1). Unfortunately we lack a good comparative collection of otoliths from Japanese fishes so we cannot identify many of these sagittae

TABLE I. Numbers and kinds of fish otoliths found in stomachs of three *Kogia simus* captured off Taiji, Japan, during July 1967.

Kind of fishes eaten	Whale identity and no. of otoliths		
	RLB 392[a]	RLB 397[a]	RLB 400[a]
Argentinidae			
Glossanodon sp.	3	5	2
Nansenia sp.		2	2
Congridae			
Congrid		2	
Gonostomatidae			
Gonostomatid			3
Ichthyococcus sp.	10		7
Macrouridae			
Coelorhynchus ?			4
Macrourid #1	3	1	7
Macrourid #2		14	9
Moridae			
Morid #1	3	14	
Morid #2		1	
Morid #3		3	
Myctophidae			
Diaphus #1			1
Diaphus #2			2
Hygophum ?			1
Lampadena #1	19	12	14
Lampadena #2			2
Symbolophorus sp.		1	3
Sternoptychidae			
Polyipnus sp.		2	1
Total	38	57	58

[a]R. L. Brownwell field numbers.

to genus and species. Cephalopod beaks and crustacean remains also were present in these stomachs.

The 38 otoliths (5 species) found in one stomach (RLB 392) were from at least 22 fish (10 *Lampadena*, 5 *Ichthyococcus*, 3 morids, 2 *Glossanodon*, and 2 macrourids), whereas the 57 and 58 otoliths in the other two stomachs represented a minimum of 37 (11 species) and 33 fish (14 species), respectively. Only 3 of the 18 species (*Lampadena*, *Glossanodon*, and a macrourid) had been eaten by all three *Kogia*, but these three species contributed 66 of the 153 otoliths. Most of the otoliths were in relatively good condition, but even those considerably eroded from digestive action were readily identified. A family-by-family discussion of the 153 otoliths in these three stomachs (Table I) follows.

Argentinidae. — Fourteen of the 153 otoliths were from two genera of argentines, *Glossanodon* and *Nansenia*, representing a minimum of eight fish. All of these otoliths were from large adults, but *Glossanodon* probably never exceeds 10 inches (250 mm) and 2 oz (60 g), whereas *Nansenia* is slightly smaller. Members of these two genera seldom approach within 650 ft (200 m) of the surface.

Congridae. — A pair of sagittae very similar to those of *Ariosoma* were found in one stomach. Congrids such as *Ariosoma* often inhabit the scattering layer and probably migrate toward the surface at night, possibly even reaching the surface. Although these sagittae were from an adult eel, it probably would have been only 10–12 inches (250–300 mm) long.

Gonostomatidae. — The 20 gonostomatid otoliths we found in these stomachs were from a minimum of 11 fish: 9 *Ichthyococcus*, and 2 unknown gonostomatids, possibly a close relative of *Vinciguerria*. Although all were from large adults, the unknown gonostomatids probably were less than 2 inches (50 mm) long, but several of the *Ichthyococcus* sagittae had come from very large individuals, probably exceeding 5 inches (125 mm). Gonostomatids usually are deeper than 1000 ft (300 m) during daylight hours, but move closer to the surface at night. *Vinciguerria* often reaches the surface at night, but *Ichthyococcus* usually remains 650–800 ft (200–250 m) down.

Macrouridae. — Macrourids usually inhabit muddy bottom areas at depths of 1500–5000 ft (450–1500 m), but some may be found as shallow as 650 ft (200 m) and others deeper than 10,000 (3000 m). Most species never leave the bottom except to move a few feet off during feeding activities. The 38 macrourid otoliths were from three species, and represented at least 22 individuals. All three *Kogia* had fed on these bottom-dwelling forms. Our use of the name *Coelorhynchus* for four of the macrourid otoliths is questionable because of the state of digestion. Most of the macrourid otoliths probably came from 10–14-inch (250–350 mm) long fish weighing as much as $\frac{1}{4}$–$\frac{1}{2}$ lb (100–225 g) each.

Moridae. — The 21 morid otoliths from at least 16 individuals also represented three species of bottom-dwelling fish. These likely were captured at depths exceeding 650 ft (200 m) but shallower than 1500 (450 m). They probably were slightly smaller than the macrourids.

Myctophidae. — Lanternfish otoliths comprised the greatest number (55) and kinds (6) found in these stomachs, but only one kind (45 sagittae from an undescribed species of *Lampadena*) was of major importance. These 45 otoliths represented a minimum of 25 lanternfish, and their sizes suggested that they had come from individuals that were 8–10 inches (200–250 mm) long. Differences between these otoliths and those of the known species of *Lampadena* were pointed out by Nafpaktitis and Paxton (1968) in a recent review of the genus. The other 10 lanternfish otoliths represented five species apparently belonging to four genera (Table I, Fig. 1). All but one of these were in the same stomach. During daylight hours members of these four genera

(*Lampadena*, *Symbolophorus*, *Diaphus*, and *Hygophum*) are found at a depth of about 1000 ft (300 m), but at night all are known to migrate into surface waters. A 4-inch (100-mm) fish would be large for an adult of most of these.

Sternoptychidae. — Three *Polyipnus* otoliths were the only representatives of this family found in these stomachs. *Polyipnus* seldom exceeds 2 inches (50 mm) when full grown, and usually remains 650–800 ft (200–250 m) beneath the surface.

FISHES EATEN BY *STENELLA GRAFFMANI* AND *S. LONGIROSTRIS*

Stenella longirostris and *S. graffmani* occur sympatrically in the eastern tropical Pacific, and frequently school together, often in the company of yellowfin tuna, *Thunnus albacares*. During 1967 and 1968, we found otoliths in the stomachs of five *S. longirostris* and one *S. graffmani* that had been netted with tuna during daylight hours in the eastern Pacific between about 12° and 20°N lat. Previous to 1967, Fitch had identified several thousand otoliths from the stomachs of two *S. graffmani* netted in the same general area during 1959 and 1963 (Table II, Fig. 2). The stomachs of these eight dolphins also contained cephalopod beaks, so there is a slight possibility that some of the otoliths had been transferred from the squid and/or octopus stomachs, particularly those of such tiny fishes as *Vinciguerria*, *Diogenichthys*, and *Bregmaceros*.

Although the five *S. longirostris* had fed upon 6–15 kinds of fish, two kinds of myctophids, *Benthosema panamense* and *Lampanyctus parvicauda*, comprised over 50% of their piscivorous diet, both individually and as a group. *Benthosema panamense* was also important to *S. graffmani*, being the only species noted in one of the three stomachs and contributing all but 3 of the 1238 otoliths found in a second. This tropical lanternfish normally undertakes a diurnal migration, but sometimes dense "balls" (schools) of *Benthosema* remain at the surface during daylight hours, where they have been observed under attack by birds, fish, and dolphins. Even though both dolphins had fed heavily upon *Benthosema*, we believe that *S. longirostris* had captured its prey several hundred feet down, whereas *S. graffmani* had fed at or near the surface.

A family-by-family discussion of the several thousand otoliths found in *Stenella* stomachs (Table II) follows.

Bathylagidae. — The *Bathylagus* otoliths in the stomachs of *Stenella longirostris* may have been those of *B. pacificus*, a species that lives more than 650 ft (200 m) beneath the surface and does not undertake a diurnal migration. They were not from *B. stilbius*, one of the commonest blacksmelt in that area and a known vertical migrator. None of the 15 otoliths in the three stomachs was from a large fish, perhaps 4 or 5 inches (100–125 mm) being about maximum.

Bregmacerotidae. — Only the stomachs of *S. longirostris* contained otoliths of *Bregmaceros bathymaster*, even though the species is known to migrate from

TABLE II. Numbers and kinds of fish otoliths found in the stomachs of two species of *Stenella* from the eastern tropical Pacific.[a]

	No. of otoliths							
	Stenella longirostris					*Stenella graffmani*		
Kinds of fishes eaten	1	2	3	4	5	1	2	3
Bathylagidae								
Bathylagus sp.	1	12			2			
Bregmacerotidae								
Bregmaceros bathymaster	11	76	6	89	61			
Centrolophidae ?								
Centrolophid ?	2	16			5	45		
Exocoetidae								
Oxyporhamphus micropterus						29		
Gonostomatidae								
Gonostomatid		7						
Vinciguerria lucetia	256	126	3	435	497		3	
Myctophidae								
Benthosema panamense	1	241	234	6147	983	4	1235	2357
Diaphus sp.	1							
Diogenichthys laternatus	949	97	60	302	15	4		
Hygophum sp.	1	35		4	1			
Lampanyctus idostigma	13	1			12			
Lampanyctus parvicauda	1916	465	30	27	282			
Myctophum aurolaternatum	2	107		7	106	8		
Symbolophorus evermanni	3	13						
Unidentifiable myctophids	78	54	13					
Paralepididae								
Stemonosudis ?					30			
Scopelarchidae								
Benthalbella sp.		1						
Unidentified		2			1	1		
Total	3234	1253	346	7011	1995	91	1238	2357

[a]Capture data (all specimens netted with tuna; M-numbers are accession numbers of Cetacean Research Laboratory, Little Company of Mary Hospital, Torrance, Calif.). *Stenella longirostris*: (1) M-57-66, about 350 miles (>550 km) SW of Manzanillo, Mex., in early 1966; (2) M-33-67, same general area as (1), but March 1967; (3) M-20-68, 12°N lat, 92°W long, February 21, 1968; (4) M-60-68, 13°N lat, 99°22′W long, April 16, 1968; (5) M-70-68, ca. 350 miles (±550 km) SW of Acapulco, Mex., April 1968. *Stenella graffmani*: (1) M-32-67, ca. 350 miles (>550 km) SW of Manzanillo, Mex., March 1967; (2) 150 miles off Acapulco, Mex., November 1959; (3) off Costa Rica, January 1963, Jules Crane data.

the depths into surface waters at night. A full-grown *B. bathymaster* would hardly exceed 3 inches (75 mm) and 1 g.

Centrolophidae ? — The most abundant otoliths in the stomach of *S. graffmani* M-32-67 were those of an unknown stromateoid. These otoliths (Fig. 2J) compare favorably with those of *Seriolella*, but according to Haedrich (1967), *Seriolella* "is restricted to cool temperate waters of the Southern

Hemisphere." We believe that they represent a species of centrolophid, but to our knowledge no member of the Centrolophidae is known from waters off tropical west Mexico. In any event, stromateoids (including Centrolophidae) are noted for their association as juveniles with floating debris, jellyfishes, and similar items, so the fishes these came from probably were eaten at or near the surface. Three of the *S. longirostris* stomachs we examined also contained otoliths of these stromateoids, but in much smaller numbers (Table II).

Exocoetidae. — Twenty-nine otoliths in the stomach of *S. graffmani* M-32-67 were from the shortwing flyingfish, *Oxyporhamphus micropterus*. This tropical species is strictly an inhabitant of surface waters. A really large individual might attain a length of 8 inches (200 mm) and a weight of 2 ounces (55 g).

Gonostomatidae. — Otoliths of *Vinciguerria lucetia* were relatively abundant in four of the five stomachs of *Stenella longirostris*, but only three were present in the *S. graffmani* stomachs. *Vinciguerria lucetia* inhabits depths of 650 to 1000 ft (200–300 m) during daytime, but migrates toward the surface at night and often can be dipnetted from under a bright light suspended above the surface. At maximum size they are just over 2 inches (50 mm) long and weigh but a few grams. Seven otoliths in one *S. longirostris* stomach (M-33-67, Fig. 2D) appear to be those of a gonostomatid but our comparative collection of gonostomatid otoliths was small, so we were unable to make a positive identification.

Myctophidae. — At least eight species of lanternfish had been fed upon by *S. longirostris* and three by *S. graffmani*. Six of these, *Symbolophorus evermanni*, *Myctophum aurolaternatum*, *Diogenichthys laternatus*, *Benthosema panamense*, *Diaphus*, and *Hygophum*, are found in surface waters at night, but the two species of *Lampanyctus* identified from the *S. longirostris* stomachs are deep (±400 m) during daylight hours and usually remain fairly deep (±200 m) at night. Three to 5 inches (75–125 mm) would be about maximum length for all of the myctophids except *Diogenichthys*, which seldom attains 2 inches when full grown.

Paralepididae. — The 30 barracudina otoliths in the stomach of *S. longirostris* M-71-68 are difficult to assign to a genus because of a shortage of comparative material. They are not from *Paralepis*, *Notolepis*, *Sudis*, or *Lestidium*, and several other genera are unknown in the eastern Pacific. Barracudinas are most abundant at depths of 1000–2500 ft (300–800 m), but range both above and below these depths. Adults of *Stemonosudis* are unknown in collections, but the otoliths in this dolphin stomach unquestionably were from adults. Based upon otoliths from other family members, these fish could have been 10–12 inches long (25–30 mm).

Scopelarchidae. — One otolith (from *S. longirostris*) was from a 4- to 5-inch (100–125 mm) *Benthalbella*, a fish that usually lives 800–1000 ft (250–300 m) down during daylight hours. Young individuals migrate into surface waters at night, but the habits of adults are unknown.

FISHES EATEN BY *LAGENORHYNCHUS OBLIQUIDENS*

We examined the stomachs of only two *L. obliquidens* that contained otoliths. One, a 5-ft 11-inch (179.5-cm) long female weighing 121 lb (55 kg), was found stranded at Long Beach, California, on August 21, 1967. Its stomach was gorged with partially digested fish (Pacific hake) and cephalopod remains (William Walker, personal communication), and yielded 57 otoliths (Table III, Fig. 3). Of the 43 hake otoliths, 29 were from fish that would have been

TABLE III. Numbers and kinds of otoliths found in the stomachs of *Lagenorhynchus obliquidens*, *Delphinus delphis*, and *Lissodelphis borealis* from southern California.

	No. of otoliths				
	Lagenorhynchus obliquidens		*Delphinus delphis*		
Kinds of fishes eaten	M-31-63[a]	M-26-67[a]	M-02-68[a]	RLB 69[b]	*Lissodelphis borealis*
Argentinidae					
Argentina sialis				3	
Bathylagidae					
Bathylagus sp.				1	
Batrachoididae					
Porichthys notatus				3	
Centrolophidae					
Icichthys lockingtoni					2
Engraulidae					
Engraulis mordax	4	12	127	14	
Melamphaidae					
Scopelogadus bispinosus					1
Merlucciidae					
Merluccius productus	3	43		97	2
Myctophidae					
Diaphus cf. *theta*			1		23
Lampadena urophaos					7
Lampanyctus cf. *ritteri*					5
Stenobrachius leucopsarus			3		
Triphoturus mexicanus					5
Unidentifiable myctophids					107
Ophidiidae					
Otophidium taylori				1	
Sciaenidae					
Seriphus politus		2			
Scomberesocidae					
Cololabis saira			2		
Total	7	57	133	119	152

[a]Accession number of Cetacean Research Laboratory, Little Company of Mary Hospital, Torrance, Calif.
[b]R. L. Brownell field number.

16–20 inches (400–500 mm) long, but the other 14 had come from smaller individuals (±8 inches or 200 mm long). These hake almost certainly had to be caught at a depth greater than 400 ft (120 m), and the anchovies may have been also, since all 12 otoliths represented large *Engraulis* (±7 inches or 180 mm), a size that typically inhabits depths of 400–650 ft (120–200 m). *Seriphus* is known only from relatively shallow (±30 m) inshore waters.

The other *L. obliquidens* (M-31-63) had stranded itself alive near the Santa Monica pier on August 29, 1963. In addition to the three *Merluccius* and four *Engraulis* otoliths, its stomach contained some cephalopod beaks and the fused pharyngeal tooth patch from a white seaperch, *Phanerodon furcatus*. *Phanerodon* is known only from relatively shallow (±30 m) inshore waters, but the hake and anchovies probably had been caught at a depth greater than 120 m.

FISHES EATEN BY *DELPHINUS DELPHIS*

Only two *D. delphis* stomachs (RLB 69, which was captured 4 miles off the San Pedro lighthouse between January and April 1964, and M-02-68, which was stranded alive at Zuma Beach on January 28, 1968) of many that Brownell examined contained otoliths. The 119 sagittae in the stomach of the 1964 dolphin represented a minimum of 63 fish belonging to six species in six families (Table III, Fig. 3). All of these fish could have been caught at or near the bottom in 650–800 ft (200–250 m) of water. None, including the hake, was very large, possibly 10 inches (250 mm) long, and weighing an ounce or so (±30 g). Five of these (*Argentina*, *Bathylagus*, *Merluccius*, *Otophidium*, and *Porichthys*) had not been reported among *D. delphis* food items previously.

The stomach of the Zuma Beach *Delphinus* contained 133 otoliths representing a minimum of 72 fish belonging to four species in three families and four genera. The Pacific saury, *Cololabis saira*, had to be taken at the surface, but the two lanternfishes and the anchovies probably were captured at least 600 ft (200 m) beneath the surface.

FISHES EATEN BY *LISSODELPHIS BOREALIS*

On November 8, 1967, an adult *L. borealis* was stranded near Imperial Beach, California, and Carl L. Hubbs, Scripps Institution of Oceanography, who examined the specimen, sent us 152 otoliths he washed out of the folds of its stomach along with "a multitude of squid beaks and lenses" (Carl L. Hubbs, personal communication). Most of these otoliths were digested considerably, but this did not preclude their identification to family, and in many cases to genus and species. Lanternfish otoliths representing at least four species (Table III, Fig. 3) comprised nearly 97% of the 152 sagittae. Less than a dozen of the 107 badly eroded otoliths that we identified only as myctophids (Table III) were of a type peculiar to *Lampanyctus*, whereas the others resembled *Diaphus* in size and configuration, but their state of digestion did not permit positive generic determinations. Possibly all seven species known to be represented had been captured and eaten at least 650 ft (200 m) beneath the surface, although $5\frac{1}{2}$-inch (135 mm) *Icichthys lockingtoni* are most abundant in the uppermost 150 ft (45 m).

FISHES EATEN BY *PHOCOENA SINUS*

The Gulf of California harbor porpoise was first described by Norris and McFarland in 1958, and to date only two entire specimens appear to have been collected (R. L. Brownell, unpublished data). Nothing has been reported in the literature on their food habits or other aspects of biology, and little is known of their behavior and distribution (Norris and Prescott, 1961). The stomach of one of the two complete specimens, an adult female (RLB 338), was examined on the beach about 20 km north of San Felipe, Baja California Norte, Mexico, on April 2, 1967. It contained 18 otoliths in rather poor condition, yet sufficient characteristics remained that the otoliths could be identified as having come from two species of shallow-water bottom-dwelling fishes. Thirteen of the otoliths (seven left and six right) were from the grunt *Orthopristis reddingi*, and five (three right and two left) were from the croaker *Bairdiella icistius* (Fig. 3). Both of these fishes are abundant throughout the upper Gulf of California; neither attains any great size, *Orthopristis* reaching perhaps 8 inches (200 mm), and *Bairdiella* possibly 10 (250 mm).

GENERAL DISCUSSION

From 18,164 otoliths found in the stomachs of 17 cetaceans, we determined that 18 kinds of fish had been eaten by *Kogia simus*, 16 kinds by *Stenella longirostris*, 9 by *Delphinus delphis*, 7 by *Stenella graffmani*, 7 by *Lissodelphis borealis*, 3 by *Lagenorhynchus obliquidens*, and 2 by *Phocoena sinus*. In addition we speculated on the sizes of fishes preyed upon as well as probable depth of feeding. On the basis of conventional recognition characters, only two fish species could have been identified: *Oxyporhamphus micropterus* in *Stenella graffmani* M-32-67, and *Merluccius productus* in *Lagenorhynchus* M-26-67.

Numerous well-documented accounts show that there is a direct relationship between otolith length and fish length for each species (Sastry, 1936; Fairbridge, 1951; Southward, 1962; Southward and Chapman, 1965; and many others). Thus, if one has accumulated this information for a given prey species, and undigested otoliths of that species have been found in the stomach of a predator, the lengths of the fishes eaten can be estimated reliably. Similarly, a reliable estimate can be obtained of the poundages ingested by utilizing a weight–length curve for the species in question.

There has been much speculation on whether the otoliths found in the folds of cetacean stomachs represent the remains of a single meal, or are an accumulation from several days of feeding. In our opinion, not even the 15,191 otoliths (representing more than 7596 fishes) reported by Schmidt (1923) represented a "full" meal for the *Delphinus* that they came from. We base this conclusion on the fact that it takes from 7000 to 10,000 adult *Benthosema panamense* to fill a gallon jar, which is a much smaller volume than a distended dolphin stomach. Thousands of *B. panamense* otoliths were found in the *Stenella* stomachs we investigated, and these lanternfishes are similar in size to the myctophids eaten by Schmidt's *Delphinus*.

In these days of extensive porpoise research, it would be a simple matter to design experiments that would produce answers as to how long cetaceans retain fish otoliths in their stomachs. Once a rate of digestion has been determined, details regarding periodicity of feeding could be worked out with a high degree of accuracy.

We feel certain that three of the cetaceans we investigated (*Kogia simus*, *Stenella longirostris*, and *Lissodelphis borealis*) had been feeding 800 ft (250 m) or more beneath the surface, and that two others (*Lagenorhynchus obliquidens* and *Delphinus delphis*) had descended to depths exceeding 400 ft (120 m). On the other hand, *Stenella graffmani* and *Phocoena sinus* appear to have fed within 100 ft (30 m) of the surface, and probably even shallower.

At least 8 of the 16 species fed upon by *Stenella longirostris* are known to ascend into (or nearly into) surface waters at night, but we doubt if the spinner dolphin did much foraging for food in these upper water layers. Trawl nets towed in these surface waters at night usually yield rather poor catches of myctophids and other mesopelagic vertical migrators, an indication that population densities are too sparse to supply either the diversity of species or numbers of individuals that made up the spinner's food. Similar logic applies to *Lissodelphis*, and the bottom-dwelling morid and macrourid remains found in the *Kogia simus* stomachs are proof of its deep-diving ability.

We have been impressed with the apparent dependency of the smaller cetaceans on mesopelagic fishes, especially myctophids, which are generally conceded as being one of the most widespread and abundant, yet least utilized, fish families in the world oceans. If the day ever arrives that man finds it economically feasible to harvest fishes from the scattering layers, uncontrolled exploitation could have a disastrous effect on our dolphin, porpoise, and whale populations.

ACKNOWLEDGMENTS

This investigation was supported in part by the National Science Foundation (NSF) through grants to Fitch (GB-1244 and GB-6490) for his research on otoliths. Brownell's work in Japan was made possible through a grant from the NSF to David K. Caldwell (GF-241) for cetacean studies, and while working at the Cetacean Research Laboratory, Little Company of Mary Hospital, Brownell received financial assistance through research grant HE 07146 from the National Heart Institute through the U.S. Public Health Service. Brownell received additional assistance from Toshio Kasuya through a grant from the Japan Society for Promotion of Science, which with the NSF co-sponsors the U.S.–Japan Cooperative Science Program.

Many individuals contributed ideas, information, specimens, or their talents, including the following: Jules Crane, Cerritos College; C. H. Fiscus,

U.S. Fish and Wildlife Service, Marine Mammal Biological Laboratory, Seattle, Wash.; John Geibel, California Department of Fish and Game, Terminal Island; Spano Giacalone, San Diego; Charles O. Handley, Jr., U.S. National Museum; Carl L. Hubbs, Scripps Institution of Oceanography; Robert J. Lavenberg, Los Angeles County Museum of Natural History; John Paxton, University of Southern California; Dale W. Rice, U.S. Fish and Wildlife Service, Seattle; Victor B. Scheffer, U.S. Fish and Wildlife Service, Seattle; Jack W. Schott, California Department of Fish and Game, Terminal Island; John Seapin, California Department of Fish and Game, Terminal Island; and William Walker, Cetacean Research Laboratory, Little Company of Mary Hospital, Torrance.

REFERENCES

Best, E. A. 1963. Contribution to the biology of the Pacific hake, *Merluccius productus* (Ayres). Cal. Coop. Ocean. Fish. Invest. Rept., **9**: 51–56.

Brown, David H., and Kenneth S. Norris. 1956. Observations of captive and wild cetaceans. J. Mammal., **37**(3): 311–326.

Cadenat, J. 1959a. Notes sur les Delphinidés ouest-africains. VI. — Le gros Dauphin gris (*Tursiops truncatus*), est-il capable de faire des plongées profondes? Bull. Inst. Franc. Afrique Noire, Ser. A, **21**(3): 1137–1141.

1959b. Rapport sur les petits Cétacés ouest-africains. Résultats des recherches entreprises sur ces animaux jusqu'au mois de mars 1959. Ibid., **21**(4): 1367–1440.

Cadenat, J., and M. Doutre. 1959. Notes sur les Delphinidés ouest-africains. V. — Sur un *Prodelphinus* à long bec capturé au large des côtes du Sénégal, *Prodelphinus longirostris* (Gray) 1828 Ibid., **21**(2): 777–798.

Daugherty, Anita E. 1965. Marine mammals of California. Calif. Dept. Fish. Game Sacramento, Calif. 87 p.

Fairbridge, W. S. 1951. The New South Wales tiger flathead, *Neoplatycephalus macrodon* (Ogilby). I. Biology and age determinations. Australian J. Marine Freshwater Res., **2**(2): 117–178.

Fiscus, Clifford H., and Karl Niggol. 1965. Observation of cetaceans off California, Oregon, and Washington. U.S. Fish Wildlife Serv., Spec. Sci. Rept. Fish., **498**: 27 p.

Frost, G. A. 1924. Fish otoliths from the stomach of a porpoise. Nature, **113**(2835): 310.

Haedrich, Richard L. 1967. The stromateoid fishes: systematics and a classification. Bull. Museum Comp. Zool., **135**(2): 31–139.

Nafpaktitis, Basil, and John Paxton. 1968. Review of the lanternfish genus *Lampadena* with a description of a new species. Los Angeles Co. Museum, Contrib. Sci., **139**: 29 p.

Norris, Kenneth S., and W. N. McFarland. 1958. A new harbor porpoise of the genus *Phocoena* from the Gulf of California. J. Mammal., **39**(1): 22–39.

Norris, Kenneth S., and John H. Prescott. 1961. Observations on Pacific cetaceans of California and Mexican waters. Univ. Calif. Publ. Zool., **63**(4): 291–402.

Sastry, N. S. 1936. The relation of the age to the length of a fish and the length of its otoliths. J. Madras Univ., **8**: 200–207.

Scheffer, Victor B., and John W. Slipp. 1948. The whales and dolphins of Washington State with a key to the cetaceans of the west coast of North America. Am. Midland Naturalist, **39**(2): 257–337.

Schmidt, Johannes. 1923. Consumption of fish by porpoises. Nature, **112**(2825): 902.

Scott, Thomas. 1903. Some further observations on the food of fishes, with a note on the food observed in the stomach of a common porpoise. Twenty-first Ann. Rept. (1902) Fish. Bd. Scotland, Pt. III, Sci. Invest. p. 218–227.

SOUTHWARD, G. MORRIS. 1962. A method of calculating body lengths from otolith measurements for Pacific halibut and its application to Portlock-Albatross grounds data between 1935 and 1957. J. Fish. Res. Bd. Canada, **19**(2): 339–362.

SOUTHWARD, G. MORRIS, AND D. G. CHAPMAN. 1965. Utilization of Pacific halibut stocks: study of Bertalanffy's growth equation. Rept. Intern. Pacific Halibut Comm., **39**: 1–33.

VLADYKOV, VADIM D. 1946. Etudes sur les mammifères aquatiques. IV. — Nourriture de Marsouin Blanc ou Béluga (*Delphinapterus leucas*) du fleuve Saint-Laurent. Contrib. Québec Dépt. Pèch., **17**: 157 p.

SECTION IV.
Faunal Surveys

Series of specimens taken during surveys of the biota of a region do far more than provide documentary evidence of the occurrence of species. Repetitive sampling at different seasons and in different years provides necessary information on many details of the lives of the species involved and on the cycles of the ecosystem itself. Detailed and thorough knowledge of this sort is needed if a region or any of its component wildlife is to be understood or interpreted properly. This principal is well expressed in the extract from Miller and Stebbin's book on Joshua Tree National Monument, a study undertaken at the behest of the National Park Service. The detailed information, of course, is provided in the species accounts that form the major portion of the book.

Leopold's *Wildlife of Mexico* is one of the more important surveys of game animals in the Americas, of interest to conservationists and sportsmen as well as scientists. The extract from the introduction to his book further indicates the importance of specimen documentation and the kinds of information that accompany it.

Many overlook the fact that specimens serve purposes beyond the original documentation. The specimens taken by Leopold in his survey formed the basis for Pitelka's taxonomic and distributional study which clarifies many points important to the management of game populations in Mexico. These faunal surveys and their associated specimens also form the basis for later studies of environmental or distributional change.

Additional Readings

Armstrong, David M. 1972. *Distribution of mammals in Colorado* (Preface, pp. v-vi, Introduction, pp. 1-12). Monograph no. 3, Museum of Natural History, University of Kansas.

Banks, Richard C. 1966. Terrestrial vertebrates of Anacapa Island, California. *Trans. San Diego Soc. Nat. Hist.* 14: 173-188.

Burleigh, Thomas D. 1958. *Georgia birds* (Preface, pp. ix-xii). University of Oklahoma Press, Norman.

Burleigh, Thomas D. 1972. *Birds of Idaho* (Preface, pp. xi-xii). Caxton Printers, Ltd., Caldwell, Idaho.

Grinnell, Joseph, H. C. Bryant, and T. I. Storer. 1918. *The game birds of California* (Preface, pp. iii-iv, and Introduction, pp. 1-5). University of California Press, Berkeley.

Lowery, George H., Jr. 1974. *The mammals of Louisiana and its adjacent waters* (Foreward by Ted O'Neil, pp. xvii-xviii, Preface, pp. xix-xx, Acknowledgements, pp. xxi-xxiii). Louisiana State University Press, Baton Rouge.

Trautman, Milton B. 1975. *The fishes of Ohio.* (Part VI, pp. 117-130). Ohio State University Press, Columbus.

THREE

PLAN AND SCOPE
OF FIELD STUDY

Alden H. Miller and Robert C. Stebbins

The need for a study of the vertebrate animal life of the Joshua Tree National Monument was felt keenly by James E. Cole during his period of service as superintendent of this unit of the national park system. Following informal discussions between Cole and Joseph S. Dixon and E. Lowell Sumner, Jr., of the research biology group of the National Park Service, the Museum of Vertebrate Zoology of the University of California, Berkeley, was urged in 1945 to undertake this task. Its purpose emerged as twofold: to conduct an original investigation of the adaptations, ecology, and distribution of the animals of this part of the Californian deserts; and to provide a source book on natural history for the park naturalists, useful to them and to the increasing number of visitors to the desert parks and monuments who as part of their intellectual recreation seriously inquire about the lives of desert animals and the problems they entail.

In May, 1945, a reconnaissance trip to the Monument was led by the senior author, and much time was spent in the field with Cole discussing the undertaking. As the plan developed, a series of trips was inaugurated by parties from the Museum directed toward becoming acquainted with all segments of biologic interest in the Monument, and scheduled to represent different seasons of the year. By spreading the field work over several years, chiefly in the period from 1945 to 1951, the annually variable conditions typical of deserts could in some measure be assessed. Basic to such a survey was extensive collecting of specimens * to determine with precision the

* Marked by asterisks in accounts of occurrences; see also p. 36.

species and subspecies represented. These documentary collections and the copious field notes on the occurrence and activities of the animals form the substance on which our report rests. Our experience in other desert areas is drawn upon in some instances to supplement our information obtained in the Monument.

The principal field trips and the personnel involved were as follows.

May 12–21, 1945: Alden H. Miller, Loye Miller, Virginia D. Miller, James E. Cole

October 12–24, 1945: A. H. Miller, Robert C. Stebbins, Ward C. Russell, Monroe D. Bryant

June 30–July 16, 1946: R. C. Stebbins, Henry G. Weston, Wade Fox, A. H. Miller

April 21–29, 1947: R. C. Stebbins, George W. Salt

August 22–September 15, 1950: A. H. Miller, W. C. Russell, Robert E. Bailey, John R. Hendrickson, Robert A. Norris, Jerry Russell

March 30–April 11, 1951: A. H. Miller, W. C. Russell

October 11–16, 1953: R. C. Stebbins, Robert Cogan

April 7–12, 1960: A. H. Miller, James D. Anderson, Gene M. Christman

The winter season, which is not represented by the Museum's field trips, was covered in very helpful manner by the continuous observations on birds by Frances Carter (1937), who was in residence at Twentynine Palms from December 30, 1933, to May 17, 1934, and from October 17, 1934, to May 30, 1935; and by Charles Sibley's work at Quail Spring from January 17 to 27 and from February 9 to 18, in 1946. Additionally a number of shorter winter trips were made by Loye Miller, and his notes with the specimens taken by him and by A. J. van Rossem, who often participated, have been made available to us.

Prior to 1945 Annie M. Alexander and Louise Kellogg had collected for the Museum in or near the Monument in February of 1935 and in April and May of 1941. Miss Alexander's direct acquaintance with the region and her enthusiasm for a concerted study of the area led to her support of the Museum's program of work begun a few years later.

The principal localities for field work are shown on the map (fig. 6). Many of these are near springs which were a convenience to the camping parties and of critical interest in observing their importance to desert animals. But at each camp we ranged widely in the vicinity. Even so, the field program could be only one of extensive sampling, as time and access did not permit us to visit every mountain sector and desert flat. Inevitably, then, our accounts must be incomplete in terms of rare and local occurrences. Certain occurrences in the Monument have been reported to us, often indirectly, without sufficient detail or substantiation to enable us to evaluate or incorporate the information in the book. Some of these items are part of continuing studies by other investigators, which are much to be desired, and which will in due course be presented in proper form in their reports. We believe that such additional data that may be forthcoming on rare or casual occurrences of animals will not materially alter the over-all picture of distribution of desert faunas which this book attempts to provide, even though they will undoubtedly improve it.

The faunal survey in the Joshua Tree Monument is a further unit in the regional surveys conducted over the years by the Museum of Vertebrate Zoology. It complements particularly those made before in nearby areas of California, namely the studies of the San Bernardino and San Jacinto mountains and, in the desert environments, the Death Valley, Providence Mountain, and lower Colorado River Valley regions (see map, fig. 20).

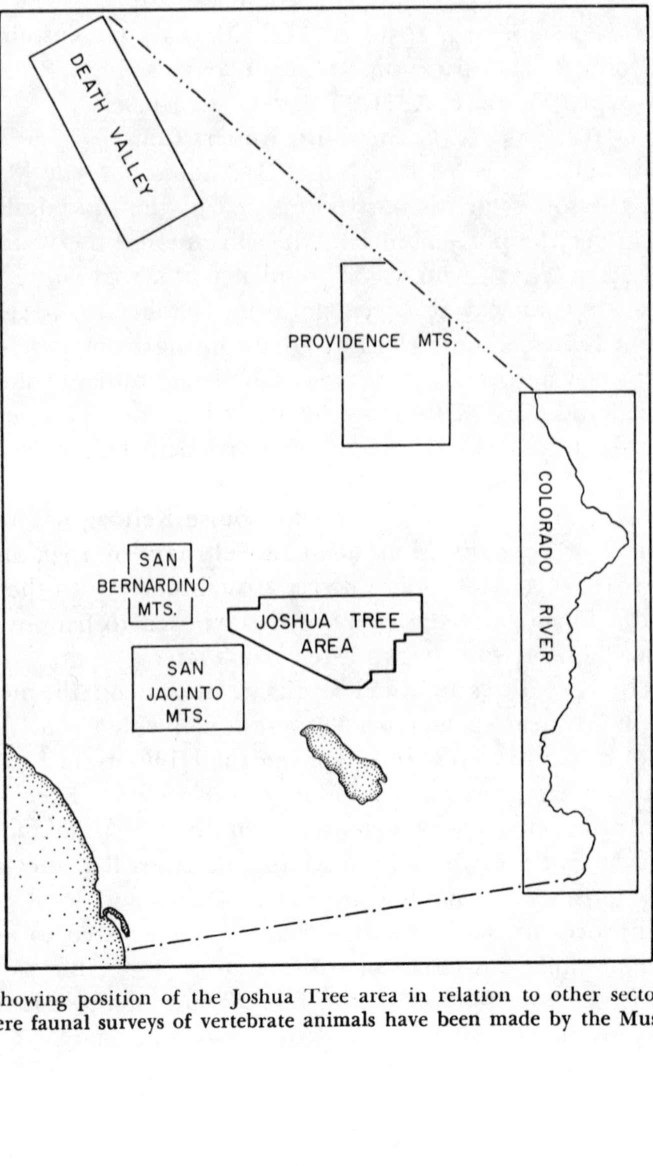

Fig. 20. **Map showing position of the Joshua Tree area in relation to other sectors of southern California where faunal surveys of vertebrate animals have been made by the Museum of Vertebrate Zoology.**

Reprinted from *Wildlife of Mexico: The Game Birds and Mammals*

Introduction

A. Starker Leopold
Illustrated by
Charles W. Schwartz

The doughty old soldier Bernal Díaz del Castillo, being the first visitor to record his impressions of Mexico, was the first to note the striking variety in altitudinal and vegetational land types of this new land. The route of the Conquest actually offered a fair cross section of the country. In the long hard struggle of 1519–1520 the Conquistadores progressed from the steaming swamps of Veracruz up jungle slopes and over the escarpment to the arid hills of Tlaxcala, thence through lovely pine uplands to the frozen shoulder of the volcano Popocatepetl, and finally into the green valley of Tenochtitlán, the heart of Montezuma's empire. Now in one day we can retrace the route in a whizzing automobile. The vegetation has been much altered, but the basic types remain in vivid contrast as Díaz described them.

Variety in Mexican topography, climate, and vegetation has given rise quite naturally to an equally varied animal life. Tropical and temperate faunas come together in unexpected combinations. For example, I recall sitting on a rimrock high in the Sierra de Tamaulipas one clear spring morning, with a level mesa of oak-grassland at my back and a lush tropical barranca at my feet. Among the oaks a wild turkey gobbled, bobwhites were whistling and Montezuma quail "buzzing," and a pair of black-bellied tree ducks explored the more venerable trees for nesting hollows. A white-tailed doe slipped over the rim seeking her bed in the cool brush below.

Down in the *monte* the monotonous whistles of tinamous and the soft cooing of red-billed pigeons and white-fronted doves echoed up to the rimrock. A handsome cock curassow sailed down the canyon on set wings, perhaps seeking his lady. And a gray fox slipped across a clearing. Where else in all North America could such an assortment of wildlife be observed from one rock?

But this animal wealth is shrinking in both variety and abundance as the Mexican landscape is cleared and put to economic uses. Overhunting likewise is contributing to the impoverishment of the fauna. Some of this loss is inevitable, of course, but not all. A strong program of wildlife conservation could guarantee for Mexico the perpetuation of much of the natural richness with which the country has been so generously endowed. Toward that end an initial step would seem to be an appraisal of the animal resources of the country, with formulation of ideas on possible ways to conserve and manage the various species and their habitats. This book is intended to be a contribution along that line.

In preparing this report, I have had in mind as my most important audience the people of Mexico who are interested—or might become so—in natural history, hunting, or conservation. There is no popular book available in Mexico that brings together what is known today of the native game animals. Such a volume, it would seem to me, is prerequisite to public understanding of the problems of wildlife conservation. To be sure, my report in its present form, being in English, will find only scant use in a Spanish-speaking country. However, a Spanish translation is planned for immediate publication in Mexico. The English edition, I hope, warrants publication for its interest to American conservationists and sportsmen.

I hasten to point out that a great many scientific papers have been written about the Mexican fauna, as the incomplete bibliography attached hereto will attest. These, in fact, were one of my principal sources of information. But most scientific writing is both unavailable and unintelligible to the average reader, and most of it concerns the problem of taxonomy, or animal classification, which is a field of science beyond the scope of all but a handful of specialists.

The idea for a survey of Mexican wildlife originated with William Vogt when he was Director of the Conservation Section of the Pan American Union. In 1944 I was employed by the Union to initiate such a survey, and I traveled in Mexico for two years, investigating local wildlife conditions in various parts of the Republic, from Sonora to Yucatán. A branch of the Mexican government gave financial support to the field work. In the fall of 1946 I joined the staff of the Museum of Vertebrate Zoölogy of the University of California, and the field program of the Museum has permitted continued work in Mexico to the extent of approximately one trip of a month or so each year. Gaps in the original survey have been filled one

Fig. 1. Travels and camps of the author in Mexico.

by one, until it now seems time to draw together the materials into the long-planned volume.

FIELD PROCEDURE

My travels in Mexico, with principal camps and study sites, are shown in figure 1. Most travel was by automobile, but to attain desirable campsites or to cross gaps in the highway system it sometimes was necessary to travel by other means of conveyance—airplane, railroad, boat, bus, chartered truck, oxcart, pack train, or afoot with native porters carrying the luggage. The surveys of migratory waterfowl were made in January, 1951, and January, 1952, in an airplane of the United States Fish and Wildlife Service, piloted by Robert H. Smith. These trips were of about a month's duration

Fig. 2. Arrival by oxcart at a camp near Galeana, Nuevo León.

and offered an opportunity to see from a low-flying airplane most of the Pacific Coast of Mexico and much of the interior.

The average length of stay at a campsite was about a week. After a camp was established I usually went first to officials of the local municipality to present my credentials and to explain my mission. This prevented much misunderstanding. Moreover, local residents could often give valuable information about the species of game in each locality and where to find

Fig. 3. A pleasant camp near Acuña, Tamaulipas. A wild turkey hangs in the oak tree awaiting preparation as a specimen.

them. Notes on hunting practices, logging of the forests, farming, grazing, fire, and many other pertinent points could be picked up in these conversations. A fair share of the data included here were obtained in this way. I owe a debt of gratitude to the many *comisarios, alcaldes,* and other local officers who supplied information and who helped along my field work in innumerable other ways.

At each camp I spent most of my time walking over the country with shotgun, field glasses, camera, and notebook, observing the range and the animals themselves and collecting specimens. I attempted to preserve at least one specimen of every kind of game bird and small game mammal in the vicinity of each camp. These served as positive locality records for the various species and also yielded information on weights, food habits, breeding condition, and general biology. In most camps no effort was made to collect deer or other large game mammals, even when they were numerous, since sight records are entirely reliable and it did not seem worth while to take the time to kill and prepare the specimens. Plants were collected which were either dominant species in the local floras or were serving as food for game. Before leaving a camp I attempted to have in my notes and in the specimen cases (1) a complete record of the game species present and the relative abundance of each; (2) information on the local flora—both its present composition and its probable composition at its climax; (3) notes

Fig. 4. Preparations at Tenosique, Tabasco, for a flight into the rain forest of eastern Chiapas. Photograph by D. L. Spencer.

on land use as related to game range; (4) notes on hunting as practiced by the local people and by visiting sportsmen; and (5) photographs of the country showing the types of range and the effects of land use on the habitat.

These field data were then supplemented with information derived from the voluminous scientific literature on Mexican fauna and flora and from some of the principal collections of Mexican vertebrates housed in museums of the United States and Mexico. The range maps for each species are based on the composite records of my own observations, museum specimens, and published records of occurrence.

Reprinted from *The Condor*

NOTES ON THE DISTRIBUTION AND TAXONOMY OF MEXICAN GAME BIRDS

By FRANK A. PITELKA

Collections of birds from Mexico recently acquired by the Museum of Vertebrate Zoology have yielded new information on the distribution and taxonomy of a number of species. It is the objective of this paper to place on record noteworthy results of critical study of one of these collections, that accumulated by A. Starker Leopold in the course of a survey of Mexican game birds. The present report is thus confined to members of the tinamiform, anseriform, galliform, gruiform, charadriiform, and columbiform groups. There were also included in this study specimens representing these groups, whether qualifying as "game birds" or not, from two other collections: first, the Milton S. Ray Collection of Guerreran birds obtained by W. W. Brown, and second, a collection from Sinaloa and Sonora obtained in October and November, 1946, by A. S. Leopold, F. A. Pitelka, and W. C. Russell. Field observations supplementing data from specimens are added from the writer's original notes on Sinaloan and Sonoran species. Topics dealt with here are range extensions of species, critical evaluations of racial characters, revisions of racial distributions, and new data on seasonal distribution and breeding season.

Crypturellus cinnamomeus. Mexican Tinamou.—The only record of this species from Campeche known to me is that of Traylor (1941:199), who described *C. c. intermedius* from the southern part of that state. A series of eight specimens taken at San Juan, Campeche (see fig. 28), in the central, coastal part of the state, also represent that race. Comparative material of other races together with an examination of data provided by previous studies of this species (especially Conover, 1933; Brodkorb, 1939; and Hellmayr and Conover, 1942) indicate that *intermedius* represents a valid race.

In describing *C. c. intermedius*, Traylor (1941:200) did not comment on presence or absence of barring on the breast of females (see Hellmayr and Conover, 1942:66-69, footnotes). Five females of that race in the Leopold Collection vary considerably in this character and form as nearly a graded series as could be expected of merely five specimens. One lacks any barring except for a faint suggestion of it on the sides of the breast; at the other extreme is a specimen prominently barred on the upper breast and rather heavily so on the sides of the breast. The other three specimens fall between these. There is some suggestion that this character may vary with age, to a partial degree at least.

Traylor (*loc. cit.*) states that in dimensions *intermedius* "falls in between" the larger races, *sallei* and *soconuscensis*, and the smaller races, *goldmani* and *cinnamomeus*. No data are given except the dimensions of the type of *intermedius*, a female (wing, 166 mm.; culmen, 25.5 mm.). Dimensions from the series of *C. c. intermedius* in the Museum of Vertebrate Zoology are given in table 1. They suggest rather that in size *intermedius* may be similar to or smaller than *cinnamomeus*.

Anser albifrons. White-fronted Goose.—But one record of this species is available from Sonora (van Rossem, 1945:44). Nine individuals were noted on a laguna at Pitahaya (40 km. SE Empalme, 100 ft.), on October 29, 1946; these plus six others were present early on October 30. None was seen again until November 3 when two individuals visited the laguna briefly.

Dendrocygna autumnalis. Black-bellied Tree Duck.—Friedmann (1947:190) has described two races, *D. a. fulgens* from extreme northeastern Mexico and southern Texas, and *D. a. lucida* from Sonora and Veracruz south at least to Costa Rica. These are said to differ in degree of blackness of the abdomen, *lucida* being the darker. I have examined ten specimens (6 ♂♂, 4 ♀♀) from these geographic ranges. Friedmann had

Table 1

Measurements of *Crypturellus cinnamomeus*

	Wing	Culmen	Tarsus
C. c. intermedius			
3 males	154 mm.	27.4 mm.	51.5 mm.
	146	30.2	46.8
	146	27.9	48.3
5 females	148	29.6	48.4
	151	28.5	46.0
	152	28.8	46.2
	147	25.6	45.2
	145	28.5	46.2
C. c. cinnamomeus			
1 male	156	31.1	47.7
1 female	163	30.2	51.2
C. c. goldmani			
1 male?[1]	152	26.3	47.7

[1] Unsexed, but plumage characters appear to be those of a male.

only eight specimens of *fulgens* (4 ♂♂, 4 ♀♀) and 11 of *lucida* (sex representation not given). While there is a slight suggestion of difference between them, this is transgressed by the sexual differences evident in our series. Thus, a male and female collected on August 4, 1945, in Tamaulipas, display the very differences used by Friedmann to distinguish *fulgens* from *lucida*. A comparable degree of sexual difference occurs between a male collected on June 29, 1932, in Costa Rica, and a female collected on July 23, 1912, in El Salvador. Friedmann did not comment on sexual differences within the races he described. In view of these facts, the status of *lucida* is probably doubtful until the characters attributed to it are re-appraised with due recognition of sexual differences.

Cairina moschata. Muscovy Duck.—No previous records from Nuevo Leon have been found. Two females were collected on July 22 and one female on July 29, 1945, at La Union (29 km. NE General Teran, 1000 ft.), in that state.

Anas platyrhynchos. Mallard.—A female was collected at Pitahaya, Sonora, on November 3, 1946. Van Rossem (1945:45) cites no Sonoran records of wintering mallards earlier than December 6.

Anas diazi. Mexican Duck.—The characters attributed to *A. d. novimexicana* by Phillips (1923:56) are "an irregular wavy barring of fulvous color on the mantle" presumably not present in *A. d. diazi* and "a tendency to a darker and more mottled breast." In seven specimens of *A. diazi* available to me, neither of these characters appears valid. With respect to the first, the "wavy barring" is produced by subterminal, more or less V-shaped fulvous marks on the mantle feathers, which, when the latter are worn, become more evident than in fresh-plumaged birds and produce the character described. But these subterminal marks are present in *A. d. diazi* as well as *novimexicana*. What is more significant is their presence in two males and their absence in five females. As to the second character, five specimens from the range of *A. d. diazi* are consistently darker on the breast than two of *A. d. novimexicana*, but there is no evidence of difference in mottling. Our material suggests that if any real differences exist, they are, in *A. d. novimexicana,* lighter ventral coloration, in particular on the abdomen, and more extensive subterminal fulvous marks on back, scapular, and flank feathers, and possibly buffier (less gray) coloration on the throat. Because of individual variation, however, the racial distinctions in this species should be reexamined further.

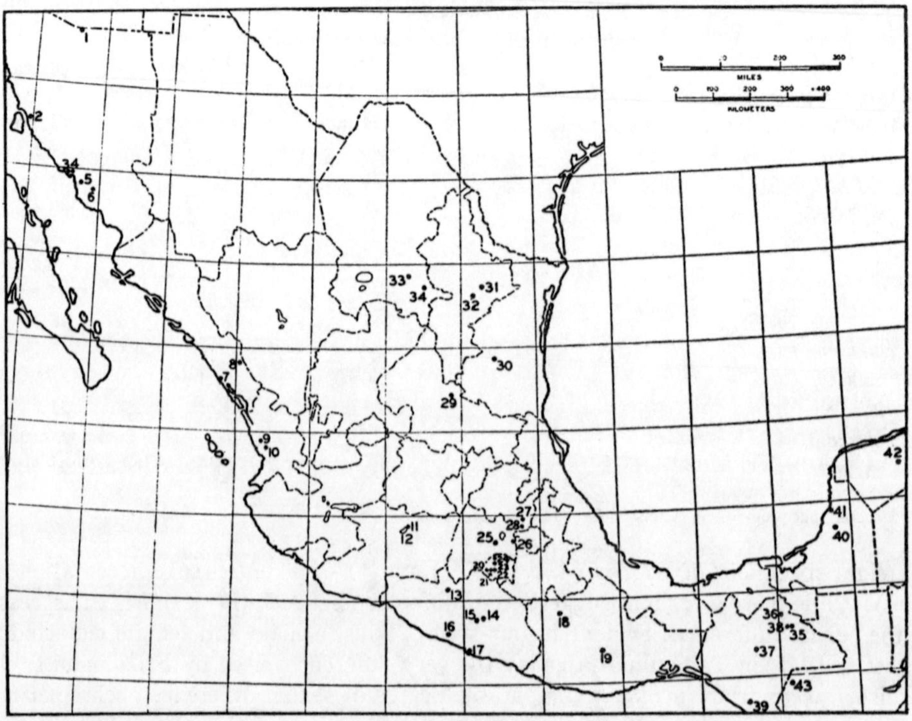

Fig. 28. Map of Mexico showing localities mentioned in the text. Cuapongo, a Guerreran locality near Omilteme and not found on reference maps, is not shown. A key follows.

SONORA
1. Nogales
2. Sierra Seri
3. Guaymas
4. Empalme
5. Pitahaya
6. Bacum

SINALOA
7. Mazatlan
8. El Batel

NAYARIT
9. Santiago Ixcuintla
10. Sauta

MICHOACAN
11. Quiroga
12. Lake Patzcuaro

GUERRERO
13. Poliutla
14. Chilpancingo
15. Omilteme
16. Atoyac
17. Acapulco

OAXACA
18. Tamazulapam
19. Tlacolula

MORELOS
20. Tetecala
21. Lake Rodeo
22. Alpuyeca
23. Cuernavaca
24. Tres Marias

DISTRITO FEDERAL
25. Pedrigal

ESTADO MEXICO
26. Rio Frio

HIDALGO
27. Lake Tecocomulco
28. Apam

SAN LUIS POTOSÍ
29. Santo Domingo

TAMAULIPAS
30. Ciudad Victoria

NUEVO LEON
31. La Union
32. General Teran

COAHUILA
33. Hipolito
34. Saltillo

CHIAPAS
35. Monte Libano
36. Palenque
37. San Cristobal
38. El Real
39. Escuintla

CAMPECHE
40. San Juan
41. Campeche

YUCATAN
42. Xocempich

GUATEMALA
43. Nenton

I find no references to the occurrence of this species in Morelos, and although thought to occur in Hidalgo by Phillips (1923:56), he had no specimens. I have examined two collected at Lake Rodeo, 4000 feet, Morelos, on March 11, 1945, and one collected at Lake Tecocomulco (25 km. N Apam, 8200 ft.), Hidalgo, on November 19, 1944.

Anas discors. Blue-winged Teal.—A male taken by W. C. Russell at Pitahaya, Sonora, on November 2, 1946, appears to represent the second record for Sonora (see van Rossem, 1945:46).

Spatula clypeata. Shoveller.—No records from Michoacan are known to me. One female was collected at Lake Patzcuaro, 6800 ft., on February 22, 1945.

Aythya valisineria. Canvasback.—Van Rossem (1945) does not list this species from Sonora. A female was collected at Pitahaya (40 km. SE Empalme, 100 ft.), on October 28, 1946.

Aythya affinis. Lesser Scaup Duck.—A male was collected by W. C. Russell at Pitahaya, Sonora, on October 30, 1946. Previously the species had not been recorded in that state earlier than November 10 (van Rossem, 1945:50).

Oxyura jamaicensis. Ruddy Duck.—But three authentic records are available from Sonora (van Rossem, 1945:51). One individual was observed at Pitahaya on October 30 and November 1, 1946.

Ortalis wagleri wagleri. Wagler Chachalaca.—Two specimens were collected at El Batel (70 km. NE Mazatlan, 5100 ft.), Sinaloa, on October 20, 1946. Altitudinal occurrence here is above that of most other records.

Dendrortyx macroura striatus. Long-tailed Partridge.—The Ray Collection contains several specimens that provide data on the breeding period of this species in Guerrero. A female collected on August 2, 1945, at Cuapongo, is marked "ovaries greatly enlarged." A half-grown juvenile from the same locality was collected on September 27, 1940. Another juvenile, in an early stage of the postjuvenal molt, was collected at Chilpancingo on August 20, 1937.

Lophortyx gambelii. Gambel Quail.—Eighteen specimens were collected at Pitahaya and one at Bacum, in Sonora, in the period from October 27 to November 2, 1946. The series includes twelve fresh-plumaged males. These display the characters attributed to *L. g. fulvipectus* by Ridgway and Friedmann (1946:296). Hellmayr and Conover (1942: 235) did not find *fulvipectus* separable from *gambelii*, but as Ridgway and Friedmann (*loc. cit.*) point out, truly fresh-plumaged specimens are necessary for comparisons.

According to van Rossem (1932:132; 1945:68) intergradation northward between *fulvipectus* and *gambelii* is confined to the regions east and northeast of Guaymas. A series of 15 Sonoran specimens additional to those mentioned above suggests, however, that this area is better assigned to the range of *fulvipectus*, and that intergradation is evident farther north and interiorward, to the Sierra Seri (opposite Tiburon Island) coastwise and to north-central Sonora interiorward. One specimen in this series, an adult male (MVZ 94933), collected approximately 37 miles south of Nogales, October 27, 1941, is indistinguishable in color from *L. g. fulvipectus*, and the remaining specimens—males and females—are intermediate to varying degree in the several characters distinguishing *fulvipectus* from the nominate race.

Nelson (1899:26) originally described *fulvipectus* as a race not only darker, but also larger in size of bill than the nominate race. The latter distinction is weak and therefore not useful in racial segregation, but it nevertheless exists, as shown by the following data on chord of the exposed culmen:

	Range	Mean
22 ♂♂ *gambelii* (Calif., Ariz., N. M.)	10.1-11.8	11.1 mm.
12 ♂♂ *fulvipectus* (southern Sonora)	11.1-12.5	11.7 mm.

With reference to Moore's (1947:28) newly described race, *L. g. friedmanni*, from northern Sinaloa, examination of specimens in the Museum of Vertebrate Zoology from Sonora and neighboring parts of Arizona and southern California suggests that most of the characters attributed to *friedmanni* are either the same as in *fulvipectus* or they are slight accentuations of them, This is not unexpected in a population to the immediate south of the present known range of *fulvipectus* if we recognize the north-south gradient over Sonora in characters separating *fulvipectus* from the nominate race (see above). If characters attributed to *friedmanni* by Moore (*loc. cit.*) in his description are checked against the series examined in this study, only two of six may prove to represent variations not already indicated in *fulvipectus*. These, in themselves variable characters, are streaking of undertail coverts (darker and "definitely" brown in males of *friedmanni*) and color of forehead (almost "solid" brown in females of *friedmanni*). I have not been able to examine any Sinaloan specimens. Moore examined only eight males and two females of "*friedmanni*." All these considerations lead me to suggest that *friedmanni* is merely an end-point in a north-south gradient already evident and recognized in *fulvipectus*.

Colinus virginianus. Bob-white.—A series of sixteen specimens from Morelos, and particularly eleven of these from Lake Rodeo (15 km. E Tetecala), clearly indicate that the locally resident population of *C. v. nigripectus* is mixed to varying degree by transplants from regions outside of the range of this race. The atypical specimens suggest the races *maculatus* or *aridus*, and it is known that birds from the vicinity of Ciudad Victoria in southern Tamaulipas have been released at Lake Rodeo (A. S. Leopold, MS).

Five males and four females collected at El Real, Chiapas, on November 29 and December 7, 1945, apparently represent the race *insignis*. This locality is outside of the known range of *insignis* and within what is generally considered the range of *minor* (see Ridgway and Friedmann, 1946:338). The type locality of the latter form, Palenque, Chiapas, is only about 45 kilometers northwest of El Real. The range of *insignis* (type locality, Nenton, Guatemala) is stated by Brodkorb (1942:3) as the "Comitan-Nenton Valley of eastern Chiapas and western Guatemala," considerably to the south of El Real. The five males from El Real agree with the characterization of *insignis* given by Ridgway and Friedmann (*loc. cit.*). In this series, the throat is consistently black, and the breast and belly are entirely chestnut. On four there is a postocular stripe, with slight touches of white above or in front of the eye, areas that are otherwise black. On the fifth, a superciliary stripe extends almost to the bill. The fifth in addition is spotted white on the malar area from a point beneath the eye posteriorly to the sides of the throat. The neighboring race *minor*, however, is white-throated. Comparative material available to me is inadequate for evaluation of characters displayed by the four females.

These data indicate that the allocation of all of northeastern Chiapas to the range of *minor* is not correct and that, therefore, there is little likelihood that *minor* also occurs in "the neighboring sections of Guatemala" (Ridgway and Friedmann, 1946:338). Continuity in the range of *insignis* between El Real and the Comitan-Nenton Valley yet remains to be established.

Odontophorus guttatus. Spotted Partridge.—Hellmayr and Conover (1942:280) recognize the race *matudae* described by Brodkorb (1941:4) from southern Chiapas, but with some reservations. Racial characters given by them are broader width of white throat stripes and tear shape of breast spots. Brodkorb had but two specimens, Conover has examined but three, and the latter author points out that because of the individual variation displayed in this species, the above characters may not be valid bases for racial distinction. A series of ten first-year and adult specimens from Monte Libano (100 km. NE San Cristobal, 2500 ft.), Chiapas, collected December 1 to 4, 1945, bear out Conover's suspicion. There is such variation in form, size, and degree of black margination of ventral spots and also in width of throat stripes that I regard *matudae* as not recognizable. This evidence is considered pertinent to the status of *matudae* in spite of the distance between Monte Libano and the geographic range ascribed to *matudae,* the vicinity of Escuintla (Hellmayr and Conover, 1942:281). Ridgway and Friedmann (1946:373) recognize no races in this species.

Cyrtonyx montezumae. Montezuma Quail.—Intergradation between *C. m. montezumae* and *C. m. sallei* in the southern part of the range of the former race is suggested by two specimens I have examined: A first-year male collected at Tres Marias (20 km. N Cuernavaca, 9800 ft.), Morelos, on January 30, 1945, departs from typical members of *montezumae* in the presence of varying amounts of chestnut on the posteriormost flank feathers. Its dorsal coloration is warmer and browner than that of any other specimen of *montezumae* examined, but there is no suggestion of *sallei* in dorsal patterning. Such is clearly present, however, in the second specimen, a first-year male collected 8 kilometers southwest of Rio Frio, 8900 feet, Estado de Mexico, on January 13, 1945. The shaft streaks of back feathers are a warm buff with or without black borders, the ground color is olive-buff, and the transverse, more or less oval spots appear in a row on either side of each shaft streak. In typical *montezumae,* however, the light buff shaft streaks are consistently black bordered and contrast with transverse black bars and lines on a reddish brown or grayish brown ground color. Thus, dorsal patterning of the specimen from Rio Frio is more similar to that of *sallei* than that of *montezumae;* but dorsal coloration is unlike either, although the general impression provided by color is perhaps closer to *montezumae* than to *sallei.*

Late breeding of this species in the highlands of Sinaloa is indicated by the record of a family group of two adults and at least three or four more than half-grown young observed on October 16, 1946, at El Batel (70 km. NE Mazatlan, 5100 ft.). Only the male was collected (left testis, 5 mm.).

Rallus limicola limicola. Virginia Rail.—A specimen of doubtful sex was taken at Alpuyeca (25 km. S Cuernavaca, 3000 ft.), Morelos, on January 17, 1946.

Charadrius vociferus vociferus. Killdeer.—This species, lacking from previous lists of Guerreran birds, is represented in the Ray Collection by a female collected on February 20, 1940, at Chilpancingo.

Eupoda montana. Mountain Plover.—No previous records from Coahuila are known to me. One female was collected at Hipolito (50 km. NW Saltillo, 5000 ft.), on February 23, 1946.

Tringa solitaria. Solitary Sandpiper.—A female collected at Cuapongo, Guerrero, on August 15, 1939, represents the race *cinnamomea* on the basis of characters of both color and size (wing, 136.0 mm.). I find no previous report of this species from that state.

Totanus melanoleucus. Greater Yellow-legs.—This species, in Sonora, is considered by van Rossem (1945:81) as "probably a fairly common transient and winter visitant, although records are few and scattered." The earliest fall date given by him from a coastal locality is November 14. Greater Yellow-legs were noted daily and repeatedly, but in small numbers, from October 27 to November 4, 1946, on a laguna at Pitahaya, Sonora.

Actitis macularia. Spotted Sandpiper.—A series of four specimens was obtained at Chilpancingo, Guerrero, between the dates of October 12 (1940) and April 26 (1939). Davis (1944:10) reports three specimens taken in the late summer (August 9-13, 1942) while Griscom (1934:373) lists one taken in January. Evidently this species occurs in Guerrero as both migrant and winter resident.

Capella gallinago delicata. Wilson Snipe.—The Ray Collection contains a female collected at Chilpancingo, Guerrero, on December 20, 1939. This species is not listed from that state by Griscom (1934) or more recent authors.

Recurvirostra americana. Avocet.—This species, in Sonora, is considered by van Rossem (1945:88) as a "transient and winter visitant of uncertain abundance." At Pitahaya, from October 27 to November 4, 1946, one flock of approximately 25 individuals as well as scattered individuals and small groups were seen daily.

Sterna albifrons. Least Tern.—Another species not previously reported from Guerrero is represented in the Ray Collection by a female collected at Acapulco on September 7, 1944, in a late stage of the postjuvenal molt. Presumably it represents the race *browni*.

Columba flavirostris. Red-billed Pigeon.—A series of 17 specimens from Texas, Mexico, and El Salvador sheds some light on the variation of this species in Mexico. According to van Rossem (1930:197) and Peters (1937:65), two races are found: *C. f. flavirostris* ranges northward through Mexico to southern Texas and southeastern Sonora; *C. f. restricta* occurs in southern Sonora and Sinaloa. *C. f. restricta* is said to be paler than the nominate race, with less red on under parts and more prominent grayish white margins on greater wing coverts. Among specimens examined in this study, only the characters of paler average coloration and ventrally reduced red coloration appear to be usable in distinguishing *restricta*. Individual variation is marked, and the magnitude of difference in average coloration is approximately comparable to that separating eastern and western races of *Zenaidura macroura*. Three adult males examined in this study together suggest a western pale race and are tentatively assigned to *restricta*: one from Sauta (12 km. S Santiago Ixcuintla, 150 ft.), Nayarit, April 18, 1946, and two from Poliutla, Guerrero, December 3, 1944. One of the latter two is ventrally the palest of specimens seen by me. These specimens would seem to bear out Hellmayr and Conover's (1942:452) postulation that the range of this race extends south beyond Nayarit. Summer occurrence of "*Columba flavirostris flavirostris*" in Guerrero has recently been recorded by Davis (1944:10).

It remains doubtful whether or not the character of more prominent white wing-covert margins actually serves in separating *restricta* from *flavirostris*. Again considerable individual variation occurs in this character among specimens seen by me. While the three specimens assigned to *restricta* do show more white wing-covert margining than *most* specimens of *flavirostris* from northeastern Mexico, they do not show nearly as much white as two from the southern part of the range of the nominate race: an adult female, San Juan, Campeche, November 23, 1945, and an adult male, Xocempich, Yucatan, November 7, 1945. Degree of wear is undoubtedly a factor in the prominence of this character; both of the specimens just cited are incompletely or recently molted, and an immature male collected at La Union, Nuevo Leon, on July 24, 1945, shows newly grown greater wing coverts with striking white margins.

Van Rossem (1930:197) notes that according to Ridgway's (1916:301) figures, birds from Texas average the largest in this species. This fact is borne out by data on wing length of adult males I have examined, as follows:

Texas, Tamaulipas, Nuevo Leon	205, 204, 200 (worn), 200 (worn), 197, 194 mm.
Nayarit, Guerrero, Vera Cruz, Yucatan	196, 196, 194, 193, 192, 192 mm.
El Salvador	187 mm.

Zenaidura macroura. Mourning Dove.—A large series of specimens from Guerrero and Oaxaca clearly indicates that populations occurring in these states do not represent the race *marginella* alone, as maintained by Ridgway (1916:349) and Peters (1937:83), but represent, in the majority of specimens, another race that is smaller and darker.

Occurrence of both breeding and wintering populations of this species in southern Mexico complicates any effort to analyze racial distribution.

Information on the breeding distribution of the Mourning Dove in southern Mexico is scanty. It supposedly breeds in Oaxaca, according to Ridgway (*loc. cit.*), and this is confirmed by four specimens from that state collected by A. S. Leopold, as follows: a male (testis 14 mm.), 6 kilometers east of Tamazulapam (6600 ft.), September 5, 1945; a male (testis 9.5 mm.) and two females (largest ova 1.5 and 4.5 mm., respectively), all from Tlacolula (5000 ft.), September 7, 1945. The dates appear to be too early for migrants from the north, and condition of gonads indicates that these specimens were obtained probably toward the end or just after the nesting season. Nesting of *Zenaidura macroura* in Guerrero, however, is apparently not known. Ridgway (*loc. cit.*) does not include that state in his statement of the breeding range, and Griscom (1934:372) lists four specimens taken at Chilpancingo in the period from October 27 to January 5. The Ray Collection contains 28 specimens from the same state representing all months except May, June, August, and September. Three of these specimens, all from Chilpancingo, were collected in April and July: female ("ovary slightly enlarged"; wing, 134 mm.), April 11, 1943; female ("ovary slightly enlarged"; wing, 140 mm.), April 12, 1944; and female ("ovary minute"; primaries undergoing molt), July 16, 1944. The latter specimens suggest the possibility that the species may breed in Guerrero, presumably at elevations above that of Chilpancingo.

In spite of the present uncertainty concerning the actual breeding range of *Z. macroura* in southern Mexico, it is possible to say that most of the specimens from Oaxaca and Guerrero examined in this study do not belong to the race *marginella*. They are clearly darker, and also smaller (see table 2), than specimens of *marginella* from southwestern United States and northwestern Mexico. The former thus suggest *Z. m. carolinensis*. Among the 36 specimens from Guerrero and Oaxaca, there are but seven actual or possible breeding specimens, already cited. The remaining 29 were obtained at times other than the presumed period of breeding and may represent, in the main, migrants and wintering birds, those from Guerrero perhaps entirely so. The probable occurrence of *Z. m. carolinensis* as a winter resident in Oaxaca and Guerrero is suggested by recently published banding data. Ridgway (1916:346) considered *carolinensis* to be a migrant along only the gulf coast of Mexico, but its presence in the interior of Mexico has been demonstrated by records of McClure (1943:407), who reports recoveries of Iowan doves in Nayarit, Jalisco, and Distrito Federal, and also in El Salvador. Two specimens collected in Michoacan on February 21, 1945, at Quiroga (6800 ft.), resemble the Guerreran and Oaxacan birds in color and thus also suggest *Z. m. carolinensis*. I do not find any breeding records of *Z. macroura* from Michoacan. McClure (letter. March 29, 1948) reports a recovery of an Iowan dove from Curimeo, Michoacan.

The western race, *Z. m. marginella,* however, does occur as a winter resident at least as far south as Guerrero. Two, not included in the total of 36 mentioned above or in table 2, collected at Chilpancingo, are here assigned to *marginella*: a male (wing, 157 mm.) taken on February 17, 1944, and a female (wing, 149 mm.) taken on November 27, 1944. Color characters of these specimens fall within the range of variation of both *carolinensis* and *marginella.*

In summary, it is shown that specimens from Guerrero and Oaxaca, which represent a breeding population in part in the latter of these two states, are distinguishable from *Z. m. marginella* by smaller size and darker coloration and are indistinguishable from *Z. m. carolinensis*. The latter race occurs in the winter in central Mexico and southward; also, most of the specimens were obtained in winter and periods of migration. These facts suggest that the majority of Guerreran and Oaxacan specimens represent wintering individuals of *carolinensis*. *Z. m. marginella* also occurs in Guerrero as a winter resident. *Z. macroura* is not known to breed in Guerrero, but there is some indication that it may do so. Foregoing data also suggest the possibility of a geographically separated population in southern Mexico that is racially indistinct from *carolinensis*; in this connection, a reexamination of breeding specimens from eastern Mexico is necessary.

Table 2

Wing Length in Three Populations of *Zenaidura macroura*

	Number of specimens	Mean with standard error	Standard deviation
Southern California, Northeastern Baja California, Nevada, Sonora, and Durango (*Z. m. marginella*)			
Males	20	148.9±0.7 mm.	2.94
Females	15	142.9±0.7	2.59
Guerrero and Oaxaca (*Z. m.* subspecies?)			
Males	17	143.9±0.7 mm.	2.76
Females	13	139.7±0.8	2.80
Ontario, New York, Michigan, Georgia, South Carolina, Wisconsin, and Missouri (*Z. m. carolinensis*)			
Males	11	143.3±1.1 mm.	3.44
Females	6	139.2±1.8	3.97

Zenaida asiatica. White-winged Dove.—Thirty-seven specimens from various parts of Mexico and adjoining parts of Texas, New Mexico, and Arizona have been examined in the light of distinctions between the races *Z. a. asiatica* and *Z. a. mearnsi* provided by Ridgway (1916:379, 383). Ridgway included Nuevo Leon and San Luis Potosí in the range of the western race (*mearnsi*), but this is now regarded as erroneous in view of the geographic gap, the northern part of the Mexican plateau, separating northern populations of the two races. Thus, Peters (1937:87) and Hellmayr and Conover (1942:500) restrict *mearnsi* to western Mexico and adjoining parts of the United States.

These authors, however, also extend the range of *mearnsi* southward and eastward to Puebla, and this I regard as open to question. Geographic relations of the two races would lead one to expect intergradation between them to occur to the south of the central plateau. This is corroborated by data from the series examined in this study. Thus, populations in the southern Mexican states east of the gulf-coast lowlands and Oaxaca are more or less intermediate in size between *asiatica* and *mearnsi*. Wing lengths of fifteen specimens from Guerrero are as follows:

Males	166, 165, 162, 161, 159, 158, 155 mm.
Females	164, 163, 160, 157, 156, 154, 153, 148 mm.

The averages, 160.8 and 156.8 mm., respectively, are below those given by Ridgway (*loc. cit.*) for *mearnsi*, but are closer to *mearnsi* than to *asiatica*. Two specimens obtained to the *west* of Puebla, however, appear to represent *asiatica*: a male (wing, 152 mm.), taken at Lake Rodeo, Morelos, on January 26, 1945, and a male (wing, 150 mm., tip slightly worn), taken at Pedrigal (5 km. S Mexico City, 7900 ft.), Distrito Federal, on July 31, 1943. In both of these specimens wing length is below the average (156.3 mm.) given by Ridgway for the race *asiatica* rather than the reverse, as might be expected from the range statements of Peters and of Hellmayr and Conover. Average color differences between the two races are slight and difficult to appraise; southern Mexican specimens, however, appear to be more or less intermediate and probably average closer to *asiatica* than to *mearnsi.*

A male taken along the east slopes of the Sierra Madre Oriental at Santa Domingo, 4500 feet, San Luis Potosí, on September 27, 1945, measures 165 mm. in wing length (to the tip of the third developed primary; the second is not fully grown). At present it seems best to regard this specimen as a large variant of *Z. a. asiatica.*

For Sonora (van Rossem, 1945:99), late fall or winter records north of latitude 29° are lacking, although the species is known to winter in southern Arizona in small numbers. A loose flock of approximately 25 individuals was observed in flight over bottomland timber of poplar and willow just south of Casita (40 km. S Nogales, 3300 ft.), on November 9, 1946.

Leptotila verreauxi. White-fronted Dove.—Guerreran specimens of this species are assigned by recent authors to *L. v. angelica* [=*brachyptera*], and seven specimens examined in this study are so placed. Characters of these specimens display an approach toward *fulviventris,* however, as vinaceous coloration of the breast and buffiness of posterior under parts both average somewhat darker than in typical *angelica* to the north. One specimen, an adult male collected 10 kilometers south of Atoyac, Guerrero, on December 12, 1944, is as richly buffy ventrally as typical *fulviventris* and represents an extreme in the entire series of *angelica* available to me.

Leptotila jamaicensis. Jamaican Dove.—This species, represented by the race *gaumeri,* is known on the Mexican mainland only in the northern part of the Yucatan Peninsula (Hellmayr and Conover, 1942:566). An adult male was obtained at San Juan, Campeche, on November 24, 1945.

Leptotila cassinii. Cassin Dove.—The known occurrence of this species within Mexico is apparently limited to southern Chiapas (Hellmayr and Conover, 1942:567). Four specimens were obtained well to the northeast of this locality, at Monte Libano (100 km. NE San Cristobal, 2500 ft.), Chiapas, December 1-5, 1945. Presumably they represent the race *cerviniventris.* It may be noted that this range extension is similar to that described under *Colinus virginianus insignis.*

Oreopelia albifacies rubida. White-faced Quail-dove.—Among 11 specimens in the Ray Collection from Cuapongo and Omilteme, Guerrero, two provide breeding data: a male collected at the latter locality on July 16, 1945, with "testes fully enlarged" and a well-grown juvenal female from the same locality collected on August 12, 1943.

ACKNOWLEDGEMENTS

Through the kindness of Robert T. Orr and Grace I. Crowe of the California Academy of Sciences, I have been able to examine Mexican specimens in the collection under their care. I am indebted also to A. H. Miller and A. S. Leopold for suggestions and comment on the contents of this paper.

LITERATURE CITED

Brodkorb, P.

1939. New subspecies of birds from the District of Soconusco, Chiapas. Occ. Papers Mus. Zool., Univ. Mich., no. 401:1-7.

1942. A new race of bob-white from interior Chiapas. Occ. Papers Mus. Zool., Univ. Mich., no. 467:1-4.

Conover, H. B.

1933. The races of the tinamou Crypturellus cinnamomeus. Proc. Biol. Soc. Wash., 46:113-118.

Davis, W. B.

1944. Notes on summer birds of Guerrero. Condor, 46:9-14.

Griscom, L.

1934. The ornithology of Guerrero, Mexico. Bull. Mus. Comp. Zool., 75:365-422.

Hellmayr, C. E., and Conover, B.

1942. Catalogue of birds of the Americas. Part I, number 1 (Chicago, Field Mus. Nat. Hist.).

McClure, H. E.

1943. Ecology and management of the mourning dove, *Zenaidura macroura* (Linn.), in Cass County, Iowa. Iowa Agric. Exp. Sta. Research Bull. 310.

Moore, R. T.

1947. New species of parrot and race of quail from Mexico. Proc. Biol. Soc. Wash., 60:27-28.

Nelson, E. W.

1899. Descriptions of new birds from Mexico. Auk, 16:25-31.

Peters, J. L.

1934. Check-list of the birds of the world. Vol. II (Cambridge, Harvard University Press).

1937. Check-list of the birds of the world. Vol. III (Cambridge, Harvard University Press).

Phillips, J. C.

1923. A natural history of the ducks. Vol. II (Boston, Houghton Mifflin Co.).

Ridgway, R.

1916. The birds of North and Middle America. Part VII (Washington, Government Printing Office).

Ridgway, R., and Friedmann, H.

1946. The birds of North and Middle America. Part X (Washington, U. S. Government Printing Office).

Traylor, M. A., Jr.

1941. Birds from the Yucatan peninsula. Field Mus. Nat. Hist., zool. ser., 24:195-225.

van Rossem, A. J.

1930. Two new subspecies of birds from Sonora. Trans. San Diego Soc. Nat. Hist., 6:197-198.

1932. The avifauna of Tiburon Island, Sonora, Mexico, with descriptions of four new races. Trans. San Diego Soc. Nat. Hist., 7:119-150.

1945. A distributional survey of the birds of Sonora, Mexico. Occ. Papers, Mus. Zool., Louisiana State Univ., no. 21:1-379.

Museum of Vertebrate Zoology, Berkeley, California, March 1, 1948.

SECTION V.
Endangered Species

The protection and preservation of endangered and threatened wildlife species has recently become one of the major concerns of wildlife management and conservation. States, nations, and international groups have produced lists of endangered species, and statutes and regulations that control human interaction with them. Because they possess a uniqueness and universality not found in vernacular names, scientific names form the basis for these lists and define the populations listed. The definition inherent in a scientific name is both inclusive and exclusive; it includes all individuals that share a common set of characteristics and excludes all that do not share those features. When used in statutes or regulations, such a name acquires a legal as well as a biological meaning. It seems obvious that the use of scientific names must be based on thorough taxonomic studies.

Taxonomic opinion does vary, however. One authority may use a single name for two or more populations while another uses two or more names for the same populations. Where there are such differences of opinion, it is important (from a legal, as well as a biological, standpoint) to know in which sense a name on a list is used. Unfortunately, lists of endangered species (or of species protected for other reasons) seldom give any indication of the authority for the names or other indication of the sense of usage. (For an exception see paper 20 in Section VI of this book). I suggest that this situation should be corrected as lists are revised.

An example of the change of status that can accompany a change in taxonomy is given in the paper by Waithman and Roest. A subspecies of Kit Fox, *Vulpes macrotis macrotis,* has been considered to be extinct. By including in that taxon other populations previously known as *V. m. devius, V. m. arsipus,* and *V. m. tenuirostris,* the subspecies *macrotis* (but not the Riverside County population) is suddenly and miraculously resurrected from extinction.

Two features of the Endangered Species Act of 1973 invite particular attention from taxonomists. The law provides that populations in a particular geographic area (e.g., part of a species or subspecies) can be considered as threatened or endangered "species," and that forms sufficiently similar to endangered species so that problems of identification might occur may receive the same protection as the endangered forms. It is conceivable that an insular population of a subspecies might be listed and protected while the mainland population of the same subspecies and all other subspecies of the species were not. Alternatively, a healthy population might be given rigorous protection merely because it differs only slightly from a critically endangered one. Precise taxonomic discrimination and characterization are obviously necessary in such situations.

Museum studies play a basic role in endangered species programs by providing names for species that may or may not already be endangered at the time of naming, and by sharpening and redefining the limits of a listed taxon. Prior to 1973, the former was essential before a population could be placed on the list; the latter will continue to be of importance as the number of listed species grows.

John Aldrich's description of the Mississippi Sandhill Crane involved naming a population already known to be endangered. The subspecies was added to the list of endangered species on June 4, 1973, soon after it formally received a name. Critical habitat was designated in an emergency rule-making in June 1975 and finally determined in August 1977. At issue in the critical habitat of this form was the placement of a major interchange on an interstate highway, the first endangered species case to reach the Supreme Court. Etnier's description of the Snail Darter from the Little Tennessee River both pointed out the endangered status of the fish and pinpointed the cause. Partly because of the existence of this fish, the completion of the Tellico Dam on that river is open to question at the time of this writing; the Supreme Court has ruled that the dam may not be completed, but the 1978 amendment to the Endangered Species Act may provide a way to accommodate both the fish and the dam.

Similarly, R. R. Miller's rediscovery of the Little Colorado Spinedace pointed out the precarious status of that species, and noted that its protection was dependent on action by the Arizona Game and Fish Department. The state placed it on a list of species needing special protection, and it is slated for proposed addition to the federal list of endangered species in the near future.

The review of the Kit Fox, already mentioned, verified the previously questioned validity of that taxon, already listed as endangered. Paradiso and Nowak's study of the Pronghorn similarly verified the taxonomic status of an endangered subspecies and provided characters by which it could be identified.

Finally, Wilbur's short paper on California Condor specimens illustrates another way in which museum collections are useful to studies of endangered species. Nearly every intensive study of individual endangered forms has begun with a query to museums throughout the world for information on their holdings and for data from the specimens labels.

Additional Readings

Aldrich, John W., and Kenard P. Baer. 1970. Status and speciation in the Mexican Duck *(Anas diazi)*. *Wilson Bull.* 82: 63-73.

Banks, Richard C. 1975. Plumage variation in the Masked Bobwhite. *Condor* 77: 486-487.

Banks, Richard C., and Roy E. Tomlinson. 1974. Taxonomic status of certain Clapper Rails of southwestern United States and northwestern Mexico. *Wilson Bull.* 86: 325-335.

Hahn, Paul. 1963. *Where is that vanished bird?* Royal Ontario Museum, Toronto.

Harris, Herbert S., Jr., and Robert S. Simons. 1976. The paleogeography and evolution of *Crotalus willardi,* with a formal description of a new subspecies from New Mexico, United States. *Bull. Md. Herpet. Soc.* 12: 1-22.

Hubbard, John P. 1977. *The biological and taxonomic status of the Mexican Duck.* Bull. no. 16, New Mexico Department of Game and Fish.

Montanucci, Richard R. 1970. Analysis of hybridization between *Crotaphytus wislizenii* and *Crotaphytus silus* (Sauria: Iguanidae) in California. *Copeia* 1970: 104-123.
Paradiso, John L., and Ronald M. Nowak. 1971. A report on the taxonomic status and distribution of the Red Wolf. *Special Scientific Report—Wildlife no. 145,* Bureau of Sport Fisheries and Wildlife.
White, Clayton M. 1968. Diagnosis and relationships of the North American tundra-inhabiting Peregrine Falcons. *Auk* 85: 179-191.

Vol. 85, No. 5, pp. 63–70 30 August 1972

PROCEEDINGS
OF THE
BIOLOGICAL SOCIETY OF WASHINGTON

A NEW SUBSPECIES OF SANDHILL CRANE FROM MISSISSIPPI

By John Aldrich
Division of Wildlife Research, Fish and Wildlife Service
U.S. Department of the Interior, Washington, D.C.

A very small resident group of sandhill cranes (*Grus canadensis*) is barely surviving in southern Mississippi. This is a remnant of a larger and more extensively distributed population of these cranes which formerly occurred along the Gulf Coastal Plain of Louisiana, Mississippi, and Alabama (Walkinshaw, 1949; Valentine and Noble, 1970).

The former status of these birds is very poorly documented, but it is probable that they always occurred in fairly widely separated pockets of suitable habitat and never in very large numbers. Breeding has been recorded for southern Louisiana by Cooke (1914), Figgins (1923), McIlhenny (1943), and Lowery (1960); for Mississippi by E. A. McIlhenny (1938), Turcotte (1947), Walkinshaw (1949), Valentine (1963), and Valentine and Noble (1970); and for Alabama by Howell (1928) and Imhof (1962).

Apparently the only place where the sandhill crane breeds on the Gulf Coast west of Florida at present is in Mississippi. There are no recent records of occurrence of these birds during the breeding season in other parts of the known former breeding range or elsewhere in that region. However, the Mississippi population has received particular attention in recent years under the stimulus of the endangered species program of the Fish and Wildlife Service. Reports by Valentine (1963) and Valentine and Noble (1970) give informative accounts of the status of these birds. These authors report that the Mississippi sandhill crane population, estimated as between 38 and 40 birds, is confined to Jackson County, generally between the Pascagoula River on the east and the Jackson-Harrison County line on the west; U.S. Highway 90 on the south and Bluff Creek on the north. The breeding grounds are near Ocean Springs and Fontainebleau and are divided into seven rather distinct nesting areas (Valentine and Noble, 1970).

The nesting habitat is described by Walkinshaw (1949), Valentine (1963), and Valentine and Noble (1970) as wet areas in semiopen pine flats. It would appear that, although savannalike in aspect, breeding areas have more trees than is usually the case with sandhill cranes (Walkinshaw, 1949). In fact, the habitat of these birds is so different from the usual wide open environment of sandhill cranes that Valentine and Noble suggest that this population should be considered a distinct ecotype.

It has been generally assumed that the Gulf Coast breeding sandhill cranes were of the same subspecies as those in Florida, *Grus canadensis pratensis*. However, as Hellmayr and Conover (1942, footnote) point out, the Gulf Coast breeding sandhills have never been critically studied and have been placed with *pratensis* only for "geographical reasons." The fact is they could not have been studied to determine their correct subspecific status since it appears that no adult specimens have ever been preserved (Walkinshaw, 1949, and personal communication).

In the course of the above mentioned endangered species program, eggs were taken beginning in 1965 from the nests of the Mississippi sandhill crane population for the purpose of propagation in captivity. This has resulted in the rearing at the Patuxent Wildlife Research Center, Laurel, Maryland, of eight Mississippi birds to adult plumage stage by 1971. One of these died on 15 September 1971, and became specimen 564841 in the National Museum of Natural History, Smithsonian Institution. Meanwhile, sandhill cranes were also reared from eggs taken in southern Florida, Okefenokee Swamp, Georgia, and Oregon.

Although slight differences were noted among individuals

in the downy stage, as soon as the first of the Mississippi cranes acquired its contour feathers, a color much darker than that of those reared from eggs taken in Florida was apparent. This dark color has been consistent in all others of this population reared subsequently and has been noticed by all persons who have viewed them critically. Subsequent to this discovery, the very dark appearance of the sandhill crane photographed beside its nest in Mississippi and shown in Figure 2, Plate 14, in Walkinshaw's (1949) monograph was noted, and Walkinshaw (personal communication) verified this condition in Mississippi cranes which he had observed.

On 19 October 1970, I made a detailed comparison of seven living, adult plumaged sandhill cranes hatched from eggs taken in Mississippi with an equal number originating in southern Florida, two from Okefenokee Swamp, Georgia, and examples of greater sandhills originating at Malheur National Wildlife Refuge, Oregon. This time of year was selected because all birds were in freshly molted plumage. These birds were intermingled in the same pen and observations were made by walking among them and examining the plumage at close range.

There was no difficulty whatever in distinguishing the Mississippi birds from all others at a glance. Compared with the south Florida and Oregon birds, they were very much darker on back, wings, neck, sides of head and breast. In these respects, the Florida birds were very similar in color to the greater sandhills (*Grus canadensis tabida*) from Oregon and to lesser sandhills (*G. c. canadensis*) taken as migrants in New Mexico. The dark color of the neck and ear patches of the Mississippi birds had the effect of making the white cheeks stand out more distinctly in contrast than was the case in cranes from other regions.

The two Okefenokee Swamp cranes appeared slightly darker than the examples from south Florida but were not nearly as dark as those from Mississippi. The significance of this is that the Okefenokee specimens represent a breeding area very close to the type-locality of *pratensis* restricted by Harper (1942) to Bronson, Levy County, Florida. Thus, the Okefenokee birds almost certainly are representative of the same population that breeds near the locality in northern Florida where Bartram (1791) described his "Great Savanna Crane" that was the basis of the formal diagnosis of *pratensis* (Meyer, 1794). Since migrant cranes from breeding populations in the central northern United States (*tabida*) are known to winter in northern Florida together with resident birds (Walkinshaw, 1960; Williams, 1970), it might not have been possible to determine which one Bartram was looking at when he provided the basis for the name *pratensis*, if he had not mentioned nesting. However, in the description of his exploration of the "Alachua Savanna" he mentioned seeing these cranes with nests and eggs. This definitely identifies his "Great Savanna Crane" with the breeding population of that area and therefore it seems certain that the name *G. c. pratensis* applies to the resident population of Florida and not the migrant birds from the north.

The differences between the birds originating in south Florida and those from Okefenokee Swamp, viewed side by side in the crane pen, were so slight that it seems justifiable to consider them the same subspecies, *Grus canadensis pratensis*. The seven specimens from Mississippi-taken eggs, however, were quite different, being consistently much darker than the Florida and Okefenokee birds. Also, they were darker than examples of greater sandhill cranes (*tabida*) and lesser sandhill cranes (*canadensis*) living in nearby pens and also specimens of the Canadian sandhill crane (*rowani*) and Cuban sandhill crane (*nesiotes*) in the National Museum of Natural History. The darker coloration considered a distinguishing characteristic of *pratensis* as compared with *tabida* (Peters, 1925) was not obvious with living birds, although it was noted in comparison of museum specimens of Great Basin breeding *tabida* and Florida *pratensis*.

So far, it has not been possible to determine the sex of the living cranes at Patuxent and thus segregate their measurements accordingly. However, comparisons of the range of measurements of the Mississippi birds of both sexes combined with the ranges known for breeding populations of previously described subspecies are indicative of major differences in overall size or in proportions. These ranges of measurements (in mm) are as follows.

Mississippi (8 live captive birds hatched from Mississippi eggs). Wing (chord), 470–493; bill (from posterior end of nostril), 82–91; tarsus, 216–258.

Florida, *pratensis* (28 live captive birds hatched from southern Florida eggs). Wing (chord) 451–506; bill (from posterior end of nostril) 82–103; tarsus, 218–274.

Greater, *tabida* (20 live captive birds hatched from Oregon eggs). Wing (chord), 479–550; bill (from posterior end of nostril), 86–111; tarsus, 215–260.

Canadian, *rowani* (10 breeding season specimens, from Walkinshaw, 1965) Wing (chord), 456–524; bill (from posterior end of nostril), 73–88; tarsus, 205–239.

Lesser, *canadensis* (50 breeding season specimens, from Walkinshaw, 1965) Wing (chord), 420–503; bill (from posterior end of nostril), 63–76; tarsus, 162–210.

From comparison of these measurements it would appear that Florida and Mississippi sandhills are quite similar in size and proportions. They both have tarsi as long or longer than greater sandhills from Oregon but are smaller in other dimensions. Measurements of lesser sandhill cranes (*canadensis*) showed them to be smaller in all respects than either Florida or Mississippi birds, particularly in bill and tarsus length. The Canadian sandhill (*rowani*) appears to have a shorter bill and tarsus but a similar or possibly longer wing than the Florida and Mississippi populations. The size of the Cuban subspecies (*nesiotes*), except for a longer bill, is apparently about the same size as the lesser sandhill according to measurements by Todd (1916) and so is definitely smaller than either Florida or Mississippi populations in wing and tarsus.

The difference in the Mississippi breeding population of sandhill cranes from all others of this species warrants its description as a distinct subspecies which may be known as:

Grus canadensis pulla new subspecies

Mississippi Sandhill Crane

Type: Adult ♀ USNM No. 564841 in the National Museum of Natural History. Hatched 15 May 1969, from egg taken 7 May 1969, about

7 miles northwest of Fontainebleau, Jackson County, Mississippi. Reared at Patuxent Wildlife Research Center, Laurel, Maryland, leg band No. 49, wing tag No. 207. Died 15 September 1971, as result of leg injury. Measurements: wing, primaries broken—unmeasurable; bill (from posterior end of nostril to tip), 91 mm; tarsus, 251 mm, middle toe (without claw), 72 mm. Plumage: worn nuptial, just beginning to molt to fresh autumn; back (worn feathers) sepia to clove brown, (fresh feathers) iron gray to dark mouse gray; breast (worn feathers) brownish olive, (fresh feathers) deep mouse gray; dorsal side of neck (worn feathers) deep neutral gray.

Subspecific characters: Similar in size to *G. c. pratensis* that breeds from southern Georgia south to southern Florida but colored portions of plumage much darker throughout. It is the darkest colored of all populations of *Grus canadensis*. From the greater sandhill crane, *G. c. tabida*, breeding in Oregon, Idaho, and Michigan, it differs on the basis of both smaller size (except tarsus) and darker color. From the Arctic breeding lesser sandhill crane, *G. c. canadensis*, the new Mississippi subspecies can be distinguished by larger size in all dimensions and darker color. From the breeding sandhills of the boreal forest region of western Canada, *G. c. rowani*, Mississippi birds can be distinguished particularly by longer tarsi and darker color. From the Cuban sandhill crane, *G. c. nesiotes*, they differ in larger size in all respects except bill length and in having darker colored plumage.

Geographical distribution: Jackson County, Mississippi, near Ocean Springs and Fontainebleau between the Pascagoula River on the east and Jackson-Harrison County line on the west; U.S. Highway 90 on the south and Bluff Creek on the north.

Examples of G. c. pulla examined: Eight in adult plumage at Patuxent Wildlife Research Center, Laurel, Maryland, hatched from eggs taken in Jackson County, Mississippi, by Jacob M. Valentine, Jr.: wing tag No. 86, hatched 3 May 1966; No. 110 hatched 3 May 1965; No. 87, hatched 16 May 1966; No. 184, hatched 26 May 1967; No. 215, hatched 4 June 1969; No. 212, hatched 16 May 1969; No. 207 (type specimen USNM 564841) hatched 15 May 1969; No. 341, hatched 30 May 1970.

Immature ♀, USNM 531663, hatched 18 May 1967, from egg taken in Jackson County, Mississippi, by Jacob M. Valentine, Jr. Died Patuxent Wildlife Research Center, Laurel, Maryland, 7 January 1968.

Downy young, sex ?, USNM 530210, hatched from egg taken 4 May 1966, in Jackson County, Mississippi, by Jacob M. Valentine, Jr., died in aviary of John Lynch, Lafayette, Louisiana, 6 July 1966.

Downy young, ♀, USNM 532029, hatched from egg taken 30 May 1968, in Jackson County, Mississippi, by Jacob M. Valentine, Jr.; died at Patuxent Wildlife Research Center, Laurel, Maryland, 12 June 1968.

Downy young, ♂?, Museum of Zoology, University of Michigan 104976, collected Ocean Springs, Mississippi, 20 April 1940, on day of hatching, by Lawrence Walkinshaw; weight 93.4 grams.

Downy young, sex?, Museum of Zoology, Louisiana State University, 2009, collected "Lake Misere, Cameron Parish, Louisiana," 7 April 1938, by E. A. McIlhenny, 3 days old. A note on the label says: "this specimen almost certainly came from southern Mississippi where E. A. McIhenny studied nesting sandhill cranes in the spring of 1938. See Auk 55, 1938: 598–602 - GHL."

Literature Cited

Bartram, W. 1791. Travels through North and South Carolina, Georgia, East and West Florida [etc.], first ed. Philadelphia: pp. xxxiv + 522, 8 pl., 1 map.

Cooke, W. W. 1914. Distribution and migration of North American rails and their allies. U.S. Dept. Agric. Bull. 128: 50 pp.

Figgins, J. D. 1923. The breeding birds of the vicinity of Black Bayou and Bird Island, Cameron Parish, Louisiana. Auk 40: 666–677.

Harper, F. 1942. William Bartram's names of birds. Proc. Rochester Acad. Sci. 8: 208–221.

Hellmayr, C. E. and B. Conover. 1942. Catalogue of birds of the Americas. Zool. Ser. Field Mus. Nat. Hist. Part 1, No. 1: 636 pp.

Howell, A. H. 1928. Birds of Alabama (2d Ed.) U.S. Dept of Agriculture, Bur. of Biol. Surv. and Dept. of Game and Fisheries of Alabama. 384 pp.

Imhof, T. A. 1962. Alabama birds. Dept. of Conservation, Game and Fish Div. by Univ. of Alabama Press. 591 pp.

Lowery, G. H. 1960. Louisiana birds. Louisana Wildlife and Fisheries Commission by Louisiana State University Press, 567 pp.

McIlhenny, E. A. 1938. Florida crane a resident of Mississippi. Auk 55: 598–602.

————. 1943. Major changes in the bird life of southern Louisiana during sixty years. Auk 60: 541–549.

Meyer, F. A. A. 1794. Zoologische Annalen 1: 412 pp (286 and 296).

Peters, J. L. 1925. Notes on the taxonomy of *Ardea canadensis* Linné. Auk 42: 120–122.

Todd, W. E. C. 1916. The birds of the Isle of Pines. Annals Carnegie Mus. 10: 146–296.

Turcotte, W. H. 1947. The sandhill crane in Mississippi. Mississippi Game and Fish. June: 8–9.

Valentine, J. M., Jr. 1963. The status of the Florida sandhill crane in Jackson County, Mississippi. Administrative Report, Bureau of Sport Fisheries and Wildlife, Atlanta, Ga. pp. 1–10.

Valentine, J. M., Jr., and R. E. Noble. 1970. A colony of sandhill cranes in Mississippi. Journ. Wildlife Mgmt. 34: 761–768.

WALKINSHAW, L. H. 1949. The sandhill cranes. Bull. 29, Cranbrook Institute of Science. 202 pp.

———. 1960. Migration of the sandhill crane east of the Mississippi River. Wilson Bull. 72: 358–384.

———. 1965. A new sandhill crane from central Canada. Canadian Field-Naturalist 79: 181–184.

WILLIAMS, L. E., JR. 1970. Spring departure of sandhill cranes from northern Florida. Auk 87: 156–157.

Vol. 88, No. 44, pp. 469–488 22 January 1976

PROCEEDINGS
OF THE
BIOLOGICAL SOCIETY OF WASHINGTON

PERCINA (*IMOSTOMA*) *TANASI*, A NEW PERCID FISH FROM THE LITTLE TENNESSEE RIVER, TENNESSEE

BY DAVID A. ETNIER
Department of Zoology, University of Tennessee, Knoxville, Tenn. 37916

The Little Tennessee River begins in Macon County, North Carolina, and flows northwest to enter the Tennessee River near Lenoir City, Loudon County, Tennessee. The flow of the Little Tennessee River is interrupted by Fontana Reservoir, a large warmwater impoundment in Swain and Graham counties, North Carolina. Tailwaters of Fontana Reservoir are consistently cold and well suited to a recreational trout fishery. The cold water habitat continues through Cheoah, Calderwood, and Chilhowee reservoirs. Below Chilhowee Reservoir a trout fishery persists for about 21 river km, ending approximately 2.5 river km above the U.S. Highway 411 bridge, Monroe County, Tennessee. Although trout occur sporadically in the lower 32 km of river, warmwater fishes are predominant.

Records indicate that the original fish fauna of the Little Tennessee River system was rich. Among the more interesting species known from this system are an undescribed darter of the subgenus *Catonotus* (R. E. Jenkins, in lit.), *Hybopsis monacha* (Cope), *Cycleptus elongatus* (Lesueur), and *Percina burtoni* Fowler. *Noturus baileyi* Taylor, a madtom that is likely extinct, is known only from a rough fish removal operation in Abrams Creek, tributary to the Little Tennessee River (Taylor, 1969). A number of reports of lake sturgeon (*Acipenser fulvescens* Rafinesque) from recent sightings and captures in the river suggest the continued existence of what is possibly a reproducing population of this vanishing species in the lower Little Tennessee River. A reasonable number of fish collection records are available from the tributaries and headwater portions of the system, and from the 21 km of river utilized as a trout fishery below Chilhowee Dam. The lower 32 km of the Little Tennessee River remained uncollected until recently. It seemed possible that (despite environmental changes that have taken place over the years) some elements of the original main channel fish fauna might persist in the lower portion of the river. An additional impoundment, Tellico Reservoir, is under construction by the Tennessee Valley Authority, and would innundate virtually all of the river remaining below Chilhowee Dam. With this in mind, Robert A. Stiles and I visited the river at Coytee Spring, River Mile 7, Loudon County, Tennessee, on 12 August 1973, in an attempt to survey the fish fauna with the aid of face masks and snorkels. Although very few fishes were seen, a single specimen of a very unusual darter was observed and "scooped up" by hand by the author. Subsequent examination of this specimen indicated that it was a new species of the genus *Percina*, subgenus *Imostoma*.

In the genus *Percina*, the subgenus *Imostoma* had until recently been envisioned as containing two species, *P. shumardi* (Girard) and *P. uranidea* (Jordan and Gilbert). Three specimens collected in Shoal Creek, tributary to the Etowah River (Mobile Bay drainage), Cherokee County, Georgia, by Don Scott and O. Tyson, were recognized by J. D. Williams in 1966 as representing a distinct species allied with *P. uranidea*. Impoundment of Alatoona Reservoir apparently eliminated the population. Stiles and I collected a specimen of this species in the Conasauga River, Polk County, Tennessee, in 1968. Approximately 60 additional specimens have now been collected from the latter stream, and the description of this species by Williams and Etnier is in preparation.

In 1970 I examined some specimens at Tulane University from the Ozark region collected by R. C. Cashner and B. A. Thompson. These distinctive specimens, collected with what was then known as *P. uranidea*, appeared to be a new species. Subsequent examination of the types of *P. uranidea* and its supposed junior synonym *P. ouachitae* (Jordan and Gilbert) by Thompson revealed that their "new" species was identical with the types of *P. uranidea*, and the types of *P. ouachitae* were identical with specimens then being referred to as *P. uranidea* (Thompson, in lit.). A paper concerning the status of *P. uranidea* and *P. ouachitae* is being prepared by Thompson and Cashner. Examination of the specimen from the Little Tennessee River suggested that its affinities were with true *P. uranidea* rather than with *P. ouachitae*, *P. shumardi*, or the undescribed species from the Conasauga River. Further studies confirmed this view and indicated that the Little Tennessee River population represented a species distinct from *P. uranidea.*

THE SUBGENUS *IMOSTOMA*

The subgenus *Imostoma* currently includes the following five species: *P. shumardi* (Girard), which lacks distinct dorsal saddles; and four species with saddles—*P. uranidea* (Jordan and Gilbert), restricted now to the White and Saline river systems, Arkansas, but formerly occurring in the Wabash River, Indiana; *P. ouachitae* (Jordan and Gilbert), widespread in the lower Mississippi River valley and in the Gulf Coastal drainages; the undescribed species from the Conasauga River; and the new species from the Little Tennessee River, described herein.

Page (1974) presented a brief diagnosis of the subgenus *Imostoma* based on *P. shumardi*, the undescribed species from the Conasauga River, and *P. uranidea* from a "ditch" in Missouri. His *P. uranidea* specimens were likely *P. ouachitae,* since *P. uranidea* is extremely rare in Missouri, and appears to be restricted to swift water in moderate size rivers. *P. ouachitae* does occur in ditch-type habitats. The inclusion of two additional species in the subgenus (*P. uranidea* and *P. tanasi*), plus examination of extensive material of the three species considered by Page, necessitates modification of the diagnosis, as follows. A "midbelly row of modified (enlarged and strongly toothed) scales" develops on individuals of all five species. These modified midventral scales are most conspicuous in large adults, and may be lacking in juveniles and small adults. In nuptial males of *P. uranidea* there is a more or less well developed band of exposed but scarcely modified ctenoid scales extending from the ventral prepectoral area transversely across the middle of the breast. In the other four species embedded cycloid and occasionally a few exposed ctenoid scales are present on the breast in many individuals. These are most conspicuous in nuptial males, and are usually confined to the posterior half of the breast.

Key to the Species of the Subgenus *Imostoma*

1. Dorsal saddles absent .... *Percina shumardi* (Girard)
   Four or five distinct dorsal saddles present 2
2. Anteriormost dorsal saddle entirely in front of dorsal fin .... *Percina* species
   Anteriormost dorsal saddle beneath dorsal fin 3
3. Posterior margin of fourth dorsal saddle over dorsal insertion of caudal fin; fifth dorsal saddle absent; anal soft rays 11–12 (extremes 10–13) 4
   Posterior margin of fourth dorsal saddle anterior to dorsal insertion of caudal fin; fifth saddle adjacent to dorsal insertion of caudal fin, moderately developed; anal soft rays 10–11 (9–12) .... *Percina ouachitae* (Jordan and Gilbert)
4. Anal rays modally 11; pectoral fins long and pointed, overlying lateral-line scale 19–24 (males) or 17–23 (females and juveniles); pelvic fins 22% of standard length or longer in males, 20% of standard length or longer in females and juveniles; nuptial males with tubercles on pelvic spine and adjacent four soft rays .... *Percina uranidea* (Jordan and Gilbert)
   Anal rays modally 12; pectoral fins shorter and more rounded, overlying lateral-line scale 17–19 (males) or 15–18 (females and juveniles); pelvic fins less than 22% of standard length in males, less than 20% of standard length in females and juveniles; nuptial males with pelvic fin tubercles confined to the four median rays .... *Percina tanasi* new species

Methods

Measurements and counts (Tables 1–8) follow the methods of Hubbs and Lagler (1958) unless defined below. Measurements were made with a needle-point divider and read to the nearest 0.1 mm. Head length was measured from the tip of the snout to the tip of the membranous extension of the operculum. Orbital diameter and interorbital width were measured between the fleshy coverings of the bones delimiting these areas. Trans-pelvic width was the measurement between the outer edges of the parallel pelvic spines. Saddle widths were measured along the dorsal midline and the saddle position from the occiput. The lateral-line scale under the tip of the pectoral fin (Table 3) was obtained by counting from the head to the fin tip when pressed against the lateral midline.

Material Studied

The number of specimens appears in parentheses after the catalog number. The following abbreviations have been used: SL (Standard length), TU (Tulane University), UT (University of Tennessee), USNM (National Museum of Natural History, Smithsonian Institution), CU (Cornell University), FSM (Florida State Museum), INHS (Illinois Natural History Survey), NLU (Northeastern Louisiana State University), UAIC (University of Alabama Ichthyological Collection), UMMZ (Museum of Zoology, University of Michigan), OAM (Oklahoma State University), HWR (Henry W. Robison collection).

Percina tanasi, new species
Snail Darter
Figure 1

Type-specimens: Holotype, TU 90858, an adult male, 59.1 mm SL, Little Tennessee River at Coytee Spring, River Mile 7, Loudon County, Tennessee, 6 Nov. 1973, G. F. Boronow, B. F. Clark, Jr., D. A. Etnier, R. L. Hensen, W. C. Starnes, B. A. Thompson. Allotype female, TU 90859, 54.6 mm SL, same data as holotype. Paratypes taken with the holotype are TU 83994 (2) and USNM 214697 (2). Additional paratypes taken at the type-locality are CU 64754 (5), FSM 20852 (5), INHS 75000 (5), NLU 32057 (5), UAIC 5020.01 (5), and UMMZ 197413 (5), all 16 Aug. 1973; USNM 214698 (5), 4 Oct. 1974; UT 91.803 (1), 12 Aug. 1973; UT 91.1074 (10), 23 April 1975. An additional paratype, OAM 9536 (1), is from River Mile 15, main channel side of Davis Island, Little Tennessee River, Loudon County, Tenn., 4 Oct. 1974.

TABLE 1. Measurements of *Percina tanasi* and *P. uranidea* expressed in thousandths of standard length. ($\bar{x}$ = mean, W = range, S_x = standard deviation, N = sample size, M = males, F = females, J = juveniles and subadults)

Characters		*P. tanasi*				*P. uranidea* (White R.)				*P. uranidea* (Saline R.)				*P. uranidea* (Wabash R.)			
		$\bar{x}$	W	S_x	N	$\bar{x}$	W	S_x	N	$\bar{x}$	W	S_x	N	$\bar{x}$	W	S_x	N
Minimum inter-	M	51.54	44–58	3.80	13	47.30	44–52	2.98	10	43.50	40–49	2.92	10	45.00	39–49	3.85	6
orbital width	F	51.18	46–61	4.38	11	45.70	41–50	3.06	10	44.11	40–49	3.18	9	41.50	40–43		2
	J	49.47	42–55	3.13	19	49.70	45–55	2.83	10	44.20	34–50	5.41	10				
Width below	M	132.46	119–147	8.77	13	116.10	107–126	6.11	10	123.10	114–133	5.43	10	119.17	114–126	4.12	6
origin of soft	F	135.91	127–149	7.24	11	124.20	108–137	8.98	10	121.10	113–129	6.26	10	117.50	112–123		2
dorsal fin	J	133.68	124–145	5.28	19	124.00	115–132	4.88	10	121.00	115–129	4.78	10				
Width below	M	95.69	76–109	8.40	13	80.70	74–88	5.06	10	90.30	84–96	4.52	10	82.57	75–88	5.09	7
origin of	F	99.91	93–113	6.02	11	87.80	81–100	7.89	10	89.00	80–100	5.54	10	82.50	80–85		2
saddle 3	J	101.37	93–109	4.30	19	95.50	90–102	4.81	10	95.10	89–102	3.84	10				
Pectoral fin	M	242.69	215–276	16.95	13	285.46	252–326	18.21	28	287.50	264–304	11.04	18	269.33	257–279	9.40	6
length	F	218.82	195–238	13.49	11	266.39	234–291	12.62	23	255.73	223–274	14.83	11	253.00	251–255		2
	J	207.79	192–224	8.85	19	248.75	223–274	10.65	24	240.90	230–253	6.12	10				
Pelvic fin	M	200.00	165–223	15.68	13	228.89	218–244	7.62	19	231.12	211–248	11.26	17	221.60	206–232	8.60	7
length	F	184.27	158–199	12.75	11	219.89	207–242	9.38	18	223.18	208–234	9.38	11	216.00	215–217		2
	J	180.42	160–194	7.90	19	210.58	187–226	10.46	24	225.80	217–232	5.88	10				
Width of first	M	64.20	52–84	9.24	10	45.40	34–57	8.02	10	57.00	42–71	9.58	10	46.67	38–58	6.47	6
saddle	F	63.00	41–82	11.85	10	48.89	31–61	10.42	9	56.10	46–65	7.08	10	31.00			1
	J	59.21	41–76	9.46	19	55.10	45–58	4.23	10	61.70	51–82	8.88	10				
Width of sec-	M	55.69	40–79	12.00	13	36.70	22–51	7.60	10	53.80	36–71	11.68	10	48.86	39–58	6.41	7
ond saddle	F	61.91	47–75	9.33	11	42.20	34–50	4.13	10	53.00	45–63	6.56	10	47.00	43–55		2
	J	56.84	40–64	6.61	19	51.10	42–60	6.45	10	61.90	56–78	6.85	10				
Width of third	M	62.15	46–81	10.42	13	30.50	19–41	6.70	10	49.70	33–67	13.69	10	49.14	37–63	8.63	7
saddle	F	64.45	50–80	9.68	11	29.40	18–38	6.67	10	49.10	36–67	10.16	10	43.00	38–48		2
	J	63.74	42–79	8.12	19	47.10	40–55	5.57	10	54.40	29–79	12.48	10				
Width of fourth	M	42.00	30–60	8.60	13	25.40	13–33	6.02	10	37.10	28–54	8.89	10	33.14	24–43	7.29	7
saddle	F	44.36	33–60	8.37	11	28.10	21–43	7.36	10	41.80	34–51	6.39	10	35.00	26–44		2
	J	46.21	24–59	8.14	19	36.20	30–45	5.29	10	48.30	41–60	5.85	10				

Table 2. Frequency of lateral-line scales in *Percina tanasi* and *P. uranidea.*

	Number of lateral-line scales																
	46	47	48	49	50	51	52	53	54	55	56	57	58	59	60	N	x̄
Percina tanasi			2	8	5	10	5	6	7	4	6	1			1	55	52.3
P. uranidea																	
White River system	1	4	2	7	18	15	16	13	8	4						88	51.3
Saline River system		1	2	4	10	9	18	19	11	9	4	5	1		1	94	52.7
Wabash River system		1	2		1	1	1	2	1							9	50.7

Table 3. Lateral-line scale beneath tip of left pectoral fin in *Percina tanasi* and *P. uranidea*

	13	14	15	16	17	18	19	20	21	22	23	24	25	N	x̄
Percina tanasi, males				1	2	8	6	2						19	18.3
P. uranidea, males															
White River system						1	1	3	5	4				14	20.7
Saline River system						1	3	9	8	5	3	2		31	20.7
Wabash River system							3	3						6	19.5
Percina tanasi, females and juveniles	2		9	9	11	5								36	16.2
P. uranidea, females and juveniles															
White River system				1	2	16	11	10	9	6	3	1	1	60	19.8
Saline River system					4	7	9	12	5	3				40	19.4
Wabash River system					1			1						2	18.5

TABLE 4. Frequency of transverse scale rows (counted from first anal spine obliquely forward to dorsal fin) and minimum scale rows around caudal peduncle for *Percina tanasi* and *P. uranidea*.

	Transverse Scale Rows								Scales Around Caudal Peduncle						
	14	15	16	17	18	19	N	$\bar{x}$	18	19	20	21	22	N	$\bar{x}$
Percina tanasi		6	22	18	8	1	55	16.6		12	32	9	2	55	20.0
P. uranidea															
White River system		21	23	13	1		58	16.2	6	18	32	4	2	58	20.2
Saline River system	1	6	18	13	6	1	45	16.4	1	9	23	12		45	20.0
Wabash River system	1	5	2	1			9	15.3	1	4	4			9	19.3

TABLE 5. Frequency of spines and soft rays in dorsal fins of *Percina tanasi* and *P. uranidea*.

	Dorsal spines						Dorsal Soft Rays							
	IX	X	XI	XII	N	$\bar{x}$	13	14	15	16	17	18	N	$\bar{x}$
Percina tanasi	1	27	22	4	54	10.5	1	10	22	17	2		52	15.2
P. uranidea														
White River system	3	25	37	2	67	10.4		12	43	6			61	14.9
Saline River system		21	21	2	44	10.6		10	26	8		1	45	15.0
Wabash River system	1	2	6		9	10.6	1	2	6				9	14.6

TABLE 6. Frequency of soft anal rays and principal caudal rays in *Percina tanasi* and *P. uranidea.*

	Anal Soft Rays						Caudal Rays				
	10	11	12	13	N	$\bar{x}$	15	16	17	N	$\bar{x}$
Percina tanasi	2	15	36	1	54	11.7	6	42	3	51	15.9
P. uranidea											
White River system	5	32	15		52	11.2	6	27	14	47	16.2
Saline River system	3	40	9	1	53	11.2	2	21	22	45	16.5
Wabash River system	3	5	1		9	10.8					

TABLE 7. Frequency of pectoral fin rays in *Percina tanasi and P. uranidea*

	Left Pectoral Rays					Right Pectoral Rays					
	13	14	15	N	$\bar{x}$	13	14	15	16	N	$\bar{x}$
Percina tanasi	10	37	4	51	13.9	10	38	2		50	13.8
P. uranidea											
White River system	9	43	2	54	13.9	12	36	6	1	55	13.9
Saline River system	5	34	6	45	14.0	5	35	5		45	14.0
Wabash River system	4	4		8	13.5	1	5			6	13.8

TABLE 8. Measurements of *Percina tanasi* expressed in thousandths of standard length. ($\bar{x}$ = mean, W = range, S_x = standard deviation, N = sample size, M = males, F = females, J = juveniles and subadults)

Characters		$\bar{x}$	W	S_x	N
Head length	M	288.00	272–304	9.40	13
	F	286.91	282–293	4.23	11
	J	289.89	278–307	7.18	19
Orbit diameter	M	71.92	66–80	3.62	13
	F	72.91	68–76	2.21	11
	J	74.32	69–78	2.52	19
Maximum body width	M	166.46	142–195	13.73	13
	F	170.73	153–200	12.27	11
	J	155.84	141–172	8.66	19
Caudal peduncle depth	M	80.76	74–86	3.30	13
	F	82.70	73–88	4.27	10
	J	83.32	78–86	2.79	19
Caudal fin length	M	200.23	177–224	12.38	13
	F	193.09	185–199	4.69	11
	J	196.74	175–206	7.71	19
Longest dorsal spine	M	119.46	100–146	15.08	13
	F	117.10	82–147	22.56	10
	J	124.72	99–154	18.25	18
Longest dorsal soft ray	M	152.08	135–173	11.02	12
	F	139.80	132–151	6.97	10
	J	138.56	127–154	7.23	18
Length of anal spine	M	72.85	61–83	6.39	13
	F	65.80	53–71	5.47	10
	J	71.29	61–78	4.12	17
Occiput to dorsal fin origin	M	131.38	117–147	9.09	13
	F	138.82	113–140	10.30	10
	J	128.70	115–138	7.53	19
Snout length	M	90.77	80–98	5.10	13
	F	86.36	80–91	3.88	11
	J	82.26	72–93	5.99	19
Trans pelvic width	M	87.31	77–104	7.67	13
	F	84.82	80–93	4.77	11
	J	82.63	74–91	4.56	19

TABLE 8. (Continued)

Characters		$\bar{x}$	W	S_x	N
Maximum body depth	M	209.17	183–240	14.67	12
	F	208.00	186–233	14.95	11
	J	186.47	167–204	9.42	19
Caudal peduncle length	M	156.30	144–171	8.18	13
	F	162.18	140–174	11.19	11
	J	169.16	154–186	11.19	19
Spinous dorsal fin length	M	269.31	236–291	15.17	13
	F	257.00	224–277	15.51	11
	J	254.53	232–275	12.01	19
Soft dorsal fin length	M	295.93	254–349	28.93	13
	F	279.91	252–326	20.63	11
	J	287.32	246–306	12.94	19
Anal fin length	M	343.23	282–414	37.32	13
	F	248.09	228–264	10.89	11
	J	266.11	238–244	18.08	19
Longest anal ray	M	179.92	134–244	35.09	13
	F	134.80	116–151	11.75	10
	J	128.21	113–140	7.93	19
Occiput to soft dorsal fin origin	M	412.69	366–443	20.98	13
	F	418.00	398–442	16.16	11
	J	406.53	391–423	7.08	19
Occiput to first saddle	M	124.20	108–137	10.34	10
	F	127.50	113–147	9.58	10
	J	122.37	99–140	11.24	19
Occiput to second saddle	M	369.92	346–386	12.80	13
	F	365.72	339–389	13.50	11
	J	360.32	333–379	11.59	19
Occiput to third saddle	M	557.38	523–574	17.60	13
	F	558.91	532–585	14.05	11
	J	543.16	513–564	12.94	19
Occiput to fourth saddle	M	702.77	650–721	21.13	13
	F	704.45	665–726	16.42	11
	J	690.84	662–710	12.56	19
Standard length	M	57.3	49.7–65.3	4.38	13
	F	54.8	50.2–63.8	4.25	11
	J	44.0	38.2–49.3	2.31	19

Additional specimens examined but not designated as types: Two specimens collected at the type-locality, 23 April 1975, have been given to the Tennessee Wildlife Resources Agency, Nashville, Tenn. Two additional specimens, currently in the possession of the Tennessee Valley Authority, Norris, Tenn., were briefly examined and included in the counts in Tables 2–7, as follows: Type-locality, 1 specimen, 10 Nov. 1974; and Little Tennessee River at River Mile 4, Greene Farm, Loudon County, Tenn., 6 Jan. 1975.

Comparative material of other species: *Percina uranidea* (Jordan and Gilbert): Arkansas: Saline River system: NLU 16831 (49 of 205, less 1 specimen reidentified as *P. ouachitae* (Jordan and Gilbert), Saline River approximately 10 km south of Johnsville, Bradley-Ashley County line, 30 July 1970; NLU 14582 (60), Saline River at end of Ark. Highway 172, Drew Co., 6 Nov. 1969; NLU 14624 (16), Saline River at Vince Bluff, west of Herbine, Cleveland Co., 18 Nov. 1969; HWR 72-23 (1), Saline River 5 km east of Poyen at U.S. Highway 270 bridge, Grant Co., 10 June 1972; HWR 71-19 (2), Saline River 12 km north of Johnsville, Bradley Co., 12 Oct. 1971; HWR 74-108 (2), Saline River southeast of Johnsville, R. 8 W., T. 15 S., S. 31, Ashley Co., 26 Oct. 1974. Arkansas: White River system: TU 66011 (10 of 46), Current River 7 km northwest of Success along Ark. Highway 211, Randolph and Clay counties, 9 Oct. 1970; TU 66806 (30), same locality as TU 66011, 21 Nov. 1970; TU 92950 (25), Strawberry River at Ark. Highway 115 bridge, 1.6 km northeast of Jessup, Lawrence Co., 28 Sept. 1974; NLU 13874 (37), White River below Lock and Dam No. 1, Batesville, Independence Co., 14 July 1969. Indiana: Wabash River system: USNM 40922 (4), Wabash River at Vincennes, Knox Co., 1888; USNM 66971 (3), same locality, 15 Sept. 1888; USNM 125337 (1), same locality, 1888; USNM 40942 (1), Wabash River at New Harmony, Posey Co., 1888.

Diagnosis: The affinities of this species are within *Imostoma* and with the saddle-backed members of the subgenus. Differing from *P. ouachitae* in having the posterior margin of the fourth saddle positioned on the anterior dorsal insertion of the caudal fin (this saddle completely anterior to the anterior dorsal insertion of the caudal fin in *P. ouachitae*). Differing from the Conasauga River form in having the anteriormost dorsal saddle positioned under the anterior three to five dorsal spines (this saddle entirely anterior to the dorsal fin in the Conasauga River form). Most closely related to *P. uranidea*, but differing from that species in having shorter paired fins (Tables 1 and 3); a more robust body (see Table 1: interorbital width, width below origin of soft dorsal fin, and width below origin of third saddle). Dorsal saddles of *P. tanasi* are wider than those of *P. uranidea* from the Saline and Wabash river systems, and much wider than those of *P. uranidea* from the White River system (Table 1.). *P. tanasi* has a modal value of 12 soft anal fin rays (11 for *P. uranidea*) and 16 caudal rays (16 or 17 for *P. uranidea*) (Table 6). Background coloration of dorsum in live speci-

FIG. 1. *Percina tanasi*, UT 91.1074 (paratypes), Little Tennessee River at Coytee Spring, River Mile 7, Loudon County, Tennessee, 23 April 1975. Top, male, 61 mm SL; middle, female, 66 mm SL; bottom, dorsal view of male, 58 mm SL.

mens brownish gray with traces of green (russet brown in *P. uranidea*, Thompson, in lit.). Nuptial males with cheek entirely black to mostly black, the posterior one-third brown with occasional gold flecks (bright metallic gold in *P. uranidea*, Thompson, in lit.). Exposed ctenoid scales entirely absent from lateral areas of breast, or rarely with one or two ctenoid scales in these areas (nuptial males in *P. uranidea* with exposed ctenoid scales usually present on lateral areas of breast, these often forming a transverse band across the middle of the breast). Nuptial males with pelvic fin tubercles confined to the median four rays (pelvic fin tubercles on pelvic spine and adjacent four rays in *P. uranidea*).

Description: *Percina tanasi* (Fig. 1) is the most robust of the saddle-backed members of the subgenus *Imostoma.* The largest specimen is a female, UT 91.1074, 66.5 mm SL. Two additional females and one male from this series, taken 23 April 1975, are over 60 mm SL. Minimum fleshy interorbital width (Table 1) greater than in *P. uranidea*, and much greater than for *P. ouachitae* or *P.* (*Imostoma*) sp. Body width (nearly 170 thousandths of SL) much greater than that of *P. ouachitae* or *P.* (*Imostoma*) sp.—both about 140 or less; the caudal peduncle tapers less rapidly than in *P. uranidea* (see width below origin of soft dorsal fin and width below origin of saddle 3, Table 1). It is so robust that the first specimen, seen underwater, was mistaken for a sculpin. Measurements of these and additional characters, expressed as thousandths of standard length, appear in Tables 1 and 8.

Lateral-line scales usually 49–56 (extremes 48–57), with 1 (0–2) pored scales posterior to hypural plate. Transverse scale rows, counted obliquely forward from first anal spine, 16–17 (15–19). Minimum caudal peduncle scale rows 20 (19–22). Frequency distributions for scale counts appear in Tables 2 and 4.

Dorsal spines X–XI (IX–XII); dorsal soft rays 14–16 (13–17). Anal spines 2; anal soft rays modally 12, often 11 (10–13). Principal caudal rays 16 (15–17). Pectoral rays modally 14, often 13, occasionally 15. Frequency distributions of fin ray counts appear in Tables 5–7.

Opercles and nape covered with exposed ctenoid scales. Cheek and breast squamation variable. Cheeks typically with embedded cycloid scales posterior to eye. In juveniles and subadults cheeks often appear naked. In adults embedded cycloid scales often more extensive on lower half of cheek; the entire cheek occasionally scaled. One adult male (USNM 214697) has about five exposed ctenoid scales on each cheek. Breast squamation variable dependent on age and apparently on sexual condition. In juveniles breast and prepectoral areas usually naked, or nearly so, with a few embedded scales in prepectoral region and/or on breast proper medial to pectoral fin base. Adults typically with embedded scales in these areas, and rarely with a transverse band of scales extending across breast between pectoral bases, as typical of *P. uranidea*. These scales occasionally ctenoid, often bearing low, rounded tubercles in nuptial males. Belly naked lateral to midventral row of modified scales.

Gill membranes weakly connected, occasionally separate. Branchiostegal rays 6-6, an occasional specimen with 7 rays on either side. Frenum varying from moderately developed to absent. Infraorbital canal complete, typically 8 pores per side. Preoperculomandibular canal with 10 pores per side. Supratemporal canal complete; supraorbital canal complete with 3 pores per side; coronal pore single.

Vertebral counts, 9 specimens, 39 (3), 40 (5), 41 (1), suggesting modal value of 40. Modal value for other saddle-backed *Imostoma* 39 (Thompson, in lit.).

Dorsum background color in life brown to brownish gray with traces of green, this pigmentation continuing ventrally to or nearly to lateral line. Dorsum crossed by four conspicuous chocolate-brown saddles. Anterior margin of saddle one at, or slightly anterior or posterior to base of first dorsal spine. Posterior margin of this saddle between bases of third to fifth dorsal spines. Saddle two under posterior end of spinous dorsal fin, extending from base of last dorsal spine to or slightly behind base of first dorsal soft ray. Anterior margin of saddle three between bases of ninth to eleventh dorsal soft rays; it extends posteriad to base of thirteenth to fourteenth soft ray. Saddle four on caudal peduncle, with its posterior margin in contact with the unpigmented middorsal line extending forward from caudal fin. This unpigmented line represents the dorsal connective tissue insertion of caudal fin. Distances of saddles from occiput appear in Table 8. All saddles narrowest on midline, extending down and forward to lateral line. Saddle one occasionally inconspicuous in both living and preserved specimens. Ventral extensions of dorsal saddles interrupted at lateral line, but otherwise continuous with series of similarly colored lateral blotches extending about one-half scale row above, and three to four scale rows below lateral line. Paler areas below lateral line, alternating with dark lateral blotches, mottled pale green to greenish yellow in life, mottled in preservative. Belly typically immaculate, or with cluster of dark chromatophores in vicinity of anus. Lower sides adjacent to anal fin and ventral portion of caudal peduncle vary from virtually immaculate to darkly mottled with continuation of midlateral pigmentation. Breast pigmentation in juveniles and subadults varies from immaculate to sparsely freckled with dark chromatophores, especially at base of pelvic spine, in prepectoral area, and near depression in middle of breast. In adults dark chromatophores typically abundant at bases of pelvic spines and often between pelvic bases; an immaculate to weakly pigmented area immediately anterior to pelvic bases; anterior portion of breast moderately to completely freckled with dark chromatophores.

Head dark olive brown dorsally. Opercles and cheeks mottled brown and yellow. In life, iris orange-yellow adjacent to pupil, becoming brownish toward periphery. A black suborbital bar, wider than pupil, extends down and slightly posteriad from eye, terminating over interoperculum. In nuptial specimens, suborbital bar expands posteriad and occupies virtually entire cheek. Anterior portion of interoperculum and rami of lower jaw irridescent pale blue in life. Ventral surface of head with scattered groups of dark chromatophores in juveniles, and liberally freckled with dark chromatophores in adults.

Dorsal fin membranes lack bright colors. In spinous dorsal fin, dark chromatophores concentrated on margin of membrane posterior to tip of each spine, forming obscure marginal dark band. Additional dark chromatophore concentrations at or near base of each membrane, more concentrated on posterior half of each membrane, and restricted to basal one-third of membrane. Membrane of spinous dorsal fin clear elsewhere. Dorsal spines pale yellow with dark chromatophores concentrated near tip and middle of each spine. Dorsal soft rays pale yellow with four to five dark marks alternating with somewhat longer clear portions of each ray. Caudal fin pigmentation similar to that of soft dorsal fin. Anal and pelvic fin pigmentation similar to that of soft dorsal fin, but dark chromatophores and pale yellow pigment on spines and rays less evident in adult males. In nuptial males these fins may be virtually immaculate. In nuptial females both anal and pelvic rays bright yellow-orange. Pectoral fins of both sexes have a yellow orange base, clear membranes, and pale yellow to orange-yellow rays, each of which marked with about five dark chromatophore concentrations. Concentrations of dark chromatophores on spines and rays of all fins neither sufficiently discrete nor aligned to form distinct bands.

Sexual dimorphism: As is typical of the subgenus *Imostoma*, the anal fin is much longer in adult males than in adult females. This character permits easy sex recognition. Anal fin length in specimens larger than 50 mm SL readily separates sexes in *P. tanasi*. Sex of *P. uranidea* specimens was easily determined in specimens of about 40 mm SL. Two male *P. uranidea* from the Saline River, HWR 74-108, collected 26 Oct. 1974, were tuberculate at 35.3 and 37.3 mm SL. Examination of the data in Tables 1 and 8 indicates that males differ from females in the following additional characters: longer snout, more slender body (probably attributable to gonad size), shorter caudal peduncle, greater length of all fins (see also Table 3), longer first anal spine, dorsal fins positioned closer to occiput, and possibly a narrower fourth saddle. *P. uranidea* exhibits similar dimorphism in these characters. *P. ouachitae* and *P.* (*Imostoma*) sp. are not measurably dimorphic in snout length, caudal peduncle length, caudal fin length, paired fin length (possibly slightly longer in male *P.* (*Imostoma*) sp.), spinous dorsal fin length, anal spine length, position of dorsal fin relative to occiput, or width of the fourth saddle.

Tubercles are conspicuous on males from early November through late April. In males with maximum development, uniserial tubercles are present on anal rays, on lower surface of the medial four pelvic rays, and on the principal caudal rays (plus two adjacent procurrent rays) on the lower half of the caudal fin. Tubercles are not present on either the pelvic or anal spines, but are present on one or both of these areas in other *Imostoma*. Although it is possible that examination of additional tuberculate specimens of *P. tanasi* might reveal the presence of tubercles on these spines, it seems rather unlikely, since the six nuptial males from UT 91.1074, collected 23 April 1975, are as tuberculate as any *Imostoma* I have examined. In several of these specimens the pelvic spines are very fleshy, but neither tubercles nor tubercle scars are present on the pelvic spine or on the adjacent pelvic ray. Tubercles are well developed on the remaining four pelvic rays. The tubercles on the anal fin had begun to slough off on these specimens, and might have already been lost from the anal spines. Branchiostegal rays have small, pointed tubercles. Single, low, rounded tubercles are present on scales on the cheek and breast, on three to four scale rows adjacent to the naked midventral area, on the modified midventral scales, on scales adjacent to the base of the anal fin, and on scales on the lower

FIG. 2. Photograph of the type-locality, Little Tennessee River at Coytee Spring, River Mile 7, Loudon County, Tennessee, as seen during minimum flow looking downstream from the gravel island which is slightly upstream from the spring outflow.

caudal peduncle and lower caudal fin base. Similar tubercles that are not borne by scales occur on the cheeks and breast, and over the interoperculum, rami of the lower jaw, and lachrymal (anterior infraorbital). Cteni on scales along the anal fin base are elongate in tuberculate males.

The female urogenital papilla is nearly cylindrical, about twice the diameter of the base of the first anal spine. Its anterior and lateral surfaces have about six deep longitudinal creases, and its posterior surface is in contact with the first anal spine. In males the urogenital papilla is a small, laterally expanded flap, also in contact with the base of the first anal spine and bearing longitudinal creases. Urogenital papillae are similar in other species in the subgenus.

Etymology: Tanasi (tan-ah'-see) was the Cherokee name for a village at River Mile 26 on the south bank of the Little Tennessee River, Monroe County, Tennessee. According to Duane King, Assistant Professor, Department of Sociology and Anthropology, University of Tennessee at Chattanooga, Tanasi served as the capital for the Cherokee Nation until 1725. The altered spelling (Tennessee) was first used by Lt. Henry Timberlake in 1762 (Williams, 1927).

Biology: *Percina tanasi* is restricted to gravel shoal areas in approximately the lower 32 km of the Little Tennessee River. Seining and underwater observation in this area indicate that fishes are not abundant. A crew of five persons, utilizing a small mesh 12-ft-long by 6-ft-deep seine, would be fortunate to collect as many as 50 fish per hour in these gravel shoal areas. At the type-locality (Coytee Spring, River Mile 7) the Little Tennessee River is about 200 m wide. At low flow a small gravel island is visible near the north bank of the river, just upstream from the spring outflow (Fig. 2). The substrate is virtually all coarse gravel interspersed with sand, with areas of silt deposition along the banks, a few bedrock outcrops, and concentrations of boulders. Even during periods of minimum flow, current is swift over the gravel areas, and depths are uniformly about 1 m. *P. tanasi* is most often found in the swiftest current, but we have encountered specimens in moderate current. According to W. C. Starnes, who continues the life history study of the species, it typically avoids areas with abundant vegetation attached to the rocks and areas where manganese deposits have blackened the rocks. Our samples indicate that *P. tanasi* is the most abundant of the gravel shoal species at the type-locality. Species collected with *P. tanasi*: *Hybopsis aestivalis* (Girard), *Nocomis micropogon* (Cope), *Phenacobius uranops* Cope, *Hypentelium nigricans* (Lesueur), *Etheostoma blennioides* Rafinesque, *Etheostoma rufilineatum* (Cope), *Percina caprodes* (Rafinesque), *Percina evides* (Jordan and Brayton), *Cottus bairdi* Girard, and *Cottus carolinae* (Gill). *Etheostoma simoterum* (Cope), *Percina shumardi* (Girard), and species of *Notropis* and *Moxostoma* inhabit the more sluggish areas near the bank.

In spite of the abundance of aquatic insects, the major dietary component of *P. tanasi* appears to be small gastropods. *Anculosa, Ferrisia, Physa, Pleurocera,* and *Viviparous* are genera tentatively identified from stomach contents. Trichoptera (mostly *Brachycentrus etowahensis* Wallace) and occasional *Ephemeroptera* are found in stomachs of specimens taken in the spring; simuliid dipteran larvae are abundant in stomachs of specimens taken in mid-summer.

Specimens taken in early November included tuberculate males. Specimens from 23 April 1975 were apparently still involved in spawning activity (sexes were segregated, one female released several mature eggs when handled, males were still tuberculate), but most specimens were spent. Water temperature on this date was 14 C (57 F). Preliminary data indicates that sexual maturity is often reached between the second and third year, and that the life span is about four years.

Phylogeny: Because of their similarity in pigmentation, meristics, morphometry, and sexual dimorphism, *P. tanasi* and *P. uranidea* are considered to be closely related species, almost certainly derived from a common ancestor. This ancestral species was presumably widespread in swift streams and rivers with clear water and firm substrates in what is now the central and lower Mississippi River drainage. These populations have disappeared (as apparently has the Wabash River population of *P. uranidea*), presumably in response to silt and loess accumulation and lowered stream gradients. The present relict distribution, higher fin ray counts, and large maximum size of *P. tanasi* and *P. uranidea* suggest that these species are modified representatives of this ancestral species. I interpret *P. ouachitae* as a more recently evolved, generalized form that has invaded areas formerly occupied by the common ancestor of *P. uranidea* and *P. tanasi*, perhaps from a population originally evolved in response to conditions in the Gulf Coastal or extreme lower Mississippi River drainages. Since *P. ouachitae* continues to occur in the main channel of the Mississippi River in west Tennessee, dispersal through the lowlands of the central portion of the Mississippi River drainage would have been easy once suitable habitats were available. Alternate interpretations, considering *P. ouachitae* as the ancestral form that has given rise to other saddle-backed *Imostoma*, appear less likely. I interpret *P.* (*Imostoma*) sp. as an early isolate in the upper Coosa River system, probably derived from the hypothetical ancestor of *P. tanasi* and *P. uranidea*, but not closely related to any existing *Imostoma*.

Taxonomic status: The population of saddle-backed *Imostoma* in the Little Tennessee River differs from known populations of *P. uranidea* in body width, paired fin length, saddle width, nuptial tubercle pattern, several aspects of pigmentation, number of anal and caudal fin rays, and probably vertebral number. I assume that genetic differences are responsible for most, if not all of this divergence. That this divergence is sufficiently large to justify recognition of the Little Tennessee River population as a distinct species is suggested by several sources of information besides the characters listed above. A useful clue to the probable taxonomic status of allopatric populations involves comparing the amount of divergence between such isolates with that between similar sympatric species in the same group. *P. uranidea* and *P. ouachitae* are sympatric in both the White and Saline river systems. The observable differences between these sympatric species (not recognized by modern ichthyologists as being distinct until 1970) are similar in magnitude to those between *P. uranidea* and *P. tanasi*. This indicates that striking differences are not prerequisite to maintenance of genetic isolation between sympatric *Imostoma*. Since the Wabash River population of *P. uranidea* does not display character states intermediate between those of Ozarkian populations and *P. tanasi* (Tables 1, 3, and 6), clinal differences are not involved.

Conservation: The entire range of *P. tanasi* would be innundated by impoundment of Tellico Reservoir, now under construction by the Tennessee Valley Authority. Although adults might persist in this reservoir for a year or two, reproduction following impoundment would no longer be possible and the species would become extinct within a few years. The possibility of finding additional populations of *P. tanasi* outside the area to be impounded appear to be extremely remote, since the remaining river and large stream habitat in this region has been well collected. *P. tanasi* is not a difficult species of darter to collect using standard seining techniques. We have not failed to collect specimens at the type-locality during periods of low flow.

Percina tanasi, the snail darter, is an endangered species in the strictest sense. It was proposed for "endangered species status" in the U.S. Federal Register, Vol. 40, No. 117, p. 25597–98, 17 June 1975, and finally listed as such in the same volume, No. 197, p. 17505–06, 9 Oct. 1975.

ACKNOWLEDGMENTS

Comparative material was provided by R. D. Suttkus, Tulane University; Neil H. Douglas, Northeastern Louisiana State University; and H. W. Robison, Southern State University, Magnolia, Ark. Specimens of *P. uranidea* from the extinct Wabash River population, housed at the U.S. National Museum, were on loan to Tulane University during this study. They were made available for my examination by Dr. Suttkus. Bruce A. Thompson, Tulane University, was extremely helpful in providing data on *P. uranidea*, vertebral counts, comments on the manuscript, and views on the subgenus *Imostoma.* Page costs were generously provided by the Tennessee Wildlife Resources Agency. The photograph of the type-locality was provided by Sam A. Venable, Jr., of the Knoxville News-Sentinel staff. Wayne C. Starnes gave valuable life history information and commented on the manuscript. I continue to be indebted to my enthusiastic students, who were eager to provide collecting assistance.

LITERATURE CITED

HUBBS, C. L., AND K. F. LAGLER. 1958. Fishes of the Great Lakes region Cranbrook Inst. Sci. Bull. 26, Rev. ed, 213 p.

PAGE, L. M. 1974 The subgenera of *Percina* (Percidae: Etheostomatinae). Copeia 1974(1):66–86.

TAYLOR, W. R. 1969. A revision of the catfish genus *Noturus* Rafinesque, with an analysis of higher groups in the Ictaluridae. U.S. National Museum Bull. 282, 315 p.

WILLIAMS, S. C. 1927. Memoirs of Lt. Henry Timberlake 1756–1765. Watauga Press, Johnson City, Tenn., 197 p.

Made in United States of America
Reprinted from COPEIA, 1963, No. 1, March 30
pp. 1–5

Distribution, Variation, and Ecology of *Lepidomeda vittata*, a Rare Cyprinid Fish Endemic to Eastern Arizona

ROBERT RUSH MILLER

THE Little Colorado spinedace, *Lepidomeda vittata* Cope, is one of the members of the compact and distinctive tribe Plagopterini (Miller and Hubbs, 1960), a group of minnows that is gradually disappearing as its habitat becomes increasingly modified by man (Miller, 1961:379–81). This species is known only from the Little Colorado River basin in Navajo, Apache, and Coconino counties of eastern Arizona. Possibly its original range also included suitable waters of western New Mexico, south of Gallup, that drain into the Zuni River (a flood tributary of the Little Colorado), but if so it no longer lives there. Although early records are few (only 4 between 1873 and 1939), there is no reason to doubt that *L. vittata* was once abundant in the main part of the upper Little Colorado River and in its several cool tributaries (Silver Creek, Showlow Creek, Chevelon Creek, and Clear Creek) that rise on the northern slopes of the White Mountains and the Mogollon Rim. It is unlikely that the species maintained permanent populations in the pristine Little Colorado River below Grand Falls, which lie about 35 miles by road northwest of Flagstaff, largely because the river habitat in that section was probably unsuited for this fish but perhaps also because this barrier (about 300 feet high, with a vertical drop on the lower falls of about 150 feet) marked the upstream limit of the large, predatory Colorado squawfish, *Ptychocheilus lucius* Girard.[1]

Between 1939 and 1960, *L. vittata* was exterminated over most of its range—in fact, over all of its range as then known. Extensive collecting by Carl L. Hubbs and R. R. Miller and family between 28 August and 5 September 1960 in the Little Colorado River (at localities where this spinedace was common in 1938–39) and in several major tributaries failed to reveal the species. Pollution, extensive and repeated use of fish poisons (toxaphene and rotenone) to rid the drainage of carp and "trash fish," introductions of exotic species, and changing stream flow probably all contributed to the disappearance of this endemic minnow. However, on 25 August 1960 a single adult was seined by R. R. and G. H. Miller from a stagnating pool in the dry channel of Clear Creek (T. 15 N, R. 13 E), south of Winslow and west of Chevelon Butte, Coconino County; this constituted the first record of the species from that stream and county and the first capture since 1939 (Miller, 1961:384).

On 15–16 July 1961 further investigation of the Clear Creek drainage, made by the R. R. Miller and Clark Hubbs families, and Teruya Uyeno, resulted in the discovery of a sizable breeding population of the species in East Clear Creek. This population was studied further on 13 August 1962. The sites of capture and the habitat of the species are described below in the ecological discussion.

It seems highly probable that *Lepidomeda vittata* is now restricted to the Clear Creek drainage and that its continued existence there may well depend in large measure on the activities of the Arizona Game and Fish Department. Thus far, Clear Creek is one of the few gamefish streams in the Little Colorado drainage that has not been chemically treated; should it be,

[1] Introduced fishes only were collected from plunge-basin pools below the falls (then dry) on 13 June 1959: *Cyprinus carpio*, *Ictalurus punctatus*, *I. natalis*, *Fundulus zebrinus*, and *Lepomis cyanellus*. However, the late Stanley Sykes (astronomer of the Lowell Observatory, Flagstaff) told me on 25 May 1950 that he caught squawfish as well as bonytails (*Gila robusta elegans*) at the base of the falls in the early 1900's.

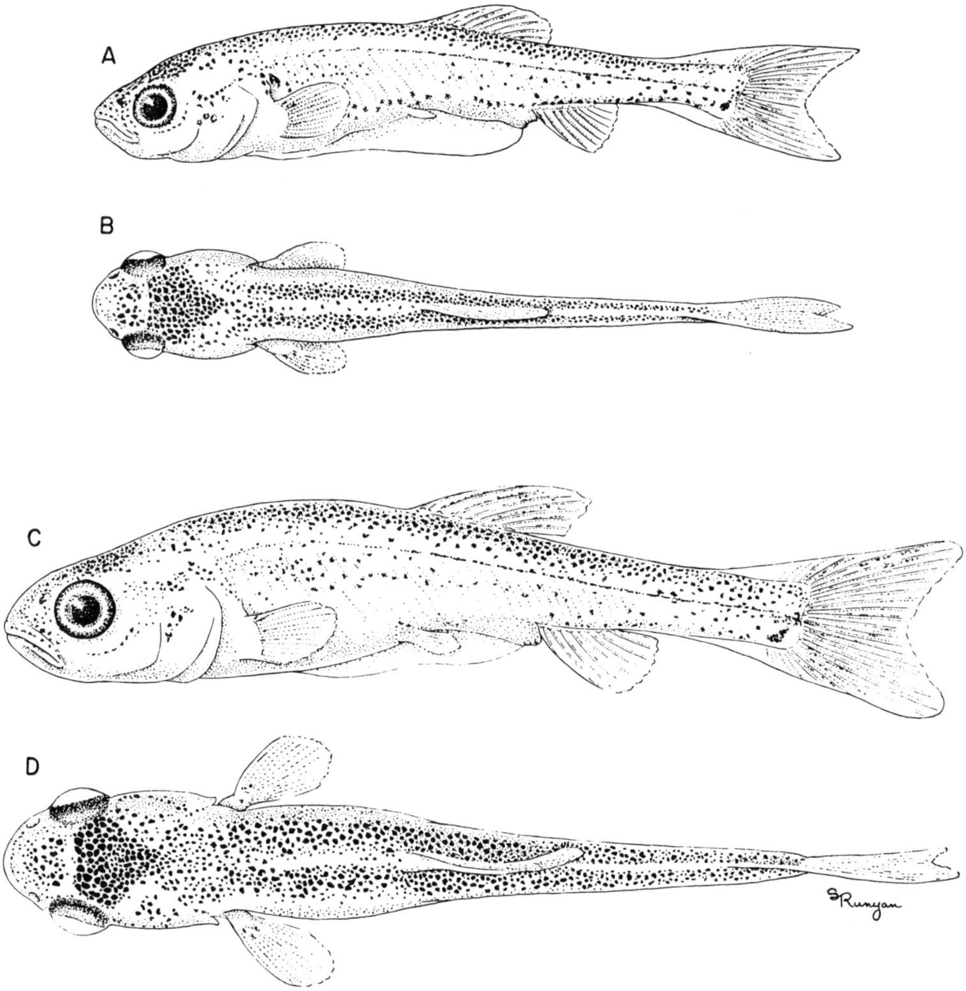

Fig. 1. Postlarval and juvenile specimens of *Lepidomeda vittata* (UMMZ 179572). A, 11.3 mm S.L.; B, same, dorsal view. C, 14.0 mm S.L.; D, same, dorsal view.

there is every reason to expect that the Little Colorado spinedace would become extinct.

In their diagnosis of this species, Miller and Hubbs (1960:29) stressed the combination of 4–4 teeth in the main row, 8 anal rays, usually more than 90 lateral-line scales, and typically 41 or 42 total vertebrae as distinguishing *Lepidomeda vittata* from the three other species of the genus. The new material from Clear Creek confirms the validity of this diagnosis and indicates a degree of variability remarkably similar to that reported for the available earlier collections, including syntypes. Counts for certain head pores and for gill rakers are reported here for the first time. Dorsal rays, 7 (1), 8 (52), 9 (7), av. 8.10; anal rays, 7 (2), 8 (57), 9 (1), av. 7.98; pectoral rays, 15 (2), 16 (17), 17 (28), 18 (12), 19 (1), av. 16.88; pelvic rays, 7–7 (53), 7–8 (2), 8–7 (1), av. 7.03; caudal rays 18 (2), 19 (26), 20 (2), av. 19.00. Lateral-line scales, 89 (1), 90 (1), 92 (3), 94 (4), 95 (7), 96 (3), 97 (2), 98 (1), 99 (2), 100 (2), 101 (1), 102 (1), 103 (2), av. 96.10. Gill rakers, 5 (1), 6 (2), 7 (8), 8 (8), 9 (9), 10 (2), av. 7.93. Vertebrae, 41 (11), 42 (16), 43 (3), av. 41.73; precaudal vertebrae 22–24 and caudal vertebrae 18–20 in six skeletons. Dental formulae, 1, 4–4, 1 (1), 1, 4–4, 2 (3), 2, 4–4, 2 (11), the teeth only occasionally missing—in contrast to the situation described by Miller and Hubbs

(1960:30). Mandibular pores, 4 (1), 5 (11), 6 (17), 7 (1), av. 5.60; preopercular pores, 6 (5), 7 (8), 8 (11), 9 (6), av. 7.60; total preoperculomandibular pores, 11 (3), 12 (7), 13 (6), 14 (10), 15 (3), 16 (1), av. 13.20. This is a somewhat greater total count than that listed for the species by Illick (1956:215–6).

The mouth is moderately oblique. The length of the depressed dorsal fin when stepped into the predorsal length varies from 1.9 to 2.35 in 32 specimens (17 males, 15 females) 53 to 69.5 mm in standard length. Head broad, its width 1.45 to 1.6 times in its length in these same specimens. These figures are very similar to those given by Miller and Hubbs (1960:15, in key), except that the range of variation is extended for the length of the depressed dorsal fin, perhaps because of the breeding condition of the males collected in 1961.

Sexual dimorphism is not marked. The most obvious morphological difference between the sexes involves the pectoral fin, which is noticeably larger in the male. This may be expressed by stepping the pectoral length into the head length: males, 1.0 to 1.1 in 17, 56.5 to 69.0 mm S.L.; females, 1.2 to 1.3 in 15, 53.2 to 69.5 mm long. The tip of the pectoral fin usually falls short of the insertion of the pelvic fin (barely reaching this point in 1 fish) in females, but consistently extends beyond the pelvic insertion in males. No mistakes were made when sexing the specimens on this basis, as verified by examination of the gonads of the 32 fish measured. Only slight sex dichromatism was noted: the ripe males have the axils of the pectoral and pelvic fins intense reddish-orange, whereas those of females are a watery reddish-orange or yellowish.

The largest specimen caught in 1961 or 1962 was a female 107 mm in standard length; a 95-mm female was made into a skeleton. The largest male taken (also skeletonized) was 75 mm long.

In discussing the relationships of *Lepidomeda*, the most generalized genus of the Plagopterini, Miller and Hubbs (1960:6, 11) remarked on its close resemblance to the genus *Gila*, especially to *Gila copei* of the adjacent Bonneville system. Further evidence of the close relationship between these genera is suggested from a comparison of the larval stages of *L. vittata* (Fig. 1) with those of *Gila* (see Winn and Miller, 1954: pl. 1, fig. C, pl. 2, fig. C). Both have a distinct oval internal pigment spot at the bases of the rays of the lower lobe of the caudal fin. (In the only available postlarval specimen of *Gila copei*, UMMZ 141677, 12.0 mm long, this pigment mark is weakly developed, and the pigmentation is generally quite different from that of *L. vittata*.) There is a striking correspondence also in dorsal pigmentation between *L. vittata* and *G. robusta intermedia*. Both have a concentration of giant melanophores in a wedge-shaped pattern just behind the eyes, and two longitudinal rows of similar melanophores extending from above the insertion of the pectoral fins to the base of the caudal fin—these rows separated by a largely unpigmented region along the midline of the back. Anterior to the wedge-shaped melanophores is an irregular, smaller patch of similar pigment between a line drawn across the top of the head between the middle of each pupil. Further comparative larval material of all of the species of *Lepidomeda* and *Gila* is needed to evaluate the significance of resemblances and differences in pigment patterns, but the data obtained thus far are not inconsistent with the idea that the Plagopterini evolved from *Gila* or a similar genus.

The stations at which *L. vittata* was taken in July and August, 1960–62, varied from the marginal habitat of permanent though stagnating pools on Clear Creek (about 7 airline miles west of Chevelon Butte) to permanent flowing sections of East Clear Creek shortly below the "96" Crossing (T. 14 N, R. 11 E) and 5.5 miles below Jones Crossing (T. 13 N, R. 10 E), at elevations between 6,000 and 6,600 feet. Water temperatures (July–August) varied from 58° to 78°F when air ranged from 62.5° at 8:45 AM to 83°F in late afternoon. Bottom consisted of sand, gravel, rocks, boulders, some silt, and bedrock. Water depths equalled or exceeded 6 feet at two of the stations and were about 4 feet at one, with depth of capture up to 4 feet. At the marginal habitats (the two lower stations) the water was greenish to greenish-brown with a green algal bloom; where the breeding population occurred (near "96" Crossing), it was clear with green algae locally dense. Here the stream channel averaged 50 feet wide, but the average width of pools was only 25 feet. Clear

Fig. 2. Isolated pool in East Clear Creek at Mack's Crossing (T. 14 N, R. 12 E), about 4 miles south of Bly Ranger Station, Coconino Co., Arizona. Six juveniles and 1 adult of *Lepidomeda vittata* were collected here on 15 July 1961.

Creek flows through a canyon with walls 300 to 400 feet high, and is shaded by alders and willows at streamside and by large stands of yellow pine beyond (Fig. 2). In the permanent sections Clear Creek is a trout stream, and both rainbow and brown trout (*Salmo gairdneri* and *S. trutta*) were taken at each of the three upper stations. Capture was made with a 12-foot "common sense" seine and a 15-foot ¼-inch (tied) seine.

The Little Colorado spinedace is trout-like in its behavior and habitat requirements. Prior to the introduction of *Salmo* into the Little Colorado drainage, before 1900 (Miller, 1961:389), it probably utilized, through the tributaries, much of the habitat that is now chiefly the domain of exotic salmonids. Its ecological niche in East Clear Creek is in clear, flowing, relatively open pools, 1 to 3 feet deep, that are cold enough for brown and rainbow trout but are generally shunned by them. It may be assumed, from the limited observations now possible, that *Lepidomeda vittata* preferred pools with large boulders that provided refuge places; when in shallower water, it presumably sought undercut banks or banks well overhung by a cover of streamside plants. The species was not taken in the large, well-shaded (hence consistently colder) sand-bottomed pools, which were occupied almost exclusively by trout. Except for trout, *L. vittata* was the most active fish seen, always moving about.

When observed on 16 July 1961, for 2 hours in the forenoon and for half an hour in midafternoon, apparent courtship activity was noted. Individuals presumed to be males were seen following apparent females and nibbling them about the vent region (observation by Clark Hubbs). No breeding was noted. The very clear water was about 2½ feet deep, but the behavior noted occurred only about 5 inches below the surface. The species evidently does not normally school but will do so rather loosely when alarmed, as by the sudden appearance of the observer or by abrupt hand clapping, etc. Individuals typically swim 4 to 6 inches below the surface and explore all parts of their pool habitat, whether shallow or deep, striking at floating debris much in the manner of trout. Young suckers (*Catostomus* and/or *Pantosteus*) avoided adults of *Lepidomeda vittata* as they swam by.

Evidently the Little Colorado spinedace had spawned earlier in the year, perhaps in May, and was again spawning in July, 1961, as suggested by the occurrence of young fish (Fig. 1) and by the courtship behavior described above.

The life colors were noted on freshly caught material in July, 1961, and, by observation of both sexes in full sunlight in a 1-gallon aquarium, in August, 1962. The sides of the body below the base of the fine, nearly vertical dark lines (that extend dorsally from the midside) shine like polished silver and the venter is white. The upper sides are olivaceous, and the back is olivaceous to bluish or lead grey. Except for pigmentation along the fin rays and on the interradial membranes near the bases of the fins, both paired and unpaired fins are largely clear. Irregularly distributed, fine, black puncticulations (giving a pepper-like effect) overlie the silvery sides from the bases of the dark vertical lines to about halfway between the lateral line and the midline of the abdomen (these markings and the vertical lines may be seen on Pl. 2, fig. A, of Miller and Hubbs, 1960). When the live fish are viewed from directly above, a conspicuous, cream-colored spot is seen at both the origin of the dorsal fin and near the bases of the terminal rays of this fin.

The native associates of *Lepidomeda vittata* in the Clear Creek drainage are two other minnows, *Gila robusta* (scarce) and *Rhinichthys osculus* (abundant), and two suckers, *Catostomus latipinnis* and *Pantosteus delphinus* (both common). The only exotics taken by us were *Salmo gairdneri* and *Salmo trutta*. *Hyla arenicola*, *Rana pipiens*, *Bufo punctatus*, and *Thamnophis elegans vagrans* were common.

The completion of this paper was made possible through financial support by the Museum of Zoology, The University of Michigan, and by a grant from the U. S. National Science Foundation (G-15914). Suzanne Runyan, staff artist of the Museum of Zoology, executed Figure 1.

Literature Cited

Illick, Helen J. 1956. A comparative study of the cephalic lateral-line system of North American Cyprinidae. *Am. Midl. Nat. 56(1): 204–23, figs. 1–39.*

Miller, Robert Rush. 1961. Man and the changing fish fauna of the American Southwest. *Pap. Mich. Acad. Sci., Arts, and Letters 46(1960):365–404.*

——— and Carl L. Hubbs. 1960. The spiny-rayed cyprinid fishes (Plagopterini) of the Colorado River system. *Misc. Publ. Mus. Zool. Univ. Mich. 115:1–39, figs. 1–2, pls. 1–3.*

Winn, Howard Elliott and Robert Rush Miller. 1954. Native postlarval fishes of the lower Colorado River basin, with a key to their identification. *Calif. Fish and Game 40(3):273–85, fig. 1, pls. 1–4.*

University Museums, The University of Michigan, Ann Arbor, Michigan.

Reprinted from *Journal of Mammalogy*

A TAXONOMIC STUDY OF THE KIT FOX, *VULPES MACROTIS*

JOHN WAITHMAN AND ARYAN ROEST

ABSTRACT.—The relationships between five subspecies of kit foxes were examined by stepwise discriminant analysis of nine cranial measurements from 104 specimens. As a result the San Joaquin kit fox, *V. m. mutica*, is recognized as a distinct subspecies and three other subspecies, *V. m. arsipus*, *V. m. devia*, and *V. m. tenuirostris*, are synonymized with *V. m. macrotis*.

The kit fox, *Vulpes macrotis*, is found throughout the more arid portions of California, Nevada, Arizona, New Mexico, parts of some adjoining states, and in Baja California and northern Mexico. Since it was first described by Merriam (1888), eight different subspecies have been recognized. *Vulpes macrotis mutica*, from the San Joaquin Valley of California, is of special interest as it is considered threatened, and rare (U.S. Bureau Sport Fisheries and Wildlife, 1973; California Dept. Fish and Game, 1974), and thus is protected by law. The protection extended to this form has economic importance because poisoned grain may not be used to control ground squirrels in areas where these foxes may find and eat dead squirrels and other rodents.

The validity of *V. m. mutica* as a distinct subspecies is not completely clear. Grinnell *et al.* (1937) commented that the type specimen of *mutica* was, relatively speaking, a giant. Furthermore, the series of skulls they studied had a much higher proportion of adult specimens than did their series of *V. m. arsipus*, with which they were comparing *mutica*. They suggested that examination of more comparable series might result in modification of the subspecies descriptions. Benson (1938) noted the similarity of all kit fox subspecies, and suggested that a revision of the group was needed. More recently, some federal and state pest control agents have told us that, in the flesh, *V. m. mutica* is indistinguishable from *V. m. arsipus* and hence should not be considered rare or endangered. This paper reports on a taxonomic analysis made to determine the relationship between these two subspecies. Three additional neighboring subspecies, *V. m. macrotis* (from Riverside County, California; now extinct), *V. m. tenuirostris* (northern Baja California), and *V. m. devia* (southern Baja California), were included in the analyses as well.

MATERIALS AND METHODS

Measurements were taken from the skulls of 127 adult kit foxes, deposited in the U.S. National Museum of Natural History (USNM), the Museum of Vertebrate Zoology, University of California, Berkeley (MVZ), and the mammal collection of California Polytechnic State University, San Luis Obispo (CPSU). Some skulls were broken, but in the final analyses complete data for nine variables were available for 104 specimens. Specimens were considered adult if the basisphenoid–basioccipital suture was closed, a feature which correlated well with other age indications (length of skull, wear to teeth, breadth of lyre) and has been used in aging other carnivores (Lensink, 1962; Orr *et al.*, 1970; Manning, 1971).

On each complete skull 22 measurements were taken; preliminary analyses of data from these unbroken skulls indicated that only nine measurements were of value in discriminating between different groups of kit foxes. These nine had F values greater than 4.0, with an alpha of 0.05, as indicated below.

Rostral breadth at canines.—Breadth across the rostrum at canine teeth, measured from outer sides of canine alveoli; F = 40.53.

Rostral breadth at first molars.—Breadth across the rostrum at first molars, measured from outer sides of most lateral portion of molar alveoli; F = 10.71.

Interorbital breadth.—Least distance dorsally between orbits; F = 7.61.

Bullar length.—From sharp anterior end of typanic bulla, near pterygoid process, diagonally to end adjacent to paraoccipital process; F = 7.08.

TABLE 1.—*Means and standard deviations of nine cranial measurements from six different populations of* Vulpes macrotis. *All measurements in millimeters, from adults. Sample size in parentheses.*

	mutica (35)		*arsipus* West (30)		*arsipus* East (22)		*devia* (6)		*tenuirostris* (6)		*macrotis* (5)	
Measurements	$\overline{X}$	S.D.	$\overline{X}$	S.D.	$\overline{X}$	S.D.	$\overline{X}$	S.D.	$\overline{X}$	S.D.	$\overline{X}$	S.D.
Condylobasal length	114.4	4.05	105.5	3.29	105.8	3.55	104.9	3.08	108.3	3.12	108.8	3.38
Zygomatic breadth	62.1	2.23	57.2	1.66	56.9	2.18	59.1	1.00	60.0	2.12	57.8	1.53
Palatal length	57.8	2.39	53.0	1.92	52.4	2.13	54.0	1.18	55.7	2.42	54.0	1.33
Breadth at first molars	33.0	1.22	30.0	0.80	30.0	1.13	31.5	0.90	31.1	0.60	31.0	0.68
Interorbital breadth	23.1	1.13	20.6	0.82	20.8	0.78	20.7	0.73	21.5	0.71	19.8	0.87
Bullar length	22.2	0.81	21.7	0.95	22.5	0.90	22.2	0.82	22.7	0.45	22.7	0.50
Postorbital breadth	21.4	1.34	21.1	0.96	21.1	1.34	20.5	1.46	20.9	0.40	22.2	1.52
Breadth at canines	18.0	0.77	16.1	0.71	15.7	0.74	16.7	0.85	15.3	0.30	15.7	0.27
Lyre breadth	7.9	2.81	15.0	3.19	13.4	3.51	10.5	2.96	12.4	1.16	12.9	2.33

Condylobasal length.—From anterior edge of premaxillae to posteriormost projection of occipital condyles; F = 6.35.

Palatal length.—From anterior edge of premaxillae to anteriormost point on posterior edge of palate; F = 8.37.

Postorbital breadth.—Least distance dorsally at constriction just posterior to postorbital processes; F = 4.36.

Lyre breadth.—Distance between temporal ridges, measured at frontal–parietal suture; F = 7.68.

Zygomatic breadth.—Greatest distance between outer margins of zygomatic processes; F = 4.18.

Body measurements were not available for many specimens, and pelage coloration is variable due to season and age, and therefore difficult to quantify (see also Egoscue, 1956; 1962); accordingly only cranial measurements were used in this study.

Because only a few specimens were available from each collecting locality, each recognized subspecies was considered a group, except for *V. m. arsipus*. Goldman (1931) described foxes from east of the Colorado River as *V. m. arizonensis*, but they were later synonymized with *V. m. arsipus* by Benson (1938). In this study specimens from east of the Colorado River were considered as one group (*arsipus* East), whereas those from west of the river comprised another group (*arsipus* West). The other groups are referred to by their subspecific names in the following discussions.

Preliminary discriminant analysis of the sexes of kit foxes in this study showed no clear sexual dimorphism in cranial features, and therefore sexes were not analyzed separately (see also Hildebrand, 1952).

Stepwise discriminant analysis of the data was performed with Biomedical Computer Program BMD07M (Dixon, 1973), which calculates means and standard deviations, evaluates the accuracy of classification, plots the first two canonical variates to develop a two-dimensional graph of the separation between groups, and determines the proportions of the total dispersion accounted for by the first two dimensions of the graph.

RESULTS

Table 1 presents means and standard deviations of cranial measurements of kit foxes. Skulls of *V. m. mutica* are distinguishable from those of the other groups on the basis of differences in condylobasal length, zygomatic breadth, palatal length, rostral breadth at first molars, interorbital breadth, rostral breadth at canines, and lyre breadth.

Table 2 shows the BMD07M program classification of individuals into groups by means of their discriminant functions. Although each individual generally classified into its proper group, *mutica* never misclassified into any other group and no member of any other group ever misclassified into the *mutica* group. This suggests a relationship between all specimens except for those in the *mutica* group, which stand out from the others.

TABLE 2.—*Discriminant classification of individual specimens to groups, based on nine skull variables.*

Specimen from this group	N	Specimen classified to this group: *mutica*	*arsipus* West	*arsipus* East	*devia*	*tenuirostris*	*macrotis*
mutica	35	35 (100%)	0	0	0	0	0
arsipus West	30	0	26 (87%)	3 (10%)	0	1 (3%)	0
arsipus East	22	0	2 (9%)	17 (77%)	0	0	3 (14%)
devia	6	0	0	0	5 (83%)	1 (17%)	0
tenuirostris	6	0	0	0	0	6 (100%)	0
macrotis	5	0	0	0	0	0	5 (100%)
Total	104						

Table 3 indicates further analysis of the manner in which individual specimens were misclassified, and indicates that in most cases these individuals were placed into a geographically adjoining group. Each individual is given a probability of belonging to each group. Probabilities usually were greater than 90 percent, but in some instances the probabilities indicated only slightly greater affinity for one group over another. In Table 3, individuals with less than a 70 percent probability of belonging to one group were separated into a composite group representing the two most probable groups of membership. In most instances such individuals were shifted into a composite group, or into one of the geographically adjoining groups. Both the *arsipus* East and *arsipus* West groups, for example, misclassify primarily into each other or into their composite group. Substantial misclassification of individuals of one group into a geographically adjacent group suggests strong similarity between those groups.

Fig. 1 is a graph developed by plotting the first and second canonical variables against each other, thereby providing a visual display of the separation between groups. The first canonical axis (X) in this study represents 67 percent of the total dispersion, whereas the second axis (Y) represents 12 percent of the total. The third and fourth axes (not graphed) represent 11 percent and 6 percent, respectively.

The first canonical axis illustrates the distinct separation of the *mutica* group from all other groups. The *mutica* centroid is a comparatively great distance from any other group centroid, and there is no overlap of individuals from this group with those of any other group. There is a great degree of overlap between the other groups, and their centroids are all about the same distance out along the X axis. This axis discriminates on the basis of size, *mutica* being larger than the other forms, which are all approxi-

TABLE 3.—*Discriminant function classification of specimens in which those with less than a 70 percent probability of belonging to one group were separated into a composite group representing the two most probable groups of membership, based on nine cranial variables. For example, if a specimen originally identified as* V. m. tenuirostris *does not clearly classify into a* tenuirostris *group, it is assigned to a group which combines features of both* arsipus *East and* tenuirostris; *this suggests close relationship between* arsipus *East and* tenuirostris *groups.*

Specimen from this group	Specimen classified to this group or combination group: *mutica*	*arsipus* West	*arsipus* East	*devia*	*tenuirostris*	*macrotis*	*mutica*/ *devia*	*arsipus* W/ *arsipus* E	*arsipus* E/ *macrotis*	*arsipus* E/ *tenuirostris*
mutica	34	0	0	0	0	0	1	0	0	0
arsipus West	0	18	3	0	1	0	0	8	0	0
arsipus East	0	2	13	0	0	3	0	3	1	0
devia	0	0	0	5	1	0	0	0	0	0
tenuirostris	0	0	0	0	5	0	0	0	0	1
macrotis	0	0	0	0	0	5	0	0	0	0

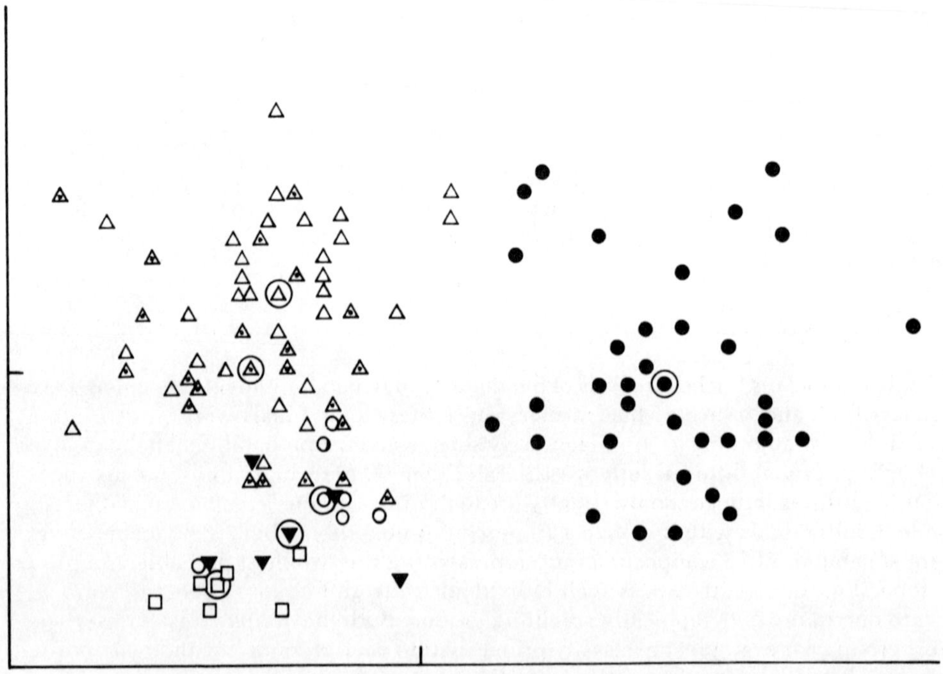

FIG. 1.—Graph of canonical analysis of discrimance in which the first canonical axis (X) is plotted against the second (Y) to illustrate the relationships between six populations of *Vulpes macrotis*. Open triangles represent *V. m. arsipus* from California (= *arsipus* West); dotted triangles represent *V. m. arsipus* from Arizona (= *arsipus* East); inverted solid triangles represent *V. m. macrotis*; open squares represent *V. m. tenuirostris*; open circles represent *V. m. devia*; solid circles represent *V. m. mutica*. Symbols surrounded by a large circle are the centroids for each group.

mately the same size. The remainder of the analysis disregards *mutica*, as it has been separated clearly at this point.

The second canonical axis (Y) shows that each of the other groups overlaps with at least one other group. The overlap of *arsipus* East and *arsipus* West is almost complete, indicating that they probably represent a single population. The vertical arrangement of the various groups suggests that differences between the groups are of a clinal nature.

DISCUSSION

The results of the analyses presented above indicate that *V. m. mutica* is clearly separable from the other kit foxes studied, and should be recognized as a distinct subspecies. The separation of *mutica* from the other forms is due primarily to its larger size. Grinnell *et al*. (1937) also reached this conclusion, but felt that because the type specimen was so large the size factor might not be as significant as it appeared to be. With the larger series examined in this study the size difference is clearly evident. In fact, two specimens from Tracy and one from Goshen, California, are larger than the "giant" type specimen.

Besides the size differences, other cranial features were observed which also distinguish *mutica* from other subspecies. Fully adult specimens of *mutica* develop a more pronounced occipital crest than do adults of the other subspecies. The lyre formed by the temporal ridges is narrower in *mutica*, and often comes close to developing into a true sagittal crest. Lyre breadths are affected by the age of the individual, younger

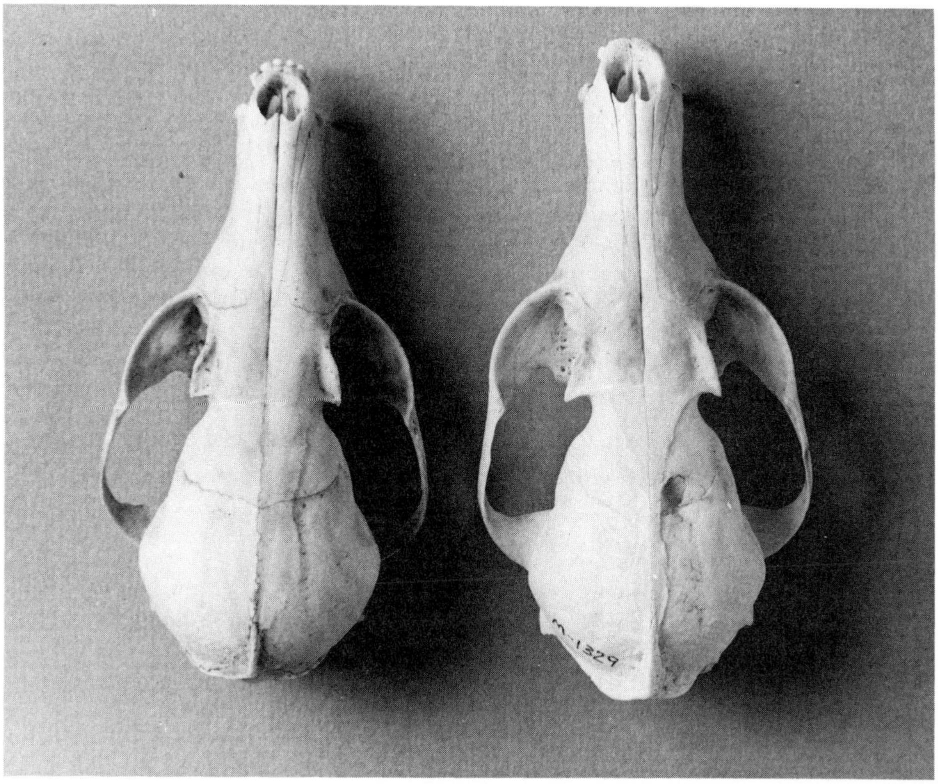

FIG. 2.—Skulls of *Vulpes macrotis macrotis* (left) and *V. m. mutica* (right), showing differences in size, sagittal and occipital crest development, and lyre width. Note occipital bump at posterior end of *macrotis* skull. *V. m. macrotis* skull from an old female taken 15 June 1970, 7.5 mi. south and 2.9 mi. east of Ocotillo Wells, Imperial Co., California; *V. m. mutica* skull from an adult male found dead 11 March 1970, near Buttonwillow, Kern Co., California.

foxes having broader lyres (Shufeldt, 1897–1901). Grinnell *et al.* (1937) did not recognize that lyre breadth was a taxonomic as well as an age feature, and hence assumed that their series of *arsipus* skulls represented younger animals than did their *mutica* series. Although lyre breadths of the two forms overlap (Table 1), most of the skulls are readily recognizable by this feature alone. A third distinguishing feature is the shape of the occipital. In *arsipus* the occipital has a bulge near its center, between the foramen magnum and the occipital crest. This projects posteriorly so far in some skulls that maximum skull length is measured to the occipital prominence rather than to the occipital crest. It can usually be seen when *arsipus* skulls are viewed from above. In *mutica*, the occipital is flat and therefore completely hidden by the larger occipital crest (Fig. 2). The other subspecies are similar to *arsipus*, showing occipital bulges when viewed dorsally.

The results also suggest that Benson (1938) was correct in synonymizing *V. m. arizonensis* (= *arsipus* East) with *V. m. arsipus* (= *arsipus* West). The canonical graph (Fig. 1) further suggests a cline involving *arsipus* and the other three subspecies studied. The close relationship of all these forms is strongly suggested by their phenetic similarity as shown by the measurements analyzed.

Because *arsipus*, *macrotis*, *devia*, and *tenuirostris* are all very similar, and because differences among them appear to be clinal in nature, we feel all these forms should be synonymized. The oldest name available for such a combined group is *macrotis*. This action would result in the recognition of only two subspecies, *V. m. macrotis* (including *arsipus*, *macrotis*, *devia*, and *tenuirostris*) and *V. m. mutica*.

This arrangement is also supported by geography. *V. m. mutica*, occurring in the San Joaquin Valley of central California, is separated from other kit foxes by the barrier of the Tehachapi Mountains, connecting the Sierra Nevada with the transverse ranges of southern California. The populations of foxes occurring in southeastern California, Arizona, and Baja California are not separated from each other by any barriers of like magnitude. The distributions of several other desert vertebrates show a similar pattern. The main geographic feature of the Colorado–Mojave Desert Region is the Colorado River, but the similarity of the *arsipus* West and *arsipus* East groups negates the river as a barrier to kit foxes.

Differences in the habitats involved lend additional support for the conclusion that only two subspecies exist. Most vegetation maps of the southern California–Arizona region indicate that the native vegetation of the San Joaquin Valley, where *V. m. mutica* is found, is grassland, whereas all other forms of kit fox occur in regions usually classified as desert or desert scrub. Even specimens of *V. m. macrotis*, the now extinct form known only from Riverside County, California, and *V. macrotis* specimens from western New Mexico, occur in fingers of desert type environment which project into these areas.

The other three currently recognized subspecies of *V. macrotis*, *neomexicana*, *nevadensis*, and *zinseri*, were not examined during this study, but inclusion of these forms in future studies should further clarify the relationships of kit foxes. It has already been suggested that *V. m. nevadensis* is not recognizably distinct from populations located further to the south (the *arsipus* East and West groups of this study) and should perhaps be included in *V. m. macrotis*, as here described (Hall, 1946).

The two subspecies of kit fox in the western portion of the range of *V. macrotis* are described below.

Subspecies Accounts

In the following descriptions the abbreviations listed earlier under Materials and Methods have been used to indicate the collections from which specimens were examined.

Vulpes macrotis macrotis Merriam

Vulpes macrotis Merriam, Proc. Biol. Soc. Washington, 4:136, 1888.

Vulpes arsipus Elliot, Field Columb. Mus., Publ. 87, Zool. Ser., 3:256, 1903.

Vulpes macrotis devius Nelson and Goldman, Proc. Biol. Soc. Washington, 22:25, 1909.

Vulpes macrotis arsipus, Grinnell, Proc. California Acad. Sci., ser. 4, 3:287, 1913.

Vulpes macrotis arizonensis Goldman, J. Washington Acad. Sci., 21:249, 1931.

Vulpes macrotis tenuirostris Nelson and Goldman, J. Mamm., 12:302, 1931.

Vulpes macrotis devia, Hall and Kelson, The Mammals of North America, 2:858, 1959.

Geographic distribution.—Desert scrub areas of southeastern California, southern Nevada, western and southern Arizona, western Sonora and Baja California , Mexico.

Comparative description.—Smaller and paler than *V. m. mutica*; skull smaller than *mutica* in all measurements except breadth of lyre formed by temporal ridges measured at the frontal–parietal suture (Table 1); occipital crest poorly developed; distinct bulge present on occipital between occipital crest and foramen magnum (Fig. 2).

Adult specimens examined (69).—ARIZONA, *Pima Co*.: La Osa, 1 (USNM); Quitobaquito, 1 (USNM). *Yuma Co*.: Aztec, 3 (USNM); Ehrenberg, 1 (USNM); Tinejas Altas, 5 (MVZ); Tule Tanks, 2 (USNM); Vicksburg, 1 (USNM); Wellton, 1 (MVZ).

CALIFORNIA, *Imperial Co.*: Brawley, 1 (USNM); Calexico, 14 mi. W., 1 (MVZ); Calexico, 18 mi E., 1 (MVZ); Glamis, 1 (MVZ); Palo Verde, 1 (MVZ); Pilot Knob, 2 (MVZ). *Inyo Co.*: Death Valley, 1 (MVZ); Kelly's Well, 1 (MVZ); Resting Spring, 1 (USNM); Triangle Spring, 2 (MVZ). *Riverside Co.*: Morongo Pass, 1 (MVZ); Palm Springs, 2 (1 MVZ, 1 USNM); Pinto Basin, 2 (MVZ); Riverside, 1 (USNM); San Jacinto Valley, 4 (USNM). *San Bernardino Co.*: Cima, 1 (MVZ); Daggett, 1 (USNM); Goffs, 1 (USNM); Hodge, 1 (USNM); Ivanpah, 2 (USNM); Kelso, 1 (MVZ); Lone Willow Spring, 3 (USNM); Victorville, 1 (MVZ). NEVADA, *Clark Co.*: Mesquite, 2 (MVZ). NEW MEXICO, *Grant Co.*: Cliff, 2 (USNM). SONORA, MEXICO: Cerro Colorado, 2 (MVZ); Puerto Kino, 1 (MVZ); Punta Penasco, 2 (MVZ). BAJA CALIFORNIA, MEXICO: Llano de Yrais, 6 (USNM); Valle de Trinidad, 6 (USNM).

Vulpes macrotis mutica Merriam

Vulpes muticus Merriam, Proc. Biol. Soc. Washington, 15:74, 1902.
Vulpes macrotis muticus, Grinnell, Proc. California Acad. Sci., ser. 4, 3:287, 1913.
Vulpes macrotis mutica, Hall and Kelson, The Mammals of North America, 2:858, 1959.

Geographic distribution.—Arid grassland of the southern and western San Joaquin Valley, the Carrizo Plain, and the Cuyama Valley, and (recently) the southern Salinas Valley, in California.

Comparative description.—Larger and slightly darker than *V. m. macrotis*; skull generally larger than *macrotis* in all dimensions except breadth of lyre formed by temporal ridges measured at frontal–parietal suture (Table 1); lyre tending to develop into sagittal crest posteriorly; occipital crest pronounced in adults; occipital flatter than in *macrotis*, without distinct bulge (Fig. 2).

Adult specimens examined (35).—CALIFORNIA, *Fresno Co.*: Coalinga, 3 (USNM); 5–30 miles west of Firebaugh, 9 (MVZ); Mercy Hot Spring, 2 (MVZ); Panoche Creek, 3 (MVZ). *Kern Co.*: Bakersfield, 1 (USNM); Blackwell's Corner, 1 (MVZ); Buttonwillow, 1 (CPSU); Keck's Corner, 1 (MVZ); Rose Station, 1 (MVZ); Snedden Ranch, 1 (CPSU). *San Joaquin Co.*: Tracy, 7 (USNM). *San Luis Obispo Co.*: Shandon, 1 (MVZ). *Tulare Co.*: Alila, 3 (USNM); Goshen, 1 (USNM).

Acknowledgments

We express our sincere appreciation to the staffs at the Museum of Vertebrate Zoology, University of California, Berkeley and at the National Museum of Natural History for their courtesies while we examined specimens in their care; to Dr. David Chesemore, California State University, Fresno for the loan of several specimens; and to Mr. Howard Leach (Nongame Wildlife Coordinator, California Department of Fish and Game) and other members of the department for sending us several specimens obtained recently as road kills. We also thank Richard A. Pimentel, Douglas D. Donaldson, Richard Nelson, Robert Brown, Archie Waterbury, Luther Bertrando, and Elmo Keller, all of California Polytechnic State University, for their advice and assistance.

Results of this study were submitted in partial fulfillment of the requirements for a Master of Science Degree in Biological Sciences, California Polytechnic State University, by the senior author.

Literature Cited

Benson, S. B. 1938. Notes on kit foxes (*Vulpes macrotis*) from Mexico. Proc. Biol. Soc. Washington, 51:7–24.

California Dept. Fish and Game. 1974. At the Crossroads, 1974; a report on California's endangered and rare fish and wildlife. Resources Agency, Sacramento, 112 pp.

Dixon, W. J. 1973. BMD Biomedical Computer Programs. Univ. California Press, Berkeley, 773 pp.

Egoscue, H. J. 1956. Preliminary studies of the kit fox in Utah. J. Mamm., 37:351–357.

———. 1962. Ecology and life history of the kit fox in Toole County, Utah. Ecology, 43: 481–497.

Elliot, D. G. 1903. Descriptions of twenty-seven apparently new species and subspecies of mammals. Field Columb. Mus., Zool. Ser., Publ. No. 87, 3:239–261.

GOLDMAN, E. A. 1931. Two new desert foxes. J. Washington Acad. Sci., 21: 249–251.

GRINNELL, J. 1913. A distribution list of the mammals of California. Proc. California Acad. Sci., 3:265–390.

GRINNELL, J., J. S. DIXON, AND J. M. LINSDALE. 1937. Fur-bearing mammals of California. Univ. California Press, Berkeley, 2:xiv + 377–777.

HALL, E. R. 1946. Mammals of Nevada. Univ. California Press, Berkeley, 710 pp.

HALL, E. R., AND K. R. KELSON. 1959. The mammals of North America. Ronald Press, New York, 2:viii + 547–1083 + 79.

HILDEBRAND, M. 1952. An analysis of body proportions in the Canidae. Amer. J. Anat., 90:217–256.

LENSINK, C. J. 1962. The history and status of sea otters in Alaska. Unpublished Ph.D. thesis, Purdue Univ. Lafayette, Indiana, 188 pp.

MANNING, T. H. 1971. Geographical variation in the polar bear *Ursus maritimus* Phipps. Canadian Wildlife Ser., 13:1–27.

MERRIAM, C. H. 1888. Description of a new fox from southern California. Proc. Biol. Soc. Washington, 4:136–138.

———. 1902. Three new foxes of the kit fox and desert fox groups. Proc. Biol. Soc. Washington, 15:73–74.

NELSON, E. W., AND E. A. GOLDMAN. 1909. Eleven new mammals from Lower California. Proc. Biol. Soc. Washington, 22:23–28.

———. 1931. New carnivores and rodents from Mexico. J. Mamm., 12:302–306.

ORR, R. T., J. SCHONEWALD, AND K. W. KENYON. 1970. The California sea lion: skull growth and a comparison of two populations. Proc. California Acad. Sci. 37:381–394.

SHUFELDT, R. W. 1897–1901. The osteology of *Vulpes macrotis*. J. Acad. Nat. Sci. Philadelphia, Ser. 2, 11:395–418.

U. S. BUREAU OF SPORT FISHERIES AND WILDLIFE. 1973. Threatened wildlife of the United States. Off. Endang. Sp. and Int. Act., Resource Publication 114 (rev. Res. Publ. 34), 289 pp.

Department of Agriculture, Veterinary Research Station, Glenfield, New South Wales, Australia, and Biological Sciences Department, California Polytechnic State University, San Luis Obispo, California 93401. *Submitted 12 April 1976. Accepted 25 November 1976.*

Reprinted from *Journal of Mammalogy*

TAXONOMIC STATUS OF THE SONORAN PRONGHORN

John L. Paradiso and Ronald M. Nowak

The Sonoran pronghorn (*Antilocapra americana sonoriensis*) was named and described by Goldman (Proc. Biol. Soc. Washington, 58:3–4, 1945) on the basis of a female from 40 mi. N Costa Rica, a ranch on the north side of the Río de Sonora, southwest of Hermosillo, Sonora, Mexico. Goldman also examined one other specimen from Fort Buchanan (now Crittenden, Santa Cruz Co., Arizona), which he referred to *A. a. sonoriensis*. He noted, however, that on geographic grounds, specimens from southeastern Arizona could be expected to exhibit morphological gradation towards *mexicana*. The range of *sonoriensis* was defined by Goldman as the desert plains of central western Sonora and north to southern Arizona.

Hall and Kelson (The mammals of North America, pp. 1023, 1959) assigned the pronghorns mentioned by Huey (Trans. San Diego Soc. Nat. Hist., 9:362–363, 1942), and Burt (Misc. Publ. Mus. Zool., Univ. Michigan, 39:73, 1938) to *sonoriensis* on geographic grounds. Hoffmeister and Goodpaster (Illinois Biol. Monogr., 24:137, 1954) stated that although the pronghorn was exterminated from the Huachuca Mountains, the native population was referrable to either *mexicana* or *sonoriensis*. They were dubious, however, about the taxonomic validity of *sonoriensis*. None of these authors reported having examined any specimens. To our knowledge, the type, and the female from Crittenden, Arizona, are the only specimens of *sonoriensis* that have been reported in the literature.

The Bird and Mammal Laboratories have recently received the skulls (without skins) of four male pronghorn from northwestern Sonora. These specimens are from within the ascribed range of *sonoriensis*, and agree closely with the holotype of that subspecies. Inasmuch as *sonoriensis* is on the Department of the Interior's list of rare and endangered wildlife, and because since it was known in the literature from only two female specimens, we felt it pertinent to discuss these new specimens and to offer comments on the taxonomic status of *sonoriensis*.

The skulls are those of one adult male, with full development of horns, and with all teeth erupted and showing considerable wear, and three juvenile males with little development of horns and third upper molars partly erupted. They were illegally killed about 1 February 1969, near the village of Caborca in northwestern Sonora, and were smuggled into Tucson, Arizona, where they were seized by United States Game Management agents. They were deposited in the mammal collection at the National Museum of Natural History (USNM 347452–347455).

The adult male (347452) differs from males of the subspecies *americana*, *mexicana*, and *peninsularis* in essentially the same characters that differentiate the female holotype of *sonoriensis* from females of those other races. The skull is markedly narrower, and more delicately structured. The frontal depression is less pronounced; the auditory bullae are smaller, flatter, and less projecting below the level of the basioccipital; and the premaxillae are less extended posteriorly along the median line. These characters were noted by Goldman in his description of *A. a. sonoriensis*. In addition, the holotype of *sonoriensis* and the recently acquired adult male from Sonora differ from comparably sexed adults of *americana*, *mexicana*, and *peninsularis* in that the molar teeth are smaller in both length and width; the rostrum is slenderer and the mastoidal breadth and greatest width at posterior border of orbits are less (a reflection of the narrower skull of *sonoriensis*). A noticeable feature that may have taxonomic significance is that in all but three of 54 *americana*, *mexicana*, and *peninsularis* that we examined, the depression of the palate posterior to the palatine foramina is bordered on both sides by a high and sharp ridge that extends from the anterior edge of the alveolus of the first premolar to the anterior tip of the maxilla. In all of the *sonoriensis* that we have seen, this ridge is lacking except in the region immediately anterior to the alveolus of the first premolar.

The three juvenile male skulls of *sonoriensis* are narrower than are comparably aged specimens of *americana*, *mexicana*, and *peninsularis*, and differ from them in the same characters listed above for the adults.

The female from Crittenden, Arizona, although probably referrable to *sonoriensis*, appears intermediate between that race and *mexicana* is some characters. It has the narrow mastoidal region, poorly developed frontal depression, and small bullae of *sonoriensis*, but is broader across the orbits and has bigger molar teeth. Another Arizona specimen, obtained in March 1971 by the Arizona Game and Fish Department, approximately 4 mi. N Parker Dam and 3½ miles east of the Colorado River in Mohave County, and sent to us for identification, shows no approach to the cranial characters of *sonoriensis*.

In Table 1, cranial measurements of the holotype of *sonoriensis*, the female from Crittenden, Arizona, and the adult male from the vicinity of Caborca, Sonora, are compared with measurements of other pronghorn populations to illustrate several of the characters discussed above.

Although there are still insufficient specimens to allow a complete appraisal of the taxonomic status of *sonoriensis*, the fact that the four recently acquired specimens from near Caborca, Sonora, show marked similarities to the holotype of *sonoriensis* and differ from specimens of *americana*, *mexicana*, and *peninsularis* in the same characteristics as does the holotype of *sonoriensis*, provides strong support for the continued recognition of *sonoriensis* as a distinct subspecies of *A. americana*. In fact, judged from the six specimens of *sonoriensis* that we examined, that taxon is more distinct from the other subspecies of pronghorn than any of them are from each other.

Specimens measured in this study (47 in National Museum of Natural History) are listed below.

Antilocapra americana americana.—MONTANA: (Buffalo Butte), 1; Capitol, Carter County, 1; Chief Mountain, Glacier County, 1; Gardner, Park County, 1; Little Dry Creek, Garfield County, 1; Poplar, Roosevelt County, 1; (Three Buttes), 1; no exact locality, 5. NORTH DAKOTA: Forbes, Dickey County, 2; Little Missouri River, McKenzie County, 1; Union, Cavalier County, 1; White Earth River, Montrail County, 1. SOUTH DAKOTA: Fort Pierre, Stanley County, 2.

Antilocapra americana mexicana.—ARIZONA: Parker Dam vicinity, Mohave County, 1. CHIHUAHUA: Presidio del Norte, 1; Whitewater, 1; Sierra en Media, 6.

Antilocapra americana peninsularis.—BAJA CALIFORNIA: 25 to 50 mi. S Callamalli, 13.

Antilocapra americana sonoriensis.—ARIZONA: Fort Buchanan (Crittenden), 1. SONORA: Caborca vicinity, 4; Costa Rica, 1 (holotype).

JOHN L. PARADISO AND RONALD M. NOWAK, *Bird and Mammal Laboratories, Bureau of Sport Fisheries and Wildlife, National Museum of Natural History, Washington, D. C., 20560, and Museum of Natural History, University of Kansas, Lawrence, 66044.* *Accepted 14 August 1971.*

TABLE 1.—*Cranial measurements of four subspecies of* Antilocapra americana. *The single male* sonoriensis *is from Caborca, Sonora. The first measurement for female* sonoriensis *is of the holotype, the second is of specimen from Crittenden, Arizona. See text and list of specimens measured for localities of other taxa. For larger samples, the mean is followed by the range (in parenthesis);* N *indicates the number in the sample and SD is the standard deviation.*

	Males			Females		
Subspecies	Mean (range)	*N*	SD	Mean (range)	*N*	SD
			Greatest length			
A. a. sonoriensis	281.0			250.8, —		
A. a. americana	284.2 (278.3–297.0)	13	4.68	281.1 (268.3–294.0)	6	8.37
A. a. mexicana	276.0, 294.0			274.0 (268.4–289.0)	5	7.96
A. a. peninsularis	285.1 (274.0–298.0)	7	6.64	270.3 (261.0–288.0)	6	9.05
			Basilar length			
A. a. sonoriensis	243.2			219.3, —		
A. a. americana	248.8 (243.0–257.9)	12	4.15	246.6 (238.5–257.8)	6	6.48
A. a. mexicana	239.4, 254.2			239.2 (232.1–250.4)	5	6.61
A. a. peninsularis	247.2 (238.0–260.1)	7	6.26	238.2 (231.5–250.5)	6	6.55
			Occipitonasal length			
A. a. sonoriensis	221.8			192.0, 208.3		
A. a. americana	224.0 (215.2–237.5)	13	6.62	221.9 (209.9–232.2)	5	8.21
A. a. mexicana	220.5, 234.2			215.7 (207.8–221.2)	6	6.64
A. a. peninsularis	224.9 (213.1–235.3)	7	7.54	211.4 (199.5–230.5)	6	9.67
			Length of maxillary toothrow			
A. a. sonoriensis	72.0			67.0, 68.1		
A. a. americana	73.1 (67.0– 80.4)	15	2.77	72.1 (68.8– 75.0)	7	1.91
A. a. mexicana	72.1, 71.7, 68.7			70.2 (67.7– 72.2)	6	1.41
A. a. peninsularis	73.6 (70.8– 80.0)	7	2.98	70.0 (65.5– 73.0)	6	2.67
			Alveolar length of M3			
A. a. sonoriensis	16.6			15.5, 18.3		
A. a. americana	18.6 (16.4– 22.0)	15	1.67	20.1 (16.6– 23.5)	7	2.27
A. a. mexicana	17.8, 19.3, 19.5			17.6 (16.0– 19.3)	6	1.44
A. a. peninsularis	18.3 (15.1– 19.9)	7	1.47	17.9 (16.3– 20.6)	6	1.58
			Breadth of rostrum across M2			
A. a. sonoriensis	65.7			66.7, 66.2		
A. a. americana	71.9 (67.0– 75.7)	14	2.74	71.0 (67.0– 78.3)	7	3.39
A. a. mexicana	67.0, 69.7, 72.0			70.0 (64.5– 71.4)	6	2.44
A. a. peninsularis	69.8 (65.2– 73.3)	7	2.53	66.5 (65.0– 67.7)	6	.94
			Mastoidal breadth			
A. a. sonoriensis	77.5			69.2, 70.2		
A. a. americana	83.6 (77.2– 87.9)	14	2.86	75.9 (72.5– 78.7)	5	2.37
A. a. mexicana	85.2, 88.0			74.6 (71.7– 77.3)	6	1.97
A. a. peninsularis	81.7 (80.0– 83.2)	7	.98	74.8 (72.3– 77.8)	6	1.63

Reprinted from *The Wilson Bulletin*

CALIFORNIA CONDOR SPECIMENS IN COLLECTIONS

Sanford R. Wilbur

As part of a study of former numbers and distribution of the California Condor (*Gymnogyps californianus*), I have been able to locate 185 mounted birds and study skins, 51 skeletons, and 55 eggs of this endangered species. All but two skins and three eggs are housed in public or scientific institutions. United States collections contain 151 skins and mounts, 50 skeletons, and 52 eggs. In the following alphabetical-by-locality list, specimen numbers are listed in order: adult (A) and immature (I) plumaged skins or mounts/skeletons/eggs. I believe most specimens now in existence are included, and would appreciate learning of any additions.

AUSTRIA: Naturhistorisches Museum (Vienna)—2A,1I/0/0; Oberosterreichisches Landmuseum (Linz)—1A/0/0. BELGIUM: Institut Royal des Sciences Naturelles (Brussels)—2A/0/0. CANADA: National Museum of Natural Sciences (Ottawa)—1I/0/0; Royal Ontario Musuem (Toronto)—1A,1I/0/0. ENGLAND: British Museum (Tring)—7A, 3I/1/3. FRANCE: Museum National d'Histoire Naturelle (Paris)—2A,1I/0/0. GERMANY: Zoologisches Museum (Berlin)—1A/0/0. MEXICO: Universidad Nacional Autonoma (Mexico City)—1A/0/0. NETHERLANDS: Rijksmuseum van Natuurlijke Histoire (Leiden)—1A/0/0. SWEDEN: Naturhistoriska Museum (Gothenburg)—2A/0/0; Naturhistoriska Rikmuseet (Stockholm)—1A/0/0. SWITZERLAND: Museum d'Histoire Naturelle (Geneva)—1A/0/0; Naturhistorisches Museum (St. Gallen)—1A,1I/0/0. U.S.S.R.: Zoological Institute-Academy of Sciences (Leningrad)—2A,1I/0/0. UNITED STATES, ARIZONA: University of Arizona, (Tucson)—1A/1/0. CALIFORNIA: Arthur Bryant Collection (Los Angeles)—1A/0/0; Bakersfield College (Bakersfield)—1I/0/0; California Academy of Sciences (San Francisco) 12A,3I/1/1; Clarke Museum (Eureka)—1A,1I/0/0; Fillmore High School (Fillmore)—0/1/0; Foster Bighorn Bar (Rio Vista)—1A/0/0; Fresno State College (Fresno)—1A/0/0; Kern County Museum (Bakersfield)—1I/0/0; Los Angeles County Museum (Los Angeles)—10A,6I/10/1; Museum of Natural History (Santa Barbara)—3A,1I/1/2; Museum of Vertebrate Zoology (Berkeley)—4A,3I/12/2; Natural History Museum (San Diego)—4A/0/1; Pioneer Museum (Ventura)—1I/0/0; Pomona College (Claremont)—1I/0/0; San Bernardino County Museum (Bloomington)—1I/0/1; Sidney Peyton Collection (Fillmore)—0/0/1; University of California (Los Angeles)—2A/11/0; Western Foundation of Vertebrate Zoology (Los Angeles)—1A/0/7. COLORADO: Museum of Natural History (Denver)—5A,1I/0/2. CONNECTICUT: Peabody Museum (New Haven)—1A/0/1. DELAWARE: Museum of Natural History (Greenville)—1I/0/2. DISTRICT OF COLUMBIA: U.S. National Museum—8A,12I/9/5. FLORIDA: Florida State Museum (Gainesville)—0/0/2. ILLINOIS: Field Museum of Natural History (Chicago)—6A,1I/0/7; University of Illinois (Urbana)—2A/0/0. IOWA: University of Iowa (Iowa City)—1A/0/0. LOUISIANA: Louisiana State University (Baton Rouge)—1A/0/0. MASSACHUSETTS: Museum of Comparative Zoology (Cambridge)—6A,8I/0/8. MICHIGAN: University of Michigan (Ann Arbor)—2A/0/0. NEBRASKA: Hastings Museum (Hastings)—1A/0/0. NEW YORK: American Museum of Natural History (New York City)—14A,7I/0/3. OHIO: Museum of Natural History (Cincinnati)—2A/0/0; Ohio State University (Columbus) 1I/0/0. PENNSYLVANIA: Academy of Natural Sciences (Philadelphia)—2A,2I/3/2; Carnegie Museum (Pittsburgh)—2A/0/1; Nelson Hoy Collection (Media)—0/0/1; Public Museum (Reading)—1A/0/0. TEXAS: Robert L. More Jr. Collection (Vernon)—0/0/1. VIRGINIA: Virginia Polytechnic Institute and State University (Blacksburg)—1A/0/0. WASHINGTON: University of Puget Sound (Tacoma)—0/0/1. WISCONSIN: Public Museum (Milwaukee) —2A,1I/1/0.—SANFORD R. WILBUR, *U.S. Bureau of Sport Fisheries and Wildlife, Patuxent Wildlife Research Center, 1190 East Ojai Avenue, Ojai, California 93023.* *Accepted 10 September 1973.*

SECTION VI.
Regulation and Protection

Comments introducing the sections on Endangered Species and Identification pointed out the importance of being able to attach an accurate, well-defined name to a population or individual. The papers in this section emphasize that the name is the key to the suite of biological characteristics of a population and that differences in such character suites form the basis for regulations that permit or restrict the taking or other utilization of a species.

The Migratory Bird Treaty Act is one of the nation's primary laws offering protection to wildlife. The importance of providing a firm basis for the names of the protected species is shown by a brief extract from the preamble to the most current set of regulations that lists species protected by this Act. These regulations also indicate that a species in a group listed in a treaty must occur in the United States to achieve federal protection. The note by White and Baird is one of many possible examples of distributional papers that provide the factual basis for the addition of a species to the protected list. The Needle-tailed Swift was placed on the protected list for the first time in the revision of November 16, 1977.

John Aldrich, whose name appears so frequently in this book, studied Mourning Doves taken in a special early hunting season in southern Florida. As a result of his finding that these birds were migrants from the population in northern states instead of, as previously thought, the West Indies subspecies, the early season was instituted throughout the state after 1951.

The contribution by R. R. Miller attempts to trace the history of occurrence of three related fishes in the Great Lakes, two of which may have been introduced or transplanted by man. These newly occurring species have had an effect on a commercially important native species and have themselves attained commercial importance. This study suggested that methods of utilizing this new resource should be devised. By the early 1960s the stock of Alewives reached commercially harvestable proportions and an industrial trawl fishery switched opportunistically to that species (J. Kutkuhn, in litt.).

Although not stated in Don Wilson's paper on the Polar Bear, his study responded to a direct request from management authorities for information on the relationships of the heavily hunted southern Alaskan population of this species to others in North America. The degree of relationship can indicate individual and genetic exchange between populations and is necessary in evaluating harvest data and the overall impact of the harvest, thus forming a basis for the setting of quotas or harvest limits.

The paper by Perrin and its earlier companions (see references) form the basis for the definition of porpoise stocks. Quotas placed on these stocks regulate the take by the Pacific tuna purse-seine fishery. Similarly, the Johnson and Stewart paper is one of a series of studies of variation in a species that attempt to define populations and their limits so that proper hunting and protective regulations can be devised.

The final selection in this section calls attention to a situation where neither systematics nor management was applied soon enough, a situation likely to be repeated often because of the failure of many to consider reptiles or amphibians as "wildlife." Although directed as a warning to laboratory biologists, this contribution should be of great significance to anyone working with live animals, in the field or in captivity. It can also be taken as a plea to systematists to provide more precise information on even "common" species and to managers to pay attention to species not traditionally within their purview.

Additional Readings

Aldrich, John W. 1963. Geographic orientation of American Tetraonidae. *Jour. Wildl. Mgt.* 27: 529-545.

Aldrich, John W., Allen J. Duvall, and Aelred D. Geis. 1958. Racial determination of origin of Mourning Doves in hunters' bags. *Jour. Wildl. Mgt.* 22: 71-75.

Boeker, Erwin L., John W. Aldrich, and William S. Huey. 1961. Study of experimental Sandhill Crane hunting season in New Mexico during January 1961. *Special Scientific Report—Wildlife no. 63,* U.S. Fish and Wildlife Service.

Chapman, Joseph A., and Raymond P. Morgan II. 1973. Systematic status of the cottontail complex in western Maryland and nearby West Virginia. *Wildl. Monog.* no. 36.

Gibbs, Robert H., Jr., and Bruce B. Collette. 1969. Comparative anatomy and systematics of the tunas, genus *Thunnus*. *Fishery Bull.* 66: 65-130.

Harrison, W. I. 1962. The first record of the Rufous-backed Robin in the United States. *Auk* 79: 271.

Levy, Seymour H. 1959. Thick-billed Kingbird in the United States. *Auk* 76: 92.

Perrin, William F. 1969. Color pattern of the eastern Pacific spotted porpoise *Stenella graffmani* Lönnberg (Cetacea, Delphinidae). *Zoologica* 54: 135-142, Pl. I-VII.

———. 1972. Color patterns of Spinner Porpoises (*Stenella* cf. *S. longirostris*) of the eastern Pacific and Hawaii, with comments on delphinid pigmentation. *Fishery Bull.* 70: 983-1003.

Saunders, George B. 1968. Seven new White-winged Doves from Mexico, Central America, and southwestern United States. *North Amer. Fauna* no. 65.

Reprinted from *Federal Register*

RULES AND REGULATIONS

Title 50—Wildlife and Fisheries

CHAPTER I—UNITED STATES FISH AND WILDLIFE SERVICE, DEPARTMENT OF THE INTERIOR

SUBCHAPTER B—TAKING, POSSESSION, TRANSPORTATION, SALE, PURCHASE, BARTER, EXPORTATION, AND IMPORTATION OF WILDLIFE

PART 10—GENERAL PROVISIONS

Revised List and Definition of Migratory Birds

AGENCY: Fish and Wildlife Service, Interior.

ACTION: Final rulemaking.

SUMMARY: This rulemaking revises the list of migratory birds contained in 50 CFR 10.13. As revised, the list contains all species covered by the treaties signed between the United States and Great Britain (on behalf of Canada), the United States and Mexico, and the United States and Japan for the preservation of migratory birds. Consequently, the revised list now includes all species protected by the Migratory Bird Treaty Act and its implementing regulations. The new list also eliminates the unnecessary distinction between game and nongame species.

To complement the revised list, the definition of "migratory bird" in 50 CFR 10.12 is amended so that the term now refers to any bird belonging to a species included in the list. The new definition also makes clear its coverage of mutations and hybrids as well as parts, nests, eggs, and products.

EFFECTIVE DATE: December 16, 1977.

FOR FURTHER INFORMATION CONTACT:

Marshall L. Stinnett, Special Agent in Charge, Regulations and Penalties, Division of Law Enforcement, United States Fish and Wildlife Service, U.S. Department of the Interior, Washington, D.C. 20240, 202-343-9237.

SUPPLEMENTARY INFORMATION: The Migratory Bird Treaty Act (16 U.S.C. 703-711) [hereinafter referred to as "the Act"] expressly protects any migratory bird included in the terms of the Convention for the Protection of Migratory Birds, August 16, 1916, United States—Great Britain (on behalf of Canada), 39 Stat. 1702, T.S. No. 628; the Convention for the Protection of Migratory Birds and Game Mammals, February 7, 1936, United States—Mexico, 50 Stat. 1311, T.S. No. 912; or the Convention for the Protection of Migratory Birds and Birds in Danger of Extinction, and Their Environment, March 4, 1972, United States—Japan, 25 U.S.T. 3329, T.I.A.S. No. 7990 (16 U.S.C. 703). Regulations implementing the Act, which are found principally in Title 50, Code of Federal Regulations, Parts 10, 20, and 21, are likewise applicable to any bird covered by one of the three treaties (16 U.S.C. 704). Accordingly, 50 CFR 10.12 currently defines "migratory birds" as "* * * all birds, whether or not raised in captivity, included in the terms of conventions between the United States and any foreign country for the protection of migratory birds * * *"

However, neither the Act nor its implementing regulations presently contain a definitive list of all the species covered by the treaties. The current list in 50 CFR 10.13 is only a reference list and does not include the birds receiving protection under the Japanese treaty. Thus, the public must constantly refer to the treaties in order to determine which species are protected by the Act and its regulations.

To eliminate the need for such cross-referencing, and to clear up other problems with the list as well, the Service proposed in November of 1976 to amend 50 CFR 10.13 so that the list would include all species covered by the treaties and hence protected by the Act and its regulations (41 FR 50010, Nov. 12, 1976). At the same time, the Service proposed a complementary amendment to 50 CFR 10.12 so that the definition of "migratory birds" would refer to the list rather than to the treaties (41 FR 50011). The proposed amendment to section 10.12 was also aimed at clarifying the status of mutations, hybrids and parts. Lastly, the Service proposed a new exception for hybrid "barnyard ducks" (41 FR 50014). Final regulations on that proposal have been reserved for a later time.

Public comments on the November

proposal focused exclusively on which birds were intended to be covered by the treaties and therefore should be included in the list. In response to those comments, and in light of other information received by the Service, several changes have been made in the proposed list. The majority of comments concerned typographical errors or inadvertent omissions or deletions from the list. These omissions or deletions, such as the chickadee, catbird, and others, have been restored to the list.

In addition, the introductory paragraph to the list has been modified for greater clarity, and the definition of "migratory bird" has been reworded to make clear its applicability to nests, eggs, and products.

Accordingly, as adopted in this final rulemaking, the list of migratory birds in 50 CFR 10.13 now contains all species covered by the three treaties and consequently all species protected by the Act and its implementing regulations. The species classification was selected because it is the smallest taxon common to all three treaties.

Generally, the species are listed alphabetically by the most widely used common name, followed by the scientific name. However, all species of ducks are listed under the heading "DUCKS" for easier reference. Scientific and English names are based on the "Check-list of North American Birds," fifth edition, published in 1957 by the American Ornithologists' Union, the "Thirty-second Supplement to the American Ornithologists' Union Check-list of North American Birds" (The Auk 90:411–419, 1973, as corrected in the Auk 90:887, 1973), and the "Thirty-third Supplement to the American Ornithologists' Union Check-list of North American Birds" (The Auk 93:875–879, 1976, as corrected). The names of species not included in the "Check-list" or supplements are based on the other current technical literature. As provided in the proposal, the list's current distinction between game and nongame species has been eliminated. Game birds are those for which open seasons are prescribed in 50 CFR Part 20. Since game birds are designated in Part 20, there is no need for a separate listing in Part 10. Also, eliminating the game bird listing in Part 10 will avoid the necessity of dual amendments when changes in game birds are made in Part 20.

Concerning the definition of "migratory bird," this rulemaking amends 50 CFR 10.12 to define that term as any bird, whatever its origin and whether or not raised in captivity, which belongs to a species listed in section 10.13. The definition also provides that "migratory bird" includes mutations or hybrids of listed species, as well as parts, nests, eggs, or products. Because of the great difficulty in distinguishing mutations and hybrids from purebreds, coverage of the former two is essential to adequate enforcement of the Act and treaties. By including parts, nests, eggs, and products, section 10.12 merely restates the coverage of the Act (16 U.S.C. 703).

This rulemaking was prepared by Marshall L. Stinnett, Special Agent in Charge, Regulations and Penalties, Division of Law Enforcement.

Reprinted from *The Auk*

FIRST NORTH AMERICAN RECORD OF THE ASIAN NEEDLE-TAILED SWIFT, *HIRUNDAPUS CAUDACUTUS*

Clayton M. White and William M. Baird

On 21 May 1974, a swift with a weak and languid wing beat was collected on Shemya Island, Aleutian Islands, 52°43′ N, 174°07′ E, as it slowly foraged along a tundra hillside. This is the first New World record for a swift of this genus and species. The pattern of flight suggested that the bird was in poor condition and under a nutritional stress. Examination showed it to have an extremely reduced pectoral muscle, but the bird, deposited in the National Museum of Natural History (No. 526402, male, right testis 5 × 2 mm) weighed 88 g, which suggests a near average condition. As we were unable to locate spring weight data for this species, weights on another swift of nearly identical size, *Apus melba* from the Aegean region, were provided by G. E. Watson. The average weight of 12 breeding males of *A. melba* was 85.8 g.

The specimen is referable to the nominate form (*H. c. caudacutus*) based on wing length and forehead color. This race breeds essentially in central and eastern Asia and migrates to winter in Australia and Tasmania. Vaurie (1969, Birds of the palearctic fauna, vol. 2, Non-Passeriformes, London, H. F. and G. Witherby, Ltd., pp. 647–648) indicates that it breeds eastward to near the mouth of the Amur River, and on Sakhalin and the Kurile Islands, westward and northward to about 85° E, 56° N (Tomsk) in western Siberia, and southward to northern China and Japan. It is a straggler to England and Finland and its breeding range is moving westward and northward (in the region of 58° N, 78° E) in Siberia (Vaurie, op. cit.). Other races occur farther south in southeastern Asia.

In view of its breeding range in the northmost part of the Kuriles, it is not too surprising that this migrant wandered to the Aleutians, especially as the spring prevailing winds are from the southwest. Shemya is about 900 miles (1440 km) directly over water from the northern Kuriles or about 1100 miles (1760 km) following along the land masses of the Kamchatka Peninsula thence out the Aleutians to Shemya. This does not seem to be an extensive distance considering the statements of Vaurie (op. cit.: 647) ". . . swifts of this genus are credited as being the most powerful and fastest flyers of all birds" and of Slater (1971, A field guide to Australian birds, non-passerines, Pennsylvania, Livingston Publishing Company, p. 400) "Entirely aerial; not known to land in Australia." The southernmost record of this species wandering beyond Tasmania is from Macquarie Island, 54°37′ S, 158°54′ E (Warham 1961, Emu 16: 189).

Several English names are currently used for the species, the commonest and preferred one being White-throated Needle-tailed Swift.—CLAYTON M. WHITE, *Department of Zoology, Brigham Young University, Provo, Utah 84602*, and WILLIAM M. BAIRD, *69 Hartwell Avenue, Littleton, Massachusetts 01460*. Accepted 24 Feb. 76.

Reprinted from *Journal of Wildlife Management*

THE SOURCE OF MIGRANT MOURNING DOVES IN SOUTHERN FLORIDA

John W. Aldrich

U. S. Fish and Wildlife Service

The Problem

There has been a separate early open season on the Mourning Dove for the southern three counties of Florida (Monroe, Dade and Broward) for a number of years. During recent years the southern Florida season has been the month of October, while the season in the rest of the state has started about two months later. The original basis for dividing Florida with respect to the open season in this way was the fact that doves usually appeared in numbers in the extreme southern counties early in the fall and were abundant for only a short time. A belief has become prevalent that this early flight of doves in southern Florida originates in the West Indies and thus the doves that are shot in Monroe, Dade, and Broward Counties are from an entirely different breeding population from that hunted in counties to the north. This theory has continued to give support for a separate open season.

Evidence for West Indian Origin

The source of the theory that the southern Florida dove population is made up largely of birds from a West Indian breeding population is obscure. However, it appears to have been based on observations of numbers of Mourning Doves migrating in fall up the Keys toward the mainland and northward through the southern counties. These observations have been described in detail in reports by Harold S. Peters (1941 and 1948) and by Frederick C. Lincoln (1945). There seems to be abundant evidence that such a northward flight does occur in the fall in southern Florida, both from observations in the field and from the banding records. Of the approximately 6,500 birds banded at Key West, mostly during the nineteen thirties, by the late Wm. W. Demeritt, there were numerous direct recoveries during the same fall migration from points to the northward in Florida and even as far as the southern part of Georgia (Fig. 1).

The fact that this northward migration was observed early in the fall season in southern Florida without any preceding southward movement having been detected, has been considered as evidence that the doves observed must have come from some breeding area to the south of Florida; and the island of Cuba is the nearest area in that direction where the Mourning Dove breeds in numbers. It seems to be the concensus of opinion that neither the Florida Keys nor the mainland of extreme southern Florida affords a breeding population sufficiently large to have been the origin of the large numbers of birds sometimes reported (Peters, 1941; and Lincoln, 1945). Peters has suggested that the hurricane which swept up into southern Florida prior to his 1948 investigations might have brought in birds from Cuba

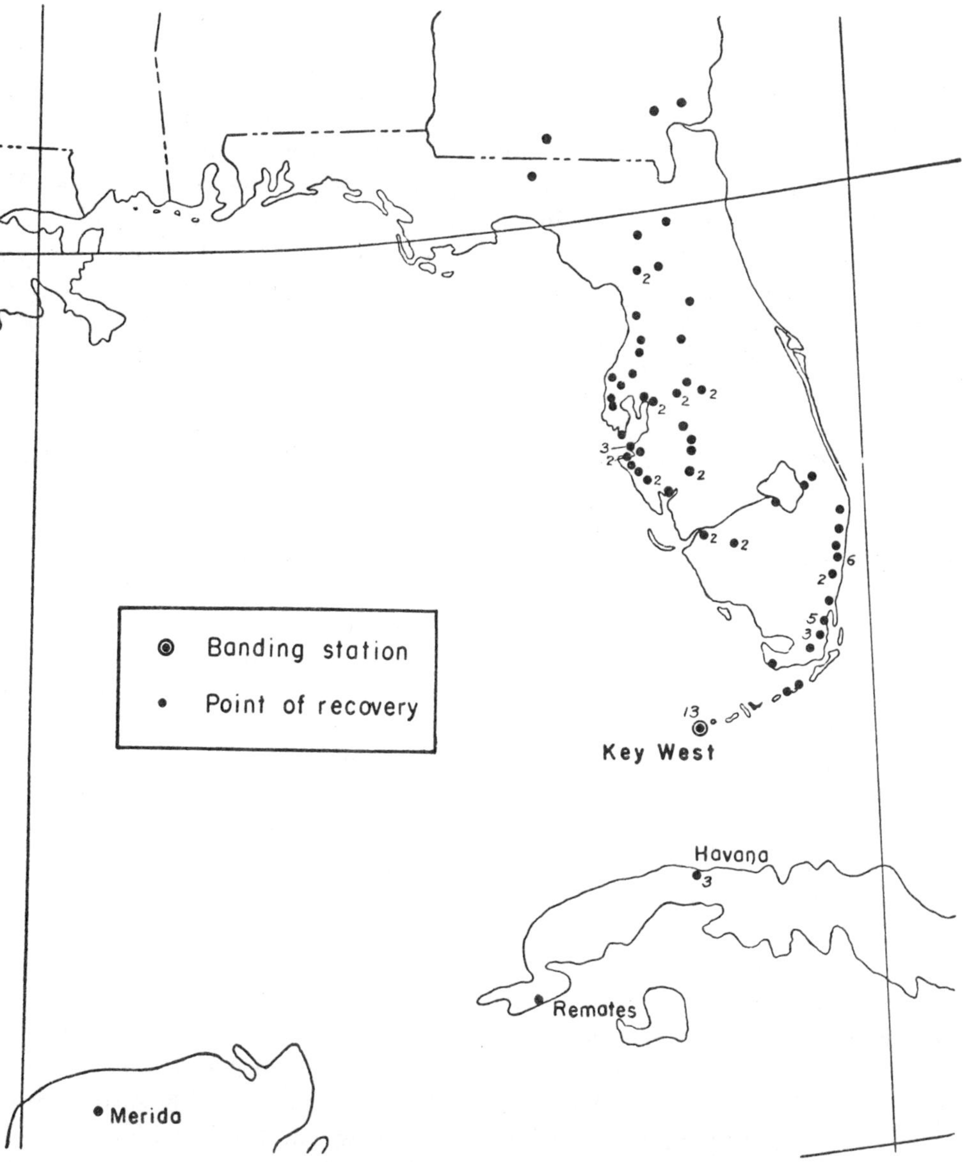

FIG. 1. Direct recoveries of birds banded at Key West, Florida.

and accounted for the large number of birds present that year. However, reports do not indicate any noticeable change in the dove population after the severe hurricane during the middle of October of 1950, and sportsmen report that there are fewer doves after hurricanes than before (*fide* F. C. Lincoln).

STUDY OF SPECIMENS

Previous to the present study the writer had identified as the eastern North American race, 6 specimens of Mourning Doves taken by Frederick C. Lincoln and one by Harold Peters, as well as 2 old specimens in the National Museum collection, all taken in southern

Florida during migration. However, up until the present investigation there has been no attempt to identify a large sample of the Mourning Doves taken during the southern Florida hunting season, to determine whether they belonged to the race which breeds in the West Indies or that of eastern North America, or even that of western North America. Because of the importance which had become attached to the theory of West Indian origin of southern Florida doves in connection with formulation of Federal hunting regulations, the writer was delegated by the Fish and Wildlife Service to investigate this matter, during the hunting season of 1950. In preparation for these investigations a study was made of 311 Mourning Dove specimens in the United States National Museum collection, and the limits of geographic variation of the two North American races and the one West Indian race were determined.

The West Indian race of Mourning Dove, *Zenaidura macroura macroura*, is distinguishable from the eastern North American race, *Zenaidura macroura carolinensis*, by having in all plumages, a deeper, more reddish buffy wash to the under parts and, in adults, by smaller size. Thirty-nine adult male specimens of West Indian Mourning Doves had wings ranging in length from 134 to 146, averaging 138.3 mm., whereas 54 male Eastern Mourning Doves had wings ranging from 140 to 157, averaging 146.7 mm.; 18 adult female West Indian birds had wings ranging from 128 to 138 and averaging 131.9 mm., whereas 24 female Eastern Mourning Doves had wings ranging from 135 to 144 and averaging 140.1 mm. Although there was considerable overlap in the measurements of these two races, the color differences were much more absolute, with only a slight overlap. By combining color and size characters it was found that a very high percentage of specimens of the West Indian race could be distinguished from specimens of the Eastern Mourning Dove. The specimens of the Western Mourning Dove were found to be different in all plumages from the other two races in the paleness of both upper and lower parts. They lacked the bright cinnamon buff of the under parts characteristic of the West Indian birds and in this respect they were more like the eastern North American race. There was no significant size difference between the eastern and western races. After working out the ranges of variation in coloration of these three subspecies, typical examples of each, including different sex and age plumages, were selected to use as standards for identifying specimens in the field.

With this preparation the writer spent the first 15 days of the hunting season (October 1 to October 15, 1950) in company with game officers in southern Florida, examining Mourning Doves in hunters' bags. During that period a total of 299 specimens were identified. The specimens were first sorted according to age in three categories. If the white feather edgings, characteristic of juvenal plumage, were present on the body plumage, the bird was grouped as a juvenile. If all of the white edgings, except the greater primary coverts, had been lost it was listed as immature. If all white-edged feathers had been molted, the specimen was classified as adult. Undoubtedly some of the earliest hatched birds of the year had already molted the last juvenal primary covert and thus were mistaken for adults. Of the total birds examined, 23.4 percent were classified as adults, 52.8 percent immatures, and 23.7 percent juveniles. Thus 76.5 percent were young of the year.

The next step was to sort the immature and mature specimens by sex on the basis of coloration of breast and head plumage. In these age groups it is possible, with experience, to distinguish fairly easily between the brighter, pinkish breast of the males as contrasted with the duller brownish breast of the females. It was not found practicable to attempt to determine the sex of the juveniles which have not yet molted their body plumage. The sorted doves were compared with specimen samples of the appropriate age and sex of the three races which I carried about with me in a flat box. The wings of all adult specimens were measured and recorded. Final identifications were made on the basis of color of young birds and combined color and measurements of adults. On this basis, of the 299 specimens examined, 4 (1.3%) were identified as the West Indian race, *macroura;* 4 (1.3%) as the western race, *marginella;* and 291 (97.3%) as the eastern race, *carolinensis.* The specimens of the West Indian and Western Mourning Dove were fairly well scattered as to locality and date. The West Indian Mourning Doves were an adult male, Perrine, Oct. 1; immature male, Golden Glades, Oct. 4; adult male, Virginia Gardens, Oct. 7; and immature male, Kendall, Oct. 8, all in Dade County, Florida. The specimens of Western Mourning Dove were an immature male, Perrine, Oct. 1; a juvenile, Perrine, Oct. 2; an immature female, Virginia Gardens, Oct. 7, in Dade County; and a juvenile, West Deerfield, Broward County, Florida, Oct. 11. No effort was made to obtain any of these specimens for preservation because it was important not to jeopardize the continued cooperation of the hunters in making these birds available for study.

The percentage of young of the year ranged from 75% on October 1 to 80% on October 7, and averaged 76.5% for the first 15 days of the season. This is considerably lower than the 95.4% found by Peters from October 16 to November 9 in the large flight of 1948, or even the 90% based on 1,481 doves examined during the entire season of 1950 (Peters, 1951). This may indicate that doves which migrate to southern Florida are practically all young birds, and that the early season birds are made up to a large extent of locally reared birds, including a higher proportion of adults. However, it should be noted that two of the four West Indian Mourning Doves identified were adults. Although all four of the Western Mourning Doves examined were young of the year, the sample was not large enough to prove that only young birds of this race ever reach southern Florida.

On the basis of my 1950 observations and specimen identifications it is concluded that the Mourning Doves which were shot in southern Florida during the first two weeks of the 1950 season did not come primarily from breeding populations in the West Indies to the south, but rather to a very large extent from breeding grounds in the eastern United States. Information is insufficient at present to determine the relative abundance of locally reared doves on the south Florida shooting grounds. However this question is being investigated by Florida State Biologist Frank Winston.

THE 1950 MIGRATION

Table 1 gives the dates on which Mourning Doves were identified and the numbers of each age group identified on these dates. From this table it is apparent that by far the largest number of doves examined was on the first day of the hunting season (Sunday, Oct. 1). From all reports this reflects the number of doves present on the shooting grounds as compared with the two subsequent Sundays (Oct. 8 and 15). There was a rapid decline in the number of doves found in hunters' bags after the first day of the season, with an increase indicated on the 7th and 8th of October. This increase was partly due to the fact that it was a week-end which would bring more hunters into the field, and also to the opening of a shooting ground which had not been shot in previously and in which the birds had remained in fair numbers. The fact that the two preceding days were very rainy, keeping the hunters at home also may have had its effect. Field observations failed to disclose indication of a general increase in doves in the region on the week-end of the 8th. The next increase in bagged doves is shown on October 15, which took place on areas that had very poor shooting for over a week previous to that time. All observations seemed to indicate that it probably represented an actual influx of doves into the region. This influx coincided with a fairly strong cold front which arrived in southern Florida the morning of October 14, on which date the writer was in the Florida Keys. The one party of hunters encountered between Key Largo and Long Key reported that doves were much more numerous that day than at any time since the opening of the season. That evening in Miami, I received reports from hunters that shooting grounds at Opa Locka had produced more doves than at any time since the first few days of the season. I, therefore, concluded that the influx of doves on the cold front of October 14 had been fairly general. Peters (1951) noted an increase of doves in the Miami area on October 27, 1950, following a cold front. This coincided with the first big flight of the year on the Keys noted by Robert Allen between October 25 and November 1, and reported by Brookfield (1951).

From various sources I obtained the information that the hunting season in 1950 had produced an average number of Mourning Doves; that the shooting had not been as good as it was in 1948;

TABLE 1.—MOURNING DOVES IDENTIFIED IN SOUTHERN FLORIDA, 1950

Date Identified	Adult	Immature	Juvenile	Total
October 1	26	54	26	106
" 2	5	0	4	9
" 3	3	4	5	12
" 4	8	9	6	23
" 7*	16	52	15	83
" 8	2	4	4	10
" 9	0	0	0	0
" 10	0	0	0	0
" 11	1	2	3	6
" 12	0	0	0	0
" 13	0	1	0	1
" 14	0	5	1	6
" 15	9	27	7	43
Totals	70	158	71	299
Percent of totals	23.4%	52.8%	23.7%	
		76.6%		

* No observations were made on Oct. 5 and 6 because of hard rains during shooting periods.

but that it was somewhat better than in 1949. It should be noted that the areas favorable for dove concentrations were quite limited in number and extent and were all hunted regularly. Therefore the observations of hunting success were probably an accurate index to dove abundance during the first two weeks of October of an average year. Thus the proportions of the three races of Mourning Doves identified in hunters' bags were probably normal for that time of year also.

EVIDENCE FROM BANDING RECORDS

The specimen evidence cited above shows that the doves present during the shooting season in southern Florida come to a large extent from breeding populations of the eastern United States but the question remains how extensive this area is. On this question we may obtain considerable information from the banding records of the U. S. Fish and Wildlife Service. Fig. 2 shows the locality at which birds banded at Key West by the late Wm. W. Demeritt have been recovered outside of Florida during the nesting and fall migration seasons. It also shows the reverse picture of doves banded to the north and recovered in southern Florida. It can readily be seen that the area from which these doves came, extends from the Atlantic seaboard to east Texas, and from Illinois to Georgia. The especially large concentration of these recoveries in the state of Illinois probably reflects the fact that this is the only northern state which had a hunting season during the nineteen thirties when banding was in progress at Key West. However, it is significant that there were more recoveries from the northern state of Illinois than for any of the southern states except Florida. This is despite the fact that birds from northern breeding grounds must pass through the southern states to reach Florida, and all of these have open seasons on doves. The reverse picture of recoveries in southern Florida of doves banded in other states, although based on few records, also suggests the importance of the more northern midwestern states as producers of southern Florida dove flocks.

If we examine all of the recoveries of doves in Florida which were banded in various states to the north (only those from southern Florida are shown on Fig. 2) we find that their points of banding cover an extensive area of the eastern United States. It should be noted that a relatively small percentage of doves banded in the north and recovered in Florida were recovered in the southern part of the state. This probably indicates a relatively high shooting pressure in the northern counties as compared with the southern. The fact that many of the fall migrant doves banded at Key West were subsequently recovered farther north during the same fall migration indicates that many of the birds shot in the northern part of the state could have made the trip to the southern tip first before being finally brought to bag in the northern section during its later hunting season. In contrast to the recoveries of migrant birds, doves banded in southern Florida during the breeding season have been recovered exclusively in the same general region. This indicates that the southern Florida breeding population is largely sedentary. It is significant that all but one of the recoveries of doves south of Florida are direct fall migration

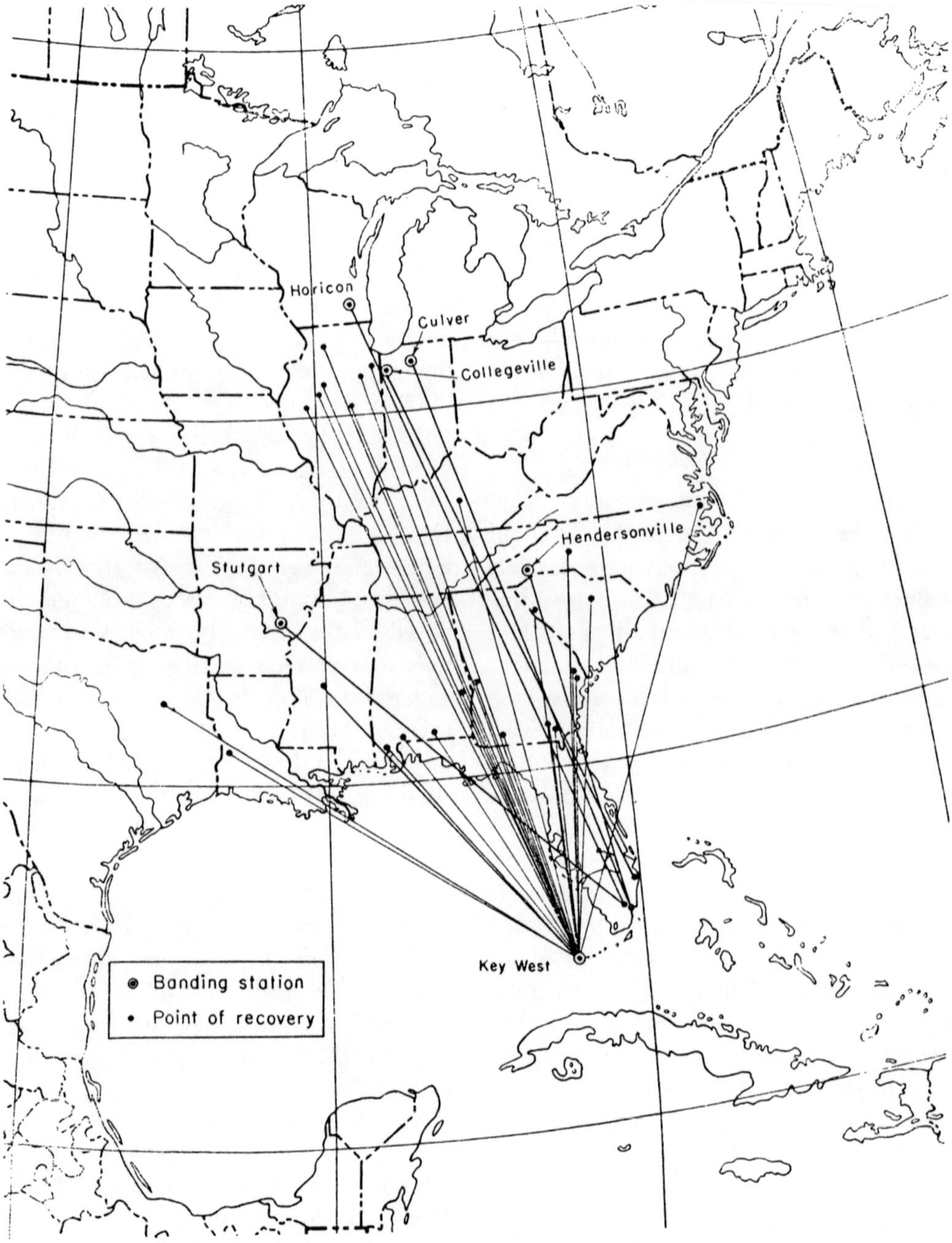

Fig. 2. Movements of banded birds to and from Southern Florida.

recoveries, indicating a northern rather than a southern origin. The one exception, April 1, 1939 at Las Martinas, Pinar del Rio Province, Cuba, is the only one that could conceivably have been of West Indies origin. Dr. Abelardo Moreno informs me that the locality is one of the best breeding areas in Cuba for Mourning Doves and April 1 is within the breeding season for doves there.

FLUCTUATION IN MIGRATION INTENSITY

There seems to be general agreement that the number of doves present in southern Florida varies greatly from year to year. Also there is considerable variation in the dates on which the big flights arrive. It is well known that southward migration of birds in general tends to coincide with the movements of masses of cold air, the advance edges of which are called "cold fronts." It was in the latter part of October 1948, when Harold Peters reported that a very large number of doves were shot in southern Florida. There is no exact record of the intensity of migration during the season of 1949 but hunters seem to agree that it was a relatively poor season. Correlated with these observations it may be noted in the weather records that there were more and stronger cold fronts reaching southern Florida during October 1948 than in the same period of 1949. In fact in 1949 no cold front of the magnitude of those of 1948 reached southern Florida until November 2. Since this was after the end of the hunting season an influx of doves at that time might not have been noticed. In 1950 doves were present in numbers before the beginning of my observations on October 1, the opening day of the shooting season. Their numbers were decimated very markedly during the first two weeks of the season and then increased noticeably after the arrival of a cold front on the 14th. Peters (1951) and Brookfield (1951) reported another increase during the last week of October, following a cold front.

The banding schedules of the late William W. Demeritt of Key West, recording the banding of approximately 6,500 doves, contain probably the most extensive and accurate information on Mourning Dove migration over about a 10-year period for any southern Florida locality. Demeritt's reports contain such notations as this: "On Nov. 9, 1930 northwester blowing, hundreds of doves at station." "Dec. 14, 1932, saw very few. On Dec. 20 about 150 with a strong cold northerly wind blowing."

HOW MIGRANT DOVES REACH SOUTH FLORIDA

The most baffling feature of the migration of Mourning Doves into southern Florida has been the arrival of large numbers of these birds in the southern counties without being noted on their way south. This has been one of the chief reasons that credence has been given to the theory of West Indies origin. One suggestion has been that these birds move down the west coast. However, up to the present time information on flights of doves in that area is lacking. Questioning of sportsmen from the southwestern coast area has failed to elicit information of noticeable flights of birds there. William D. Wood, manager of the Sanibel National Wildlife Refuge on Sanibel Island, southwest of Fort Meyer, reported 56 Mourning Doves there from September to December 1950, but there is no way of knowing whether these birds were headed north or south. Florida state biologist Frank Winston is now investigating this problem through banding and coloring doves trapped in the northwestern part of the state.

That few Mourning Doves reared in North America get south of the Florida Keys is attested by the fact that so few banded birds have been reported from the West Indies, where it has been noted by Dr. Abelardo Moreno of the University of Havana that Mourning Doves are shot in large numbers. Only 6 doves banded in the United States have been recovered in the West Indies, while one was recovered to the south of Cuba in Yucatan. This is even more remarkable when we realize that more doves have been banded at Key West, the nearest point in the United States to Cuba, than at any other station in North America, and no less than 168 of these have been recovered.

CONCLUSIONS

The general conclusions which can be drawn from the results of both banding and specimen evidence are that Mourning Doves which are shot in southern Florida are largely migrants of the breeding population from a very extensive area east of the Mississippi River; that they include an especially high percentage of young birds; that a large percentage of these birds appear to come from breeding grounds in the extreme northern part of the species range; that they reach the southern tip of Florida in varying numbers from late September to early November, accompanying movements of large masses of cold air to the southern tip of Florida during the migration season. The number and intensity of cold fronts during that period may control, to some extent, the number of migrant doves that will reach southern Florida. The reproductive success of doves in the northern states during that year presumably would also have a considerable effect on the numbers of birds reaching southern Florida, since such a large percentage of this population is composed of young of the year. During periods when no large migrations arrive from the north the doves shot in southern Florida presumably consist to a greater extent of the relatively sedentary local population. At such times it might be expected that a higher percentage of adult birds would be present.

The period beginning in late September or early October, when doves normally begin to appear in numbers in southern Florida, is preceded by a concentration of doves in the southeastern states to the north of Florida in late August and early September. These concentrations appear to decrease just prior to the time that southern Florida populations normally build up (Peters, 1943). During the course of flights of doves into Florida, it appears that upon reaching the southern tip of the peninsula the majority of these birds stop their southward progress. Thus most of the birds land either in the Keys or the southern tip of the mainland, move northward along the east coast to the latitude of Lake Okeechobee; thence they appear to cross the peninsula and move northward along the west side, gradually spreading out into the northern part of the state and south Georgia (Fig. 1). A few occasionally proceed across the straits to the West Indies. This migration appears usually to diminish noticeably during the latter part of November.

ACKNOWLEDGMENTS

For valuable assistance while in the field in Florida I am greatly indebted to

U. S. Game Management Agent Jacob Wolfley, to Florida Game and Fresh Water Fish Commission Officers I. P. Hodges, W. P. Cooper, Jack Spaulding, and particularly to I. H. Schwartz, who took me with them on their patrols of the dove shooting fields; also to an interested group of sportsmen, the dove hunters who cooperated in a most cordial and helpful manner in allowing me to handle and carefully study the doves which they had shot. I am especially indebted to my colleague, Harold S. Peters, who has had considerably more field experience in southern Florida than I have had, for reading the manuscript and offering a number of helpful suggestions.

References Cited

Brookfield, Charles M. 1951. Region Reports, Fall Migration, Florida Region. Audubon Field Notes, 5: 10–12.

Demeritt, William W. 1931–1940. Bird-banding Schedules. (in files of Fish and Wildlife Service).

Lincoln, Frederick C. 1945. Report on Trip to South Florida to Investigate the Status of Mourning Doves, September 26-October 6, 1945 (unpublished report in files of the Fish and Wildlife Service).

Peters, Harold S. 1941. Southern Florida Mourning Dove Investigations, December 2–16, 1941 (unpublished report in files of Fish and Wildlife Service).

———. 1943. Dove Investigations in Mississippi and adjoining States, September 27-October 24, 1943 (unpublished report in files of Fish and Wildlife Service).

———. 1948. Mourning Dove Investigations in Southern Florida, October 21 to November 10, 1948 (unpublished report in files of Fish and Wildlife Service).

———. 1951. Mourning Dove Investigations in Southern Florida, October 1950 (unpublished report in files of Fish and Wildlife Service).

Accepted for publication Dec. 3, 1951.

ORIGIN AND DISPERSAL OF THE ALEWIFE, *ALOSA PSEUDOHARENGUS,* AND THE GIZZARD SHAD, *DOROSOMA CEPEDIANUM,* IN THE GREAT LAKES

ROBERT RUSH MILLER

Museum of Zoology, University of Michigan
Ann Arbor, Michigan

ABSTRACT

It is possible that the alewife is native to Lake Ontario and the gizzard shad to Lake Erie, although conclusive evidence for this is lacking. There appear to be no records of the alewife from Lake Ontario prior to 1873, after the lake had been stocked with American shad. That species may still occur in Lake Ontario occasionally. A tabular comparison of these species is given to facilitate identification. The alewife is referred to the genus *Alosa* (rather than *Pomolobus)* because no reliable characteristics are available to distinguish the species of these two groups.

Wilmot recorded a herring-like fish (not the American shad) from Lake Ontario about 1837 which probably was the gizzard shad, though it may have been the alewife. The first record of gizzard shad in Lake Ontario is for 1913, but the species was reported from Lake Erie in 1848, 18 years after the completion of the first canal to connect Lake Erie and the Ohio River. It is hypothesized that the gizzard shad entered Lake Erie in pre-Columbian times during a warm-dry period.

A chronological record of the appearance of alewives in the upper Great Lakes shows the first specimen reported from Lake Erie was taken in 1931, Lake Huron 1933, Lake Michigan 1949, and Lake Superior 1954; actual dates of first appearance in each lake are unknown, but the species obviously moved from Lake Erie to the other lakes via Detroit River and Lake St. Clair. The alewife did not become abundant in Lake Erie or Lake Huron until around 1950 or in Lake Michigan until about 1956; it is widely distributed but not yet abundant in Lake Superior. The phenomenal increase in abundance and spreading of the alewife may be correlated with the upset in population balance by the sea lamprey; catch statistics for the alewife and shallow water cisco *(Coregonus artedii)* from South Bay, Lake Huron, suggest a direct or indirect causal relationship during 1954-1956.

The gizzard shad does not occur north of lower New York Harbor (and there only rarely) on the Atlantic slope, although it has been reported from New Brunswick, where it was probably misidentified for the alewife. Its inland range has been extended by canal connections and perhaps it reached Lake Erie in this way; it definitely entered Lake Michigan through the Chicago River canal. A reference in 1874 to the "saw-belly" from the Great Lakes undoubtedly represents the gizzard shad, also known by that name. The species has evidently spread to Lake Ontario and the vicinity of Quebec by moving through the Welland Canal, and to Lake Huron via the Detroit River and Lake St. Clair. The first specimen known to be taken in Lake Michigan was caught on October 13, 1953, west of Muskegon, Michigan; another was taken near Gary, Indiana, on November 2, 1953. The gizzard shad has not been reported from Lake Superior where cold, deep water may prohibit its establishment. It has on occasion gained some commercial importance in Lakes Erie and Huron, where

Reprinted from Volume 86 (1956), Transactions of the American Fisheries Society.
Published 1957. Printed in U.S.A.

31,600 pounds valued at $930 were taken in 1949. The gizzard shad may have been aided in expanding northward by the warming climatic trend of the past half century.

The two species, particularly the alewife, are steadily increasing in abundance and methods of utilizing this new resource should be devised.

Introduction

Three species of the herring family, Clupeidae, which is predominantly marine, have been reported from the Great Lakes drainage. The American shad, *Alosa sapidissima* Wilson, has been listed as native to Lake Ontario (*e.g.*, by Hubbs and Lagler, 1947, p. 34), although others claim that it first appeared there when planted by the U.S. Fish Commission in the 1870's (Smith, 1892, pp. 193-194; Evermann and Kendall, 1902, p. 213). The species was last recorded from Lake Ontario in 1931, but it occurs in the lower Ottawa River and stragglers may occasionally reach the lake (Radforth, 1944, pp. 94-95). As detailed below, the alewife, *Alosa pseudoharengus* (Wilson), may have penetrated the Great Lakes drainage from the St. Lawrence River as far as Lake Ontario. It has been able to surmount the barrier of Niagara Falls by passing through the Welland Canal into Lake Erie, from which it has spread throughout the Great Lakes. The gizzard shad, *Dorosoma cepedianum* (LeSueur), probably entered the Great Lakes drainage naturally from the Mississippi River basin via Lake Erie and then moved eastward into Lake Ontario. Its recent occurrence in Lake Michigan, on the other hand, seems clearly the result of immigration through the Chicago River canal; it is unknown in Lake Superior or in the northern parts of Lake Huron and Lake Michigan.

Because fishery biologists and others have difficulty in distinguishing these three species, some of the distinctive external characters are indicated (Table 1) in order to facilitate identification, particularly of young fish. Leim (1924, p. 227, figs. 28-31) gave additional characters that distinguish shad from alewives, including photographs of early life-history stages. An adult gizzard shad (unless injured) is easily recognized by the greatly lengthened last dorsal ray, as well as by the small, inferior mouth and deep body. The position of the dorsal fin, relative length of the anal fin, and comparative size of the upper jaw provide field characters for distinguishing smaller specimens of the three species. Vertebral counts do not overlap between the alewife and the shad or between the gizzard shad and the shad. As Graham (1956, p. 20) has shown, populations of the alewife from Lake Ontario have fewer vertebrae than do those of Atlantic coastal streams.

TABLE 1.—*Comparison between three species of herrings reported from the Great Lakes drainage*

Character	American shad[1]	Alewife[2]	Gizzard shad[3]
Last ray of dorsal fin...	Not modified	Not modified	Modified into a long filament (inconspicuous or absent in young)
Origin of dorsal fin.....	Well before insertion of pelvic fins	Well before insertion of pelvic fins	Over or behind insertion of pelvic fins
Anal fin..............	Relatively short, with 18-24 rays	Short, with 15-20 rays	Relatively long, with 25-36 rays
Ventral scutes.........	Typically 22 + 16 = 38 (range, 34-39)	Typically 19 + 14 = 33 (range, 32-37)	Typically 18 + 12 = 30 (range, 27-32)
Longitudinal scale rows.	52-64	45-54	52-70
Vertebrae (including hypural)............	52-60	47-50	48-51
Upper jaw..............	Long, slender, its tip reaching posteriorly to middle of pupil in young (25 mm. T.L.) and to beyond eye in adult; no notch	Shorter and wider, its tip reaching to front of pupil in young and to rear of pupil in adult; no notch	Short and moderately wide, its tip scarcely reaching front of eye in young and front of pupil in adult; a slight notch (indentation) along ventral margin in young, becoming a deep notch in adult
Lower jaw..............	Strong, longer than upper jaw, not fitting into notch in upper jaw	Strong, projecting beyond upper jaw, not fitting into notch in upper jaw	Relatively weak, equal to (young) or shorter than upper jaw, fitting into notch in upper jaw
Mouth................	Terminal	Terminal	Subterminal to inferior
Gillrakers............	Long, slender, about 25-75 on lower limb in young to adult	Long, slender, about 20-40 on lower limb in young to adult	Very long, fine, close-set, about 90-275 on lower limb in young to adult

[1]Vertebral counts taken from Vladykov and Wallace (1938); number of anal rays in part from Leim (1924, p. 224); gillraker counts approximated from Hildebrand and Schroeder (1928, p. 94).

[2]Counts original, based on Great Lakes and Ontario material only, except for gillraker number which is modified from Hildebrand and Schroeder (1928, p. 89). Leim (1924, p. 206) recorded a variation of 46 to 51 vertebrae in Atlantic Coast alewives.

[3]All counts taken from Miller (1950) except number of gillrakers, which is original.

Alewife

The scientific name of the alewife is generally given as *Pomolobus pseudoharengus* (Wilson). However, Hildebrand and Schroeder (1928, p. 93) indicated that *Pomolobus* and *Alosa* are scarcely distinct and Bailey, Winn and Smith (1954, p. 118) gave further reasons why it seems best to unite them. No clear-cut morphological characters have been proposed to distinguish as a group the species usually placed in *Pomolobus* from those assigned to *Alosa.* Moreover, the name *Pomolobus* probably rests on uncertain nomencatural grounds, for it may be questioned seriously whether Rafinesque's description of *P. chrysochloris* (type species of *Pomolobus*) is based on this genus and species or on the Ohio shad, *Alosa alabamae ohiensis* Evermann (S. F. Hildebrand, unpublished manuscript on the Clupeidae of the Western North Atlantic). Use of *Alosa,* the older of the two generic names, therefore seems preferable.

There is no conclusive evidence that the alewife occurs naturally in Lake Ontario. Bean (1884, p. 589) and others believed it was introduced there by mistake in the 1870's when shad were planted by the U.S. Fish Commission. This has been denied (see Smith, 1892, p. 188), but evidence for nativity is weak since there appear to be neither specimens nor records of the species from Lake Ontario prior to 1873. The possibility that alewives did reach this lake from the St. Lawrence River should not be dismissed, however. Radforth (1944, p. 96) felt that they may have entered Lake Ontario during the invasion of the Champlain Sea, in the later stages of the glacial history of the Great Lakes. Toner (1934) pointed out that its presence in certain lakes whose outlets pass over barrier falls in Frontenac and Leeds counties, eastern Ontario, disproves the claim that the alewife was planted in Lake Ontario. Possibly the species reached its limits of toleration in Lake Ontario and in favorable seasons was able to establish sizable populations which were decimated during inimical years, when presumably only the hardiest individuals persisted. Since around 1890, the alewife has intermittently appeared in Lake Ontario in almost incredible numbers, as noted during periodic summer die-offs (Graham, 1956, p. 21). Whether such comparative abundance reflects recency of introduction is problematical.

The historical appearance of the alewife in the upper Great Lakes, as evidenced by captures known to the author, is given in chronological order (Table 2). The first specimen was taken in Lake Erie in 1931, in Lake Huron in 1933, in Lake Michigan in 1949 and in Lake Superior in 1954. This clearly establishes the sequence of its invasion of these lakes but the dates of first appearance in each lake are not known. As Dr. M. B. Trautman pointed out *(in litt.),* there must have been thousands of alewives in Lake Erie in 1931 before one could be caught in a net. Nevertheless, the species apparently did not really become abundant there until about 1950, although commercial fishermen caught alewives infrequently, in Ohio waters, after 1940, and there were similar reports from the Cana-

TABLE 2.—*Spread of the alewife in the upper Great Lakes*

Date	Locality	Remarks[1]
Sept. 23, 1931....	L. Erie, at E. end off Nanticoke, Ont.	Adult, 7¾ in. long (Dymond, 1932).
Mar. 28, 1933....	L. Huron, 12 mi. E. of Outer Duck Id., Ont.	Adult, from about 70 fathoms; others reported (MacKay, 1934).
April 15, 1935....	L. Huron, 40 mi. from Rogers City, Mich.	Adult, 188 mm. S. L., from 70 fathoms; UMMZ No. 109109 (Van Oosten, 1935).
Dec. 26-31, 1940..	L. Erie, off South Bass Id., Ohio	Adult, 9 in. T. L. taken in gillnet by K. H. Doan (Trautman, 1957).
Dec. 27, 1942....	L. Erie, near Vermillion and Lorain, Ohio	Two adults taken in gillnet by fishermen (Trautman, 1957).
May 5, 1949.....	L. Michigan, W. of S. Manitou Id., Mich.	Adult, 176 mm. S. L., from 60 fathoms, Charles Anderson; UMMZ No. 157215.
Mar. 20, 1951[2]...	L. Michigan, off Whitehall, 14 mi. SW. of Montague, Mich.	Adult, 197 mm. S. L., from 60 fathoms, James Grover; UMMZ No. 160969 (see The Fisherman, Vol. 19, No. 5, 1951, p. 6).
Mar. 12, 1952....	L. Michigan, 10 mi. ENE. of Milwaukee, Wisc.	Adult, 177 mm. S. L., Frank Miller; UMMZ No. 162861.
July 13, 1953.....	L. Erie, on beach W. of Colchester, Ont.	Young-of-the-year taken with gizzard shad (Langlois, 1954, p. 207).
July 14, 1953.....	L. Erie, South Bass Id., Ohio	Young-of-the year (Langlois, 1954, p. 207).
Dec. 22, 1953[3]....	L. Michigan, 12 mi. WNW. of South Haven, Mich.	Adults, 180 and 188 mm. S. L., from 32 fathoms, Chris Jensen; UMMZ No. 166514.
Sept. 25, 1954....	L. Superior, off Pendill's Cr., Chippewa Co., Mich.	Adult, 168 mm. S. L., Robert Gordon; UMMZ No. 167872.
Nov., 1954.......	L. Superior, Chequamegon Bay, off Long Id., Wisc.	Adult, 164 mm. S. L., obtained by Russell Daly; UMMZ No. 167945.
June 1 & 14, 1955.	L. Superior, Whitefish Bay, about 1 mi. N. of mouth of Tahquamenon R., Mich.	Adults, 162 and 175 mm. S. L., from pound net, Ernest L. Pomeroy.
June 3, 1955.....	L. Huron drainage: Ocqueoc R. Weir near Rogers City, Mich.	Adults, 179 and 182 mm. S. L., Albert E. Hall, Jr.; UMMZ No. 169062.
June 10, 1955....	L. Superior, Whitefish Bay, Mich.	Adult, 197 mm. S. L., sent in by L. R. Anderson.
June 11, 1955....	L. Michigan, East Bay of Grand Traverse Bay, Mich.	19 juveniles, 55-76 mm. S. L., bait dealer for Stanley Lievense; UMMZ No. 170945.
Sept. 14, 1955....	L. Huron, along shore in the Les Cheneaux Channels, Mackinac Co., Mich.	Fingerlings, about 2 in. T. L., bait dealer; ident. by L. R. Anderson.
Oct. 17 and Nov. 12, 1955..	L. Superior, Gros Cap area near Sault St. Marie, Ont.	Adult male (210 mm. T. L.) and adult female (273 mm. T. L.), ident. by K. H. Loftus and W. B. Scott; ROM No. 18157 (male).
Feb. 13, 1956.....	L. Superior, off Two Harbors, Minn.	Adult from herring net, Adolph Ojard; ident. by John G. Hale and Samuel Eddy.

TABLE 2—Continued

Date	Locality	Remarks[1]
March, 1956	L. Superior, off Knife R., Minn.	Adult taken by Ivar Pederson; ident. by John G. Hale.
June 6, 1956	L. Superior, at mouth of Blind Sucker R., Luce Co., Mich.	Adult, 9½ in. long, caught by George McGary; 2 other adults also taken. Ident. by L. R. Anderson.
July 31, 1956	L. Superior, Split Rock R. area, Minn.	Adult, 161 mm. S. L., from herring net, Ragnivald Sve.
Aug. 31, 1956	L. Superior, off Shaganance Id. (near entrance to Black Bay), Ont.	Adult male, 166 mm. S. L., taken by Melville Gerow; ident. by R. A. Ryder.
Sept. 10, 1956[2] . . .	L. Michigan, East Bay of Grand Traverse Bay at O-at-ka Beach, Mich.	341 postlarvae to young, 19-40 mm. S. L., taken in minnow seine for Stanley Lievense; UMMZ No. 171308.
Nov. 5, 1956	L. Superior, Keweenaw Bay, 1 mi. off town of same name, Mich.	Adult, 181 mm. S. L., from surface gillnet over 17 fathoms, taken by Uno Weideman.

[1]Abbreviations are: S.L. standard length; T.L., total length; UMMZ, University of Michigan Museum of Zoology; ROM, Royal Ontario Museum.

[2]Approximate date.

[3]On April 24, May 10, and Oct. 20, 1954, 3 more adults were taken in 32-50 fathoms by Mr. Jensen from the same general area; UMMZ No. 167702.

dian waters of Lake Erie. "After 1950 the young and adults [alewives] were sufficiently numerous so that they were distributed rather regularly throughout Lake Erie. In December, 1954, they clogged the gill and fyke nets in the vicinity of Conneaut, Ohio, following a bad storm" (Trautman, *in litt.,* Dec. 20, 1956). This information is supported by Dr. W. B. Scott who wrote *(in litt.,* March 5, 1954) that ". . . in the last four or five years the alewife has definitely increased in numbers in Lake Erie." The Welland Canal has been available for this highly migratory species since 1829, and yet we have no knowledge of its occurrence in Lake Erie until 1931 (or of the equally migratory sea lamprey before 1921).

The rather sensational increase in abundance and spread of the alewife during the past decade has probably resulted from a number of changes particularly favorable to this species. Important modifications in the complex interrelationships of the native fauna have been brought about by the sea lamprey, *Petromyzon marinus,* and these disturbances may have influenced the change in alewife abundance. We do not yet have, however, sufficient biological information to be able to draw conclusions on this problem. The lake herring and alewife may compete for food and space at some stage in their life cycles since the adults rely heavily on animal plankton. Locally, as at South Bay (Table 3), Saginaw Bay and Green Bay (information from Stanford H. Smith), the alewife populations have recently increased greatly while the lake herring stocks have decreased. Possibly, newly metamorphosed lampreys migrating across these bays to deeper water have helped to reduce the lake herrings; or, the decrease (particularly noteworthy in these bays in 1954, 1955, and 1956) may be a result of normal population fluctuation. That it has occurred simultaneously in the three bays, however, suggests some degree of causal relationship, direct or indirect, between the alewife, lake herring and lamprey (and possibly other species).

The alewife is now abundant in many parts of Lake Huron and Lake Michigan. During the period May 6, 1954, to March 10, 1955, when the U. S. Fish and Wildlife Service research vessel M/V CISCO made thirteen cruises in the southern part of Lake Michigan (mostly south of a line joining Milwaukee and Grand Haven, but north to a line joining Manitowoc and Ludington), no alewives were caught. On the other hand, during nine similar cruises in the northern half of Lake Michigan, over the period May 10 to November 12, 1955, 21 alewives were taken from 10 stations. Nevertheless, by late 1953 the alewife had moved as far southward in Lake Michigan as the vicinity of South Haven (Table 2, item 11). Moreover, in a memorandum dated January 2, 1957, Max A. Hunt (District Fisheries Supervisor, Plainwell, Michigan) reported that "Commercial fishermen, fishing out of Saugatuck, report the alewife is being caught in their "chub" nets in up to 43 fathoms of water. There have been 200-300 fish taken in a single lift. The fishermen are complaining about the alewife fouling their nets." Saugatuck is about 20 miles north of South Haven and well within the southern half of Lake Michigan.

The author is indebted to Leonard S. Joeris for the following information on the alewife in the Green Bay area of Lake Michigan. The first indication of spawning was noted in October, 1953, when two or three schools of small alewives were present in the carp-holding ponds near Duck Creek, Wisconsin (at the southern tip of Green Bay). Additional reliable reports of schools of young were received in the early fall of 1954 and

TABLE 3.—*Catch of alewife and lake herring in South Bay, Lake Huron, Ontario* (all fish taken in 1½ inch mesh five-pound nets fished in same location)

Year	Pounds alewife	Pounds herring	Number of lifts
1951	0.2 (1 fish)	36,018	175
1952	1.0 (5 fish)	14,119	163
1953	6.0 (51 fish)	12,625	65
1954	2,316	8,731	97
1955	16,302	4,841	86
1956[1]	22,428	3,476	93

[1]Figures for 1956 are provisional, subject to checking, but final figures could vary only a few pounds from those given.

1955. The number of adults has increased yearly from a single individual in 1952 to common occurrence by 1955 in all herring pound nets—yet not in sufficient numbers to cause difficulty for the fishermen. By the spring of 1956, however, fishermen in the Ellison Bay and Gills Rock area (on the eastern side of Green Bay) were becoming quite concerned by the continuing increase of the alewife. A steady increase in numbers has also occurred for the gillnet fishery along the northern part of Green Bay and Lake Michigan.

Alewives are now abundant in the Saginaw Bay area of Lake Huron. For example, during operations of the CISCO over the period May 3 to October 25, 1953, 6 of 9 cruises caught this species. While trawling in a very shallow area near the southwest end of the Bay, about 6,000 alewives, 1.5 to 3.0 inches long, were taken in a 35-foot trawl in one 5-minute bottom tow, and 8,500 were caught in a 5-minute tow just off the bottom. Over 50 percent of all fishes caught in trawling operations by the CISCO were alewives (information from Stanford H. Smith).

Records up to the fall of 1956 suggest that the alewife may not yet have established itself in numbers in Lake Superior although it has rapidly dispersed there. The CISCO ran nine cruises on the lake during May 3 to October 25, 1953, and took no alewives, although their nets were fished in some of the regions from which the species was obtained soon thereafter (Table 2.)

Van Oosten (1935) felt that the alewife probably entered Lake Huron through Georgian Bay directly from Lake Ontario via the Trent River-Trent Canal-Lake Simcoe-Severn River waterway, rather than by way of the Detroit River and Lake St. Clair, since the species had not been reported from the last two areas. In response to my query regarding this possibility, Dr. W. B. Scott wrote (Jan. 2, 1957) that there are no records of this species (or of *Petromyzon, Osmerus* or *Anguilla)* from Lake Simcoe, or any other major portion of the waterway, even though the habitat there seems to be excellent for all of these fishes. The proposed route is a very involved, even tortuous one, and the water flows through hydroelectric plants at two marine railways at the western end of the system. Failure to take alewives in Detroit River is not surprising, since organized collecting there (as by the Michigan Institute for Fisheries Research) has been carried out chiefly in shallower waters than those which the alewife habitually seeks. Radforth (1944, pp. 95-96) also concluded that the alewife reached Lake Huron via the Detroit River and Lake St. Clair.

It seems evident that the alewife will become increasingly abundant in the Great Lakes and that utilization of this new resource should be encouraged. The species may well establish itself, by distribution in bait[1] buckets, in such inland lakes as Crystal Lake, Benzie County, Michigan.

[1]On July 6, 1956, an alewife 139 mm. S.L. was taken by I. A. Rodeheffer and J. K. Day in Mullet Lake, Cheboygan Co., Michigan (specimen in UMMZ collection).

According to Robert Heffner, bait dealer from Beulah, it occurs nearby in Betsie River, near Frankfort, and at the mouth of Platte River; he has observed dense schools of alewives at sizes up to 4 inches at Frankfort. Although they do not survive well when collected in large numbers, a few do live and it is difficult to sort them all out from bait species. Alewives have been used in Lake Ontario as a ready source of cheap food as well as for fertilizer, and on the Atlantic coast they are marketed fresh, smoked or pickled.

Gizzard Shad

Discussion regarding the northern limit of this species, both for the Atlantic seaboard and for the Great Lakes-St. Lawrence basins, has been frequently published. Early errors in range statements and evidently in identification have led to subsequent confusion. Some of the controversy, however, also hinges on the question of whether *Dorosoma* is native in the Great Lakes drainage.

At least as early as the 1870's, Gill (1873, p. 811) erred by including Cape Cod in the distribution of this species. Subsequently, Jordan and Gilbert (1883, p. 271) and others perpetuated the error. The author has been unable to discover the basis for listing Cape Cod as the northern limit of the gizzard shad on the Atlantic seaboard, but the locality has persisted in the literature almost to the present (e.g., see Vladykov, 1945, p. 36). *D. cepedianum* is uncommon north of latitude 40° N. in New Jersey and southeastern Pennsylvania, and reaches its northern coastal limit in New York Harbor, usually called the Lower Bay or Sandy Hook Bay (Breder and Nigrelli, 1934, p. 194; Breder, 1938, pp. 23, 28). Halkett (1913, p. 46) stated that this species has been recorded from St. John River, Fredericton, New Brunswick, and this formed the basis of the inclusion of New Brunswick in the range of *Dorosoma cepedianum* by Hubbs and Lagler (1947, p. 34) and by Radforth (1944, p. 58). As pointed out by Vladykov *(op. cit.)*, the New Brunswick record is based upon the remarks of an unknown author, who, in a review of Nash's Check List of the Fishes of Ontario, wrote of *D. cepedianum:* " . . .it is a native Canadian fish, and was recorded by the late Edward Jack on the St. John River, at Fredericton, N. B. . . ." The St. John River is a tributary of the Bay of Fundy and its mouth lies more than 500 miles northeast of the northernmost known coastal limit of *Dorosoma.* This record is therefore unacceptable; the fish thought to have been a gizzard shad very likely was the alewife, which occurs as far north as Newfoundland and has been confused with the gizzard shad.

Kirtland (1850, p. 2) appears to have been the first to suggest that *Dorosoma* entered the Great Lakes (i.e., Lake Erie) via canal connections, and this belief has been variously accepted and restated by subsequent authors during the succeeding century. The problem is complicated by the very different physical conditions of Lake Michigan and Lake Erie, the only two of the Great Lakes which offered direct access to the gizzard shad in glacial or postglacial times. More than a dozen species, representing three families, show a distributional pattern in Michigan that clearly indicates penetration from the south into southeastern Michigan. These fishes invaded the Lake Erie basin by utilizing the former outlet (now Maumee River) of the glacial precursor of Lake Erie, or by way of other natural connections across the very low divide between that basin and the Mississippi River system. Yet not one of these species is known from the adjacent drainage of Lake Michigan. It seems evident that the colder, deeper water of that lake has served as an effective barrier to these fishes, as it evidently has, until very recently, to the gizzard shad. Thus, although this species was reported by Nelson (1876, p. 44) from the Chicago River canal and Lake Michigan, there is no objective evidence to show that the gizzard shad established itself in the lake by migration from the Illinois River until recent years. There are no specimens of this species from the southern tip of Lake Michigan in the Chicago Natural History Museum (or any other collection), and Forbes and Richardson (1919, Map 8) gave no records nearer Lake Michigan than the junction of Fox and Illinois Rivers in La Salle County, Illinois; furthermore, Gerking (1945, Map 8) did not find it in the Lake Michigan drainage of Indiana.

Whether *Dorosoma* is native to Lake Erie cannot be solved conclusively because, to the author's knowledge, there are neither specimens nor records of gizzard shad from this lake prior to the completion in 1830 of the first canal connecting the Ohio River with Lake Erie. The first specimens of *Dorosoma* to be reported from the Great Lakes were four fish recorded by Kirtland *(loc. cit.)* in November, 1848, from near the mouth of the Cuyahoga River at Cleveland. He was certain that this species did not occur in Lake Erie or its tributaries in 1840, and believed the specimens captured in 1848 represented recent invaders from the south. Kirtland was perhaps overimpressed by the winterkill of many gizzard shad during a hard freeze in the Dayton Canal, since he cited this as evidence that *Dorosoma* "was a native of a warmer climate." He was also impressed with the failure of fishermen to recognize the gizzard shad near Cincinnati much before 1840 and with their stated belief that it was an emigrant from the south. Klippert (in Jordan, 1882, pp. 870-871) also believed that the gizzard shad was not native to Lake Erie; he wrote that "before the construction of the Ohio Canal it was not found anywhere north of the divide, or water shed between the lake [Erie] and the Ohio River." According to Jordan *(op. cit.,* p. 742), Mr John H. Klippert was "a close observer and excellent amateur ichthyologist."

In November, 1877, Samuel Wilmot (1879) collected a specimen of *D. cepedianum* from Sarnia, Ontario, at the southern tip of Lake Huron. It was identified by Baird as *Dorosoma cepedianum heterurum* and the author has verified that it is a gizzard shad (USNM No. 20836). In transmitting this specimen, Wilmot wrote that "Fish very similar in appearance to this one have been known in Lake Ontario and other of our waters for many years; I recollect them forty years ago." This would have been about 1837. Wilmot was a pioneer fish culturist and supervisor of the Canadian salmon hatchery at Newcastle on Lake Ontario (Milner, 1874, pp. 25, 58, 228, 231), and he surely knew the American shad; his testimony strongly suggests that the gizzard shad inhabited the Lake Ontario basin as early as 1837, although it is possible that the species he referred to may have been the alewife. The first published record of the gizzard shad in Lake Ontario that I have noted is for 1913, when Nash reported the species to be very rare (see Dymond, Hart and Pritchard, 1929, p. 14). Since it is now known from an actual specimen that *Dorosoma* occurred in Lake Huron in the 1870's, it seems obvious that the reference by Milner (1874, p. 36) to the "saw-belly," identified as *Pomolobus chrysochloris* Rafinesque, actually represents one of the early records of *Dorosoma cepedianum* from the Great Lakes (which lake was not specified). This conclusion was first expressed by Hubbs and Lagler (1947, footnote, p. 34).

Although the possibility that Dorosoma first entered Lake Erie through canals cannot be ruled out, it is perhaps more plausible to regard this species as a postglacial migrant into Lake Erie. This view is favored by Trautman (1957). On the northern fringe of its range *Dorosoma* is near the limit of its thermal tolerance, forging northward during a series of favorable years and then, under adverse conditions, becoming almost eliminated (see, for example, Cox, 1896, p. 610). When numbers are killed and fishermen see the species for the first time, they fail to recognize it and conclude that it is a recent invader from a more southern clime. Dr. Trautman wrote me that this has happened innumerable times in Ohio since Kirtland's day. Perhaps the gizzard shad gained access to Lake Erie during a warm-dry period in pre-Columbian times and then, with the retrun to present conditions, it retreated but left relict populations. *Dorosoma* is favored to establish outlying colonies because spawning adults enter small streams or ditches, provided they are of low gradient, and the young are later abundant there. In early historic time, there were many places along the divide between the Ohio and Lake Erie watersheds (e.g., the close approach of the Maumee and Wabash drainages near Ft. Wayne, (Indiana) where *Dorosoma* could readily have crossed from one drainage to the other. It is also possible, as Gerking (1945, p. 33) suggested, that the gizzard shad may have entered the glacial precursor of Lake Erie through the Ft. Wayne outlet (Maumee River), but this view seems less acceptable because *Dorosoma* does not appear to be adapted to very cold water, such as prevailed during the melting of the continental glaciers in the evolution of the Great Lakes. The species occurs in the Great Lakes drainage of Indiana only in the Maumee River system (Gerking, *op. cit.*, Map 8). There appears to be no need for nor support of the hypothesis (Radforth, *loc. cit.)* that *Dorosoma* utilized the Mohawk-Hudson glacial outlet (or the Chicago-Ubly outlet) to enter Ontario; if it uilized glacial waterways at all, the Maumee outlet seems to be the only possible one that it might have entered. Vladykov *(op. cit.)* concluded that this species reached the environs of Quebec by descending the St. Lawrence River from the Great Lakes.

Thus the gizzard shad, as perhaps with the alewife in Lake Ontario, apparently neared the limit of its tolerance in Lake Erie. Its increase and spread in the lower Lakes may well reflect the known warming climatic trend (Baum and Havens, 1956) or may be a response to changing biotic conditions of the Great Lakes in recent decades—or both phenomena may be operative. That the species has entered canals and thus extended its range seems well founded (e.g., see Wright, 1918, p. 544; Greeley, 1928, p. 95).

The first valid records (based on identified, preserved specimens) of *Dorsoma cepedianum* from Lake Michigan are the following: UMMZ No. 165645, an adult, 255 mm. S. L., taken by the McNab brothers, on October 13, 1953, in a trap net in 14 feet of water about 1 mile south of the channel connecting Lake Michigan and Muskegon Lake, west of Muskegon, Michigan. A second specimen, an immature female in poor condition identified by the author and then discarded, was obtained by D. C. Erlandson from a pound net set in 25 feet of water, on November 2, 1953, off Miller, 5 miles east of Gary, Indiana. The evidence clearly points to a southern origin for these stocks, since no specimens have been reported in the northern half of Lake Michigan or in northwestern Lake Huron. *Dorosoma* has thus gradually extended its range into Lake Huron by emigration from Lake Erie and into Lake Michigan through the Chicago River Canal. The gizzard shad has not been reported from Lake Superior, where low temperature and the absence of shallow water may preclude its establishment.

This species has had a long history of fluctuating abundance in Lake Erie but it evidently has rather recently established itself in numbers in certain parts of Lake Huron. For example, Bensley (1915) did not record it from Georgian Bay, and when Walter Koelz collected a specimen there late in 1919 (UMMZ No. 55632), fishermen informed him that the gizzard shad had rather recently appeared and that all those taken were small. Today the species is fairly common in suitable parts of Georgian Bay. It was so abundant in Detroit River and Lake St. Clair in the fall of 1932 that unemployed persons caught them for food with dipnets and hook and line. By October 7, 1933, the gizzard shad had moved inland up the Saginaw River drainage as far as Pine River at St. Louis, Gratiot Co., Michigan (UMMZ No. 101607, 1 adult); this is one of the few records of its occurrence well inland from the Great Lakes and populations have not established themselves at appreciable distances inland. An attempt was made to establish the species in Whitmore Lake, Washtenaw and Livingston Counties, Michigan, in August, 1935, when the Michigan Institute for Fisheries Research stocked 124 individuals (1.5 to 15.0 inches S. L.)

from Lake St. Mary's, Ohio. It is estimated that 100 adult shad survived after one week, but if so, successful reproduction did not occur and the introduction failed (IFR Report No. 315; see also Cooper and Schafer, 1954).

Abundance in lakes Erie and Huron since about 1940 is indicated by the commercial catch statistics (U. S. Fish and Wildlife Service and Michigan Department of Conservation) which show that a total of approximately 116,000 pounds valued at $2,300 were taken during the period from 1942 to 1954. The largest catch for a single year, 31,600 pounds valued at $930, was taken in 1949. Occasionally, sudden mortality of vast numbers (similar to that of the alewife in Lake Ontario) creates a public nuisance, as it did at Erie, Pennsylvania, during January-February, 1953. (This and other aspects of the biology of the gizzard shad are summarized by me in a paper to be published by the U. S. Fish and Wildlife Service entitled: "The clupeid fishes of the subfamily Dorosomatinae, with special reference to the gizzard shad, *Dorosoma cepedianum*.")

The gizzard shad is not as widespread in the Great Lakes as the alewife, probably because its preferred habitat is less extensive, and, except locally, it is not likey to attain or exceed the popuation size of that species. It has been utilized in the Great Lakes as a source of food for trout, swine and cattle, as fertilizer, and it has been steamed and pressed for oil.

Acknowledgments

Splendid cooperation from many sources has made this paper posible. I have tried to recognize individually the valuable help given by fishermen by mentioning them in the text. The Institute of Fisheries Research of the Michigan Department of Conservation has kept me informed of new records; Gerald P. Cooper, Kiyoshi G. Fukano, Walter R. Crowe, Stanley Lievense, Leland R. Anderson, and T. H. Durling have been especially helpful. The staff of the U.S. Fish and Wildlife Service's Great Lake Laboratory, particularly James W. Moffett, Ralph Hile, Stanford H. Smith, and Leonard S. Joeris, have made available pertinent records and have discussed critical aspects of the manuscript. Raymond E. Johnson and John G. Hale have supplied Minnesota records. John Van Oosten has been of great help with literature references. Milton B. Trautman has given me the benefit of his long experience with the alewife and gizzard shad in Lake Erie and has generously allowed me to draw from material in press. My colleague, Reeve M. Bailey, has critically read the manuscript and offered valuable suggestions. A number of Canadian workers, especially W. B. Scott, F. E. J. Fry, John Budd, J. Murray Speirs, Richard A. Ryder, and Kenneth H. Loftus, have filled in records for the Canadian portion of the Great Lakes, and Dr. Scott has provided stimulating discussion and useful references relating to the origin and dispersal of the alewife. All of these people have my grateful thanks.

Literature Cited

Bailey, Reeve M., Howard Elliott Winn, and C. Lavett Smith
1954. Fishes from the Escambia River, Alabama and Florida, with ecologic and taxonomic notes. Proc. Acad. Nat. Sci. Phila., Vol. 106, pp. 109-164.

BAUM, WERNER A., and JAMES M. HAVENS
1956. Recent climatic fluctuations in maritime provinces. Trans. 21st N. Amer. Wildlife Conf., pp. 436-453.
BEAN, TARLETON H.
1884. On the occurrence of the branch alewife in certain lakes of New York. In: Goode, George Brown. The fisheries and fishery industries of the United States. Sec. I, text, pp. 588-593.
BENSLEY, B. A.
1915. The fishes of Georgian Bay. Contrib. Canad. Biol., 1911-1914, Fasc. 2, pp. 1-51.
BREDER, CHARLES M., Jr.
1938. The species of fish in New York harbor. Bull. N.Y. Zool. Soc., Vol. 41, pp. 23-29, 3 figs.
BREDER, C. M., Jr., and ROSS NIGRELLI
1934. Fish notes for 1933 and 1934 from Sandy Hook Bay and other New York localities. Copeia, 1934, No. 4, pp. 193-195.
COOPER, GERALD P., and ROBERT N. SCHAFER
1954. Studies on the population of legal-size fish in Whitmore Lake, Washtenaw and Livingston Counties, Michigan. Trans. 19th N. Amer. Wildlife Conf., pp 239-259, fig. 1
COX, ULYSSES O.
1896. A report on the fishes of southwestern Minnesota. Rept. U.S. Comm. Fish and Fish., Pt. 20 (1894), pp. 605-616.
DYMOND, JOHN R.
1932. Records of the alewife and steelhead (rainbow) trout from Lake Erie. Copeia, 1932, No. 1, p. 32.
DYMOND, J. R., J. L.HART, and A. L. PRITCHARD
1929. The fishes of the Canadian waters of Lake Ontario. Univ. Toronto Studies, Publ. Ontario Fish. Res. Lab., No. 37, pp. 1-35.
EVERMANN, BARTON WARREN, and WILLIAM C. KENDALL
1902. Notes on the fishes of Lake Ontario. Rept. U.S. Comm. Fish., 1901, pp 209-216.
FORBES, STEPHEN ALFRED, and ROBERT EARL RICHARDSON
1919. Maps showing distribution of Illinois fishes to accompany a report on the fishes of Illinois. Ed. 2. Ill. Nat. Hist. Surv., Atlas, 103 maps.
GERKING, SHELBY D.
1945. The distribution of the fishes of Indiana. Inv. Ind. Lakes and Streams, Vol. 3, No. 1, pp. 1-137, Maps 1-113.
GILL, THEODORE
1873. Catalogue of the fishes of the east coast of North America. Rept. U.S. Comm. Fish and Fish., 1871-72, Pt. 1, pp. 779-822.
GRAHAM, J. J.
1956. Observations on the alewife, *Pomolobus pseudoharengus* (Wilson), in fresh water. Publ. Ontario Fish. Res. Lab., No. 74 (Univ. Toronto Biol. Ser. No. 62), pp. 1-43, figs. 1-20.
GREELEY, J. R.
1928. Fishes of the Oswego watershed. N. Y. Conserv. Dept., Suppl. 17th Ann. Rept., 1927, pp. 84-107, col. pls. 1-12.
HALKETT, ANDREW
1913. Check list of the fishes of the Dominion of Canada and Newfoundland. Ottawa, pp. 1-138, pls. 1-14.
HILDEBRAND, SAMUEL F., and WILLIAM C. SCHROEDER
1928. Fishes of Chesapeake Bay. Bull. U.S. Bur. Fish., Vol. 43 (1), pp. 1-366, illus.

HUBBS, CARL L., and KARL F. LAGLER
1947. Fishes of the Great Lakes region. Cranbrook Inst. Sci., Bull. 26, pp. xi + 186, illus.
JORDAN, DAVID S.
1882. Report on the fishes of Ohio. Ohio Geol. Surv., Vol. 4, Pt. 1, pp. 737-1,002.
JORDAN, DAVID S., and CHARLES H. GILBERT
1883. Synopsis of the fishes of North America. U.S. Nat. Mus., Bull 16, pp. i-lvi, 1-1,018.
KIRTLAND, J. P.
1850. Fishes of Ohio. Fragments of natural history. The Family Visitor, Cleveland [and Hudson], Vol. 1, No. 1, pp. 1-2 (Jan. 3, 1850).
LANGLOIS, THOMAS H.
1954. The western end of Lake Erie and its ecology. J. W. Edwards, Inc., Ann Arbor, Mich. 479 pp., illus.
LEIM, A. H.
1924. The life history of the shad *Alosa sapidissima* (Wilson), with special reference to the factors limiting its abundance. Contr. Canadian Biol., n. s., Vol. 2, pp. 161-284.
MACKAY, H. H.
1934. Record of the alewife from Lake Huron. Copeia, 1934, No. 2, p. 97.
MILLER, ROBERT RUSH
1950. A review of the American clupeid fishes of the genus *Dorosoma.* Proc. U.S. Nat. Mus., Vol. 100, pp. 387-410.
MILNER, JAMES W.
1874. Report on the fisheries of the Great Lakes; the results of inquiries prosecuted in 1871 and 1872. Rept. U.S. Comm. Fish and Fish., 1872-73, App. A, pp. 1-75.
NELSON, E. W.
1876. A partial catalogue of the fishes of Illinois. Bull. Ill. Mus. Nat. Hist., Vol. 1, pp. 33-52.
RADFORTH, ISOBEL
1944. Some considerations on the distribution of fishes in Ontario. Contrib. Roy. Ontario Mus. Zool., No. 25, pp. 1-116, fig. 1-32.
SMITH, HUGH M.
1892. Report on an investigation of the fisheries of Lake Ontario. Bull. U.S. Fish Comm., Vol. 10 (1890), pp. 177-215 pls. 21-50.
TONER, G. C.
1934. Notes on the alewife. Canadian Field-Nat., Vol. 48, pp. 51-52.
TRAUTMAN, MILTON B.
1957. The fishes of Ohio. Ohio State Univ. (in press).
VAN OOSTEN, JOHN
1935. First record of the alewife, *Pomolobus pseudo-harengus,* for the state of Michigan. Copeia, 1935, No. 4, pp. 194-195.
VLADYKOV, VADIM D.
1945. Trois poissons nouveaux pour la province de Quebec. Canadian Nat., Vol. 72, pp. 27-39, 5 figs.
VLADYKOV, V. D., and D. H. WALLACE
1938. Populations of the shad *(Alosa sapidissima)* along the Atlantic Coast region. Trans. Amer. Fish. Soc., Vol. 67 (1937), pp. 52-66.
WILMOT, SAMUEL
1879. Notes on the western gizzard shad, *Dorosoma cepedianum heterurum* (Raf.) Jordan. Proc. U.S. Nat. Mus., Vol. 1, pp. 263-264.
WRIGHT, ALBERT HAZEN
1918. Fish succession in some Lake Ontario tributaries. Sci. Mon., Vol. 7, pp. 535-544, maps 1-3.

Reprinted from the *Third International Conference on Bears: Their Biology and Management*

Paper 44

Cranial Variation in Polar Bears

DON E. WILSON
National Fish and Wildlife Laboratory, National Museum of Natural History, Washington, D.C. 20560

INTRODUCTION

Various attempts at understanding geographic variation in Polar bears have been made since Knotterus-Meyer (1908) began to confuse the issue by recognizing seven forms of what had previously been considered a monotypic circumpolar species. The most recent and noteworthy contribution was by Manning (1971), who outlined the earlier work.

Briefly summarized, Manning demonstrated a cline of increasing size from East Greenland across Canada to the Bering Strait. He suggested that the population of largest bears, from the Bering Strait area and southward, could be considered subspecifically distinct, but left it unnamed pending further investigation.

The present study was initiated in order to examine more closely the extent and kind of geographic variation in Alaskan polar bears. A variety of multivariate analyses were conducted on Manning's data, and on additional specimens obtained since the completion of Manning's work.

MATERIALS AND METHODS

A total of 295 skulls of Alaskan polar bears was examined. The seventeen skull measurements used and described by Manning (1971) were taken with calipers. The measurements were: (1) Condylobasal length (CBL); (2) Molar-premaxilla length (MPL); (3) Mastoid breadth (MB); (4) Zygomatic breadth (ZB); (5) Supra-orbital breadth (SB); (6) Cranial length (CL); (7) Facial length (FL); (8) Maxilla-supraorbital height (MSH); (9) Least cranial breadth (LCB); (10) Interorbital breadth (IB); (11) Breadth at canines (BC); (12) Palatal breadth (PB); (13) Length P4 to M2 (LP4-M2); (14) Crown length of M2 (LM2); (15) Crown length of M1 (LM1); (16) Coronoid height (CH); (17) Condylopalatal length (CPL).

Basic statistical analyses were performed with a univariate program (D-STST) in use at the Smithsonian Institution Information Systems Division. Ratio diagrams modified from Simpson (1941), as used by Anderson (1972), were used as a graphic method of comparing measurements and proportions between groups of specimens. The data were also analysed with subroutines contained in the Numerical Taxonomy System of Multivariate Statistical Programs (NTSYS), developed by F. J. Rohlf and associates of the State University of New York at Stony Brook.

Although the original analyses utilized individual specimens, many of the results presented here are based on group means generated by the earlier computer runs. Limitations on sample sizes owing to variation in age and sex, as well as incomplete specimen data, necessitated this approach. Only the results of analyses, on adult animals are reported here.

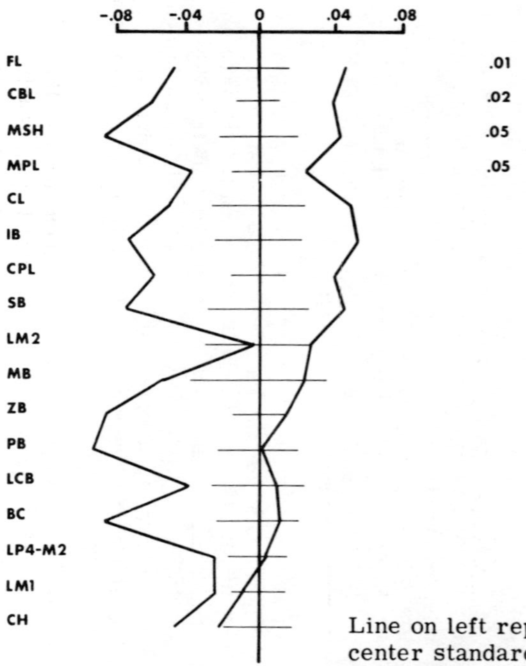

Fig. 1. Ratio diagram for males.

Line on left represents East Greenland; center standard Alaska North; line on right Alaska South. Numbers on right show significance level of T-test differences between the two Alaska populations. Scale across top represents units of difference in common logarithms among the two sample means (East Greenland and Alaska South) and that of the reference sample (Alaska North). Horizontal bars on reference sample represent confidence limits of two standard errors on either side of the reference sample mean.

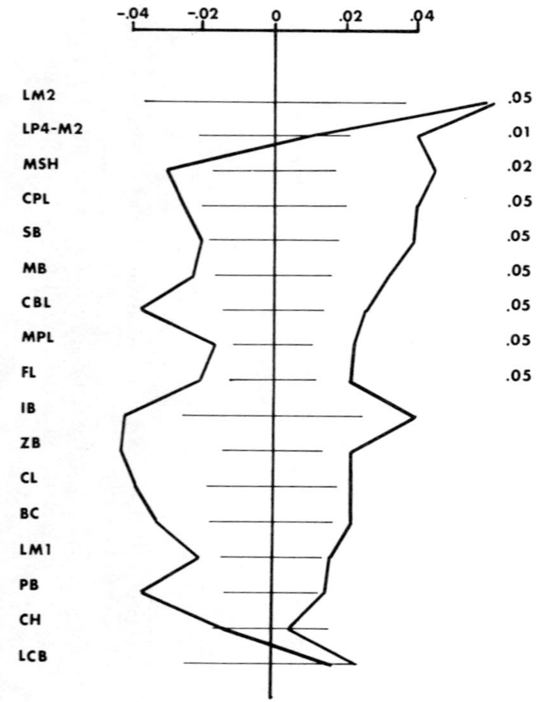

Fig. 2. Ratio diagram for females.

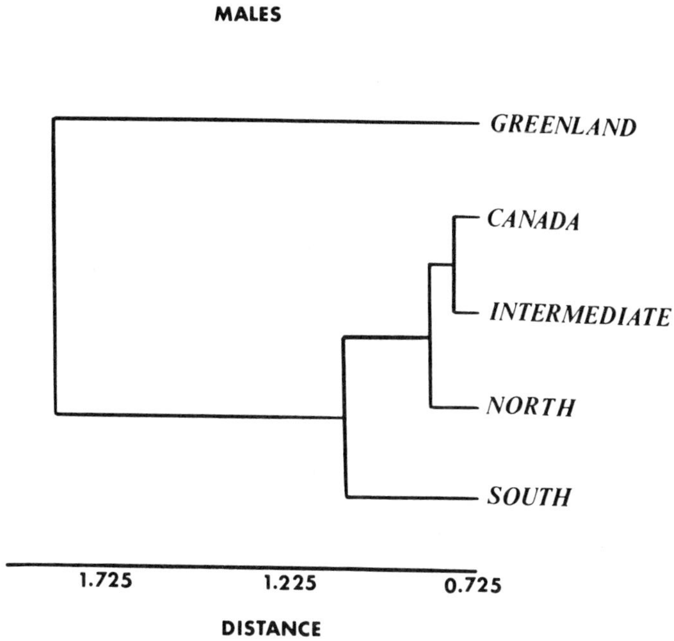

Fig. 3. Distance phenogram for males. The cophenetic correlation coefficient is 0. 907.

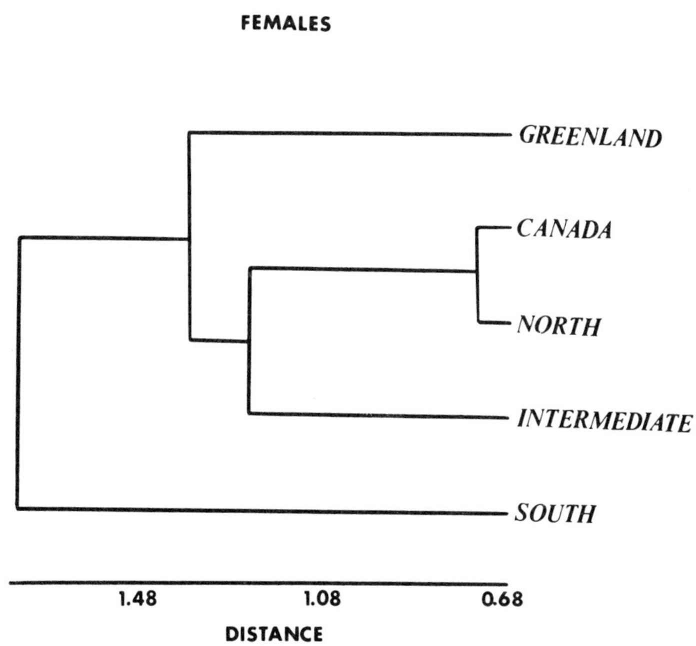

Fig. 4 Distance phenogram for females. The cophenetic correlation coefficient is 0. 770.

RESULTS AND DISCUSSION

The Alaskan material was grouped into three samples based on the earlier results of Manning (1971). The Alaska North sample includes animals mainly from areas to the east of Wainwright, expecially Pt. Barrow, Colville and Barter Island. The Intermediate sample was drawn from Pt. Lay, Wainwright, Icy Cape and Franklin Point. Animals from Cape Lisburne south were included in the Alaska South sample.

In the ratio diagrams, the animals from the Intermediate sample were allocated to the Alaska North and South populations by using Pt. Lay as the dividing line. Figure 1 is a ratio diagram comparing means of males of the East Greenland and Alaska South samples to those of Alaska North. As can be seen, there are four characters which show significant differences between the two Alaska populations. These characters (FL, CBL, MSH, MPL) reflect the greater skull length and height of male bears from the Alaska South population. Figure 2, for females, shows nine significant differences between the two Alaska populations. That the Alaska South population averages large in all 17 characters in both males and females is obvious from Figures 1 and 2.

Results of a cluster analysis on a matrix of Euclidean distance coefficients support the above results. Phenograms in Figures 3 and 4 illustrate the clinal nature of the variation across North America with the small Greenland animals at the top of the phenograms and larger animals at the bottom. The phenogram for males (Fig. 3) shows the close relationship of the Canada, Alaska North and Alaska Intermediate populations; whereas the phenogram for females (Fig. 4) appears to stress the distinctiveness of the Alaska South population, as in the ratio diagrams.

The results of the principal components analyses on character correlation matrices are shown in Figures 5 and 6. In both cases, component number one accounts for much of the variation (77% in males, 69% in females). For males, the three characters contributing most to component number one are zygomatic breadth, supraorbital breadth and interorbital breadth. For females, the correspondingly important characters are breadth at canines, condylobasal length and supraorbital breadth. The second principal component for males which accounts for an additional 12 percent of the variation, is heavily loaded for the three toothrow measurements: length of M1, length of P4-M2, and length of M2. The second principal component for females, which accounts for 19 percent of the variation, is heavily loaded for interorbital breadth, cranial length and palatal breadth. Not unexpectedly these results indicate the major source of variation in these bears to be size, with more subtle shape variations as shown by Operational Taxonomic Units (OTU) coordinates on the second Principal Component.

To more clearly define the relationships among the various populations three dimensional plots of the results from Kruskal's nonmetric multidimensional scaling (MDSCAL) are presented (Figures 7 and 8). While not greatly different from the Principal Components Analyses, the MDSCAL plots may be a better representation of the relative phenetic differences between populations (Rohlf, 1972). The results of MDSCAL confirm the distinctiveness of the Alaska South population, and also show the phenetic similarities among the northern or interior groups of bears.

The patterns suggested here agree well with current and ice flow patterns in the Arctic region. The East Greenland population occurs mainly in the south flowing Greenland current. The Alaska South population is found mainly in the northward moving currents from the Bering Strait. The interior populations

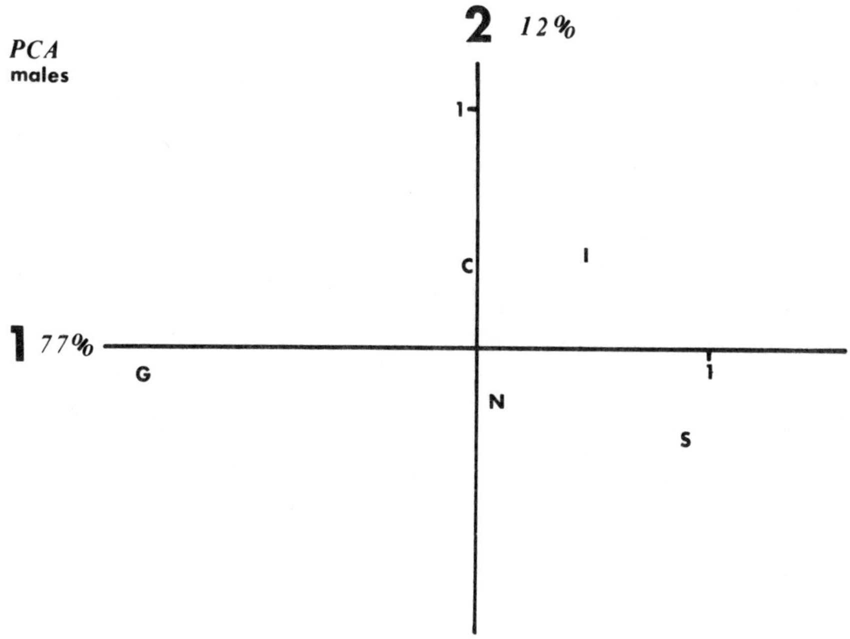

Fig. 5 Principal components analysis for males. G—Greenland; C—Canada; N—Alaska North; I—Intermediate; S—Alaska South.

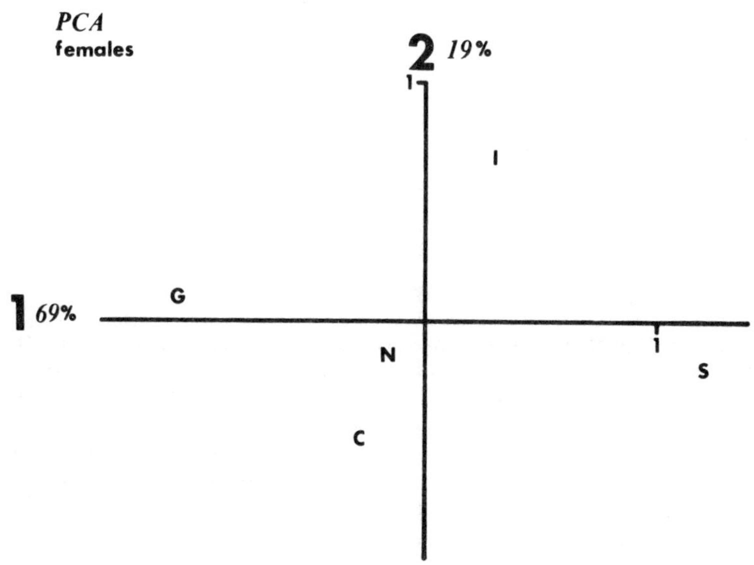

Fig. 6. Principal components analysis for females. Abbreviations as in Fig. 5.

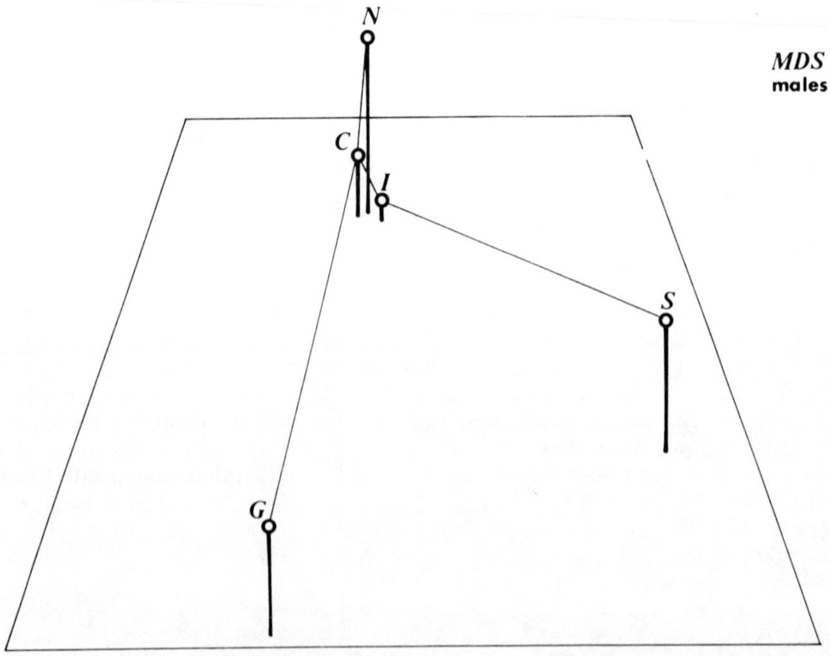

Fig. 7. Three dimensional representation of MDSCAL for males. The stress value is zero.

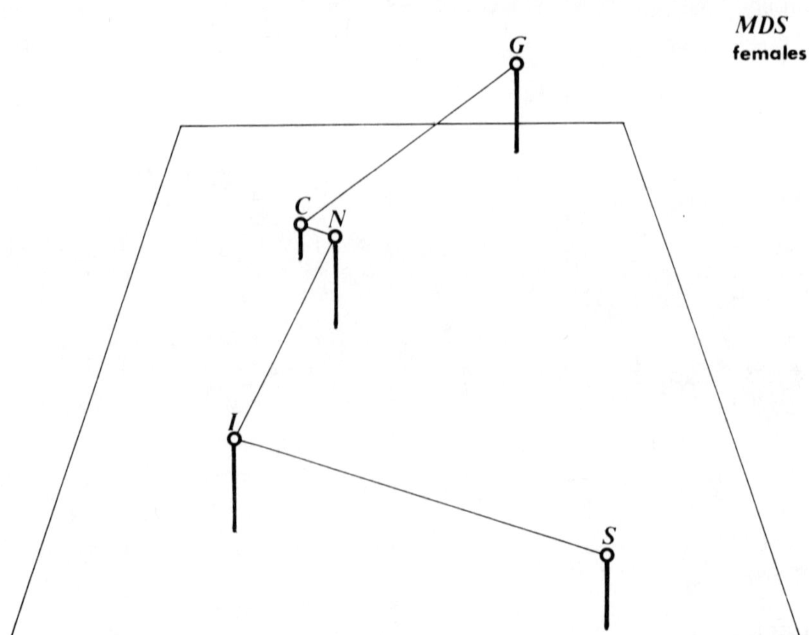

Fig. 8. Three dimensional representation of MDSCAL for females. The stress value is zero.

are affected by a variety of east or west-flowing currents across the top of North America, which may facilitate gene flow among these populations.

If subspecies were to be recognized, the East Greenland bears should be allocated to the nominate race, *Ursus maritimus maritimas* Phipps. The type locality for *U. maritimus* is Spitzbergen (Phipps, 1774). The population extending from West Greenland through Northern Alaska would take the name *U. m. labradorensis* Knotterus-Meyer. As Manning (1971) has pointed out, however, these interior Nearctic bears appear to be quite similar to the Palearctic populations which would assume the name *U. m. marinus* Pallas. In either event, there is no name currently available for the Alaska South population.

Although these results are tentative, pending the examination of a few additional specimens, I think some general conclusions are possible. The generally clinal nature of the variation originally pointed out by Manning (1971). is verified. There are definite steps in the cline at the extremes, such that the East Greenland population and the Alaska South population can be separated readily from an interior group extending from West Greenland to Pt. Lay, Alaska. Although these differences may be of sufficient magnitude to warrant subspecific recognition, I think nomenclature stability will best be served by considering *Ursus maritimus* monotypic, at least until the Old World populations have been similarly analysed.

ACKNOWLEDGEMENTS

Clyde Jones, Sandra Husar and Robert Fisher aided in measuring specimens. Jack Lentfer made available most of the specimens used in this study, and Pat Liba provided valuable help in locating some of the data. Michael Bogan provided computer assistance, and both he and Henry Setzer greatly improved the manuscript with their critical comments.

REFERENCES

ANDERSON, S. 1972. Mammals of Chihuahua Taxonomy and Distribution. *Bull. Amer. Mus. Nat. Hist.* **148**: 149-410.

KNOTTERUS-MEYER, T. 1908. Uber den Eisbären und seine geographischen Formen. *Sitzungsberichte der Gesellschaft Naturforschender Freunde zu Berlin.* Pp. 170-187.

MANNING, T. H. 1971. Geographical variation in the polar bear *Ursus maritimus. Canadian Wildlife Service Rept.* Series **13**: 1-37.

PHIPPS, C. J. 1774. *A voyage towards the North Pole*. J. Nourse, London. 253 pp.

ROHLF, F. J. 1972. An empirical comparison of three ordination techniques in numerical taxonomy. *Systematic Zool.* **21**: 271-280.

SIMPSON, G. G. 1941. Large Pleistocene felines of North America. *Amer. Mus. Novitates* **1136**: 1-27.

Reprinted from *Journal of the Fisheries Research Board of Canada*

Distribution and Differentiation of Populations of Dolphins of the Genus *Stenella* in the Eastern Tropical Pacific[1]

WILLIAM F. PERRIN

National Oceanic and Atmospheric Administration, National Marine Fisheries Service, Southwest Fisheries Center, La Jolla, California 92037, USA

PERRIN, W. F. 1975. Distribution and differentiation of populations of dolphins of the genus *Stenella* in the eastern tropical Pacific. J. Fish. Res. Board Can. 32: 1059–1067.

There are at least two races of spotted dolphin, *Stenella attenuata*, in the eastern Pacific, based on length, color pattern, and skeletal differences. The coastal form is provisionally called *S. attenuata graffmani*. The offshore form is distinct from a third, Hawaiian race. Likewise, at least three unnamed races of spinner dolphin, *S. longirostris*, occur in the eastern Pacific. Length, shape of adult male, color pattern, and cranial differences are used to distinguish a Costa Rican, an eastern, and a whitebelly race. An Hawaiian race is distinct from these. A distributional break may indicate two stocks of the striped dolphin, *S. coeruleoalba*.

PERRIN, W. F. 1975. Distribution and differentiation of populations of dolphins of the genus *Stenella* in the eastern tropical Pacific. J. Fish. Res. Board Can. 32: 1059–1067.

Il existe au moins deux races de *Stenella attenuata* dans le Pacifique oriental, se distinguant par la longueur, le motif de coloration et le squelette. Nous nommons provisoirement la forme côtière *S. attenuata graffmani*. La forme du large est distincte d'une troisième, la race hawaïenne. Semblablement, au moins trois races sans noms de *S. longirostris* se rencontrent dans le Pacifique oriental. La longueur, la forme des mâles adultes, le motif de coloration et les différences craniennes servent à distinguer une race de Costa Rica, une race orientale et une race à ventre blanc. Une race hawaïenne est différente de celles-ci. Une discontinuité dans la répartition peut indiquer l'existence de deux stocks de *S. coeruleoalba*.

Accepted March 14, 1975 — Accepté le 14 mars 1975

THE purse seine fishery for tuna in the eastern tropical Pacific employs three species of *Stenella* (Perrin 1968, 1969; NOAA 1972). The purpose of this paper is to describe the distribution of the subpopulations of these three species and the grounds for separating them.

The involved species of dolphin are *Stenella attenuata* (Gray, 1846), called "spotted porpoise" by the American tuna fishermen, *S. longirostris* (Gray, 1828), called "spinner porpoise," and *S. coeruleoalba* (Meyen, 1833), called "streaker porpoise." The taxonomy of the genus is unsettled, and the nomenclature used here is provisional. *Stenella attenuata* is the most important of the three species to the fishery (Table 1). *Stenella longirostris* is next most important. Only a very few net sets are known to have been made on *S. coeruleoalba*. Other cetaceans involved in the fishery are the delphinids *Delphinus delphis* L., *Tursiops* sp., *Lagenodelphis hosei* Fraser, 1956, and *Steno bredanensis* (Lesson, 1828), the latter three to very minor extent.

[1]Prepared for the meeting of the "Subcommittee on Small Cetaceans, Scientific Committee, International Whaling Commission," held at Montreal, Quebec, Canada, April 1–11, 1974.

Printed in Canada (J3652)
Imprimé au Canada (J3652)

Materials and Methods

Most of the distribution data and the specimens used for morphological studies were collected by scientific observers aboard commercial tuna seiners during the period 1968–74. I also surveyed available U.S. museum collections and the literature. Complete details concerning the specimens and methods used in the analyses of morphology have been previously presented (Perrin 1972a, 1973).

In tagging dolphins to delineate their ranges and migrations, several tagging methods and types of tags were used. Animals were tagged on the deck after being extricated from the net or separated from the catch of tuna, in the net during the "backing down" rescue operation (Perrin 1969; Coe and Sousa 1972), or as they rode the bow-wave of the vessel. Dart-type spaghetti tags of two types were used. Initially, an all-plastic tag was used (Mather 1963). Since 1971, a tag with a metal dart (Mather 1963) has been used. Animals on the deck or in the net were tagged with a short wand. Bow-riding animals were tagged with a long wand with a quick-release head (Beckett 1968) or with a crossbow (Evans et al. 1972). Some

TABLE 1. Observed net sets on delphinids by tuna purse seiners, 1971–74.

Stock	Number of observed net sets (% of year's observed sets in parentheses)			
	1971	1972	1973	1974[a]
Stenella attenuata	64 (94.1)	219 (90.5)	616 (81.5)	681 (84.4)
Coastal race	0	0	7 (1.0)	0
Offshore race	64 (94.1)	219 (90.5)	609 (80.6)	666 (82.5)
Unidentified to race	0	0	0	15 (1.9)
S. longirostris	40 (58.8)	138 (57.0)	311 (41.1)	313 (38.8)
Eastern race	28 (41.1)	70 (28.9)	147 (19.4)	286 (35.4)
Whitebelly race	12 (17.6)	28 (11.6)	43 (5.7)	49 (6.1)
Unidentified to race	11 (16.2)	6 (2.5)	117 (15.5)	1 (0.1)
S. coeruleoalba[b]	0	0	2 (0.3)	2 (0.2)
Delphinus delphis	2 (2.9)	26 (10.7)	97 (12.8)	110 (13.6)
Northern race[c]	0	0	15 (2.0)	0
Southern race[c]	2 (2.9)	26 (10.7)	82 (10.8)	110 (13.6)
Tursiops sp.[b]	3 (4.4)	7 (2.3)	5 (0.7)	34 (4.2)
Lagenodelphis hosei	1 (1.5)	0	0	0
Steno bredanensis	0	0	0	2 (0.2)
Total sets observed[d]	68	242	756	807

[a]Partial data, as of March 25, 1974.
[b]May be more than one stock.
[c]Races *fide* W. E. Evans personal communication.
[d]Do not equal sum of species totals; some sets are on mixed herds.

TABLE 2. Races of *Stenella attenuata* in eastern Pacific and Hawaiian waters. Length data previously unpublished; color pattern and skeleton data from Perrin (1972b). Sample sizes in parentheses.

Race	Adult length (see Fig. 2) (*cm*)		Adult color patterns	Skeleton		
	Males	Females		Adult condylobasal length (*mm*)	Adult tooth diameter (*mm*)	No. vertebrae
Coastal	203–257 (37)	184–227 (76)	Dorsal spotting light to heavy, but medium to heavy in most. Dorso-ventral contrast low.	419–460 (18)	3.4–5.7 (9)	77–79 (9)
Offshore	179–224 (332)	167–206 (1042)	Dorsal spotting light to heavy, but medium to light in most. Far off-shore (> 800 km) animals lightly spotted.	356–411 (68)	2.7–4.0 (65)	77–83 (61)
Hawaiian	182–202 (6)	189–203 (3)	Dorsal and ventral spotting so obscure as to appear absent. Adult resembles neonate of offshore race (Perrin 1970).	395–440 (12)	—	79 (2)

plastic dorsal-fin tags (Norris and Pryor 1970) were also used.

Species Accounts

Stenella attenuata — There are at least two races of spotted dolphin in the eastern Pacific. They differ in several features (Table 2). A coastal form is relatively large and robust, has heavy jaws and teeth, is heavily spotted, and is restricted to onshore waters from the Gulf of California to northern Peru (Fig. 1). An offshore race is relatively small and slender, has lightly built jaws and teeth, is spotted to varying degree, and occurs in offshore waters west to about 145°

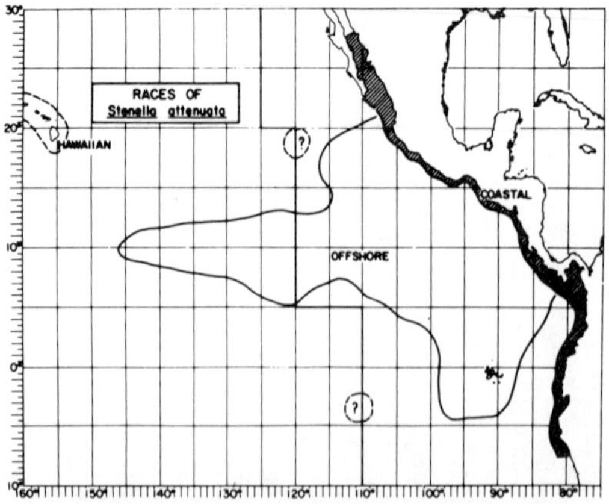

FIG. 1. Known ranges of races of *Stenella attenuata* in the eastern Pacific, based on collections and sightings by National Marine Fisheries Service observers aboard commercial tuna seiners and research vessels, museum collections and literature records. Boundaries are approximate only, and fluctuate seasonally and from year to year.

W long. The ranges of the two forms may overlap, or the boundary between them may move seasonally or from year to year. They have not been observed or captured together. In the central latitudes of the range the coastal form has been collected as far offshore as 50 km and the offshore form as far inshore as 20 km from the coast. The offshore form predominates in the fishery (Table 1). Another race inhabits the waters around Hawaii and may extend into the South Pacific and to the west. This form is also small and slender, but is less spotted.

The existence of two forms was first suspected upon comparison of a large series of skeletal specimens from the tuna fishery with the material available in museums, which consists mostly of animals harpooned on the coast and includes the type of *Stenella graffmani* (Lönnberg, 1934). Both series included obviously very young and very old individuals. The skulls in the coastal series were longer, broader, and had proportionately larger posttemporal fossae and teeth than those in the series from the tuna fishery (Perrin 1972a). The skulls of the two forms differ mainly in those features that have to do with feeding. The coastal form appears to be adapted to feeding on larger, more substantial prey than does the offshore form. The offshore form feeds primarily on epipelagic squids and fishes (Perrin et al. 1973). No data are available on the diet of the coastal form. The two series of specimens are statistically separable on the basis of aspects of the skull related to feeding (Perrin 1972a). The Hawaiian form is also separable on the basis of skull measurements. It has a longer, broader skull than does the offshore form but does not have the more massive feeding apparatus like that of the coastal race. Within the offshore race, an east–west cline may exist. Specimens from farthest west tend to be most Hawaiian-like.

The coastal form also has a larger, more massive postcranial skeleton than does the offshore form and, on the average, fewer vertebrae (Perrin 1972a). The range of vertebral counts for the coastal series is 77–79 (nine specimens), as opposed to 79–83 (48 specimens) for offshore animals taken from adjacent waters. Offshore animals from more than 800 km offshore tend again to have lower vertebral counts (77–82 in 13 specimens), supporting the concept of an east–west cline within the offshore race.

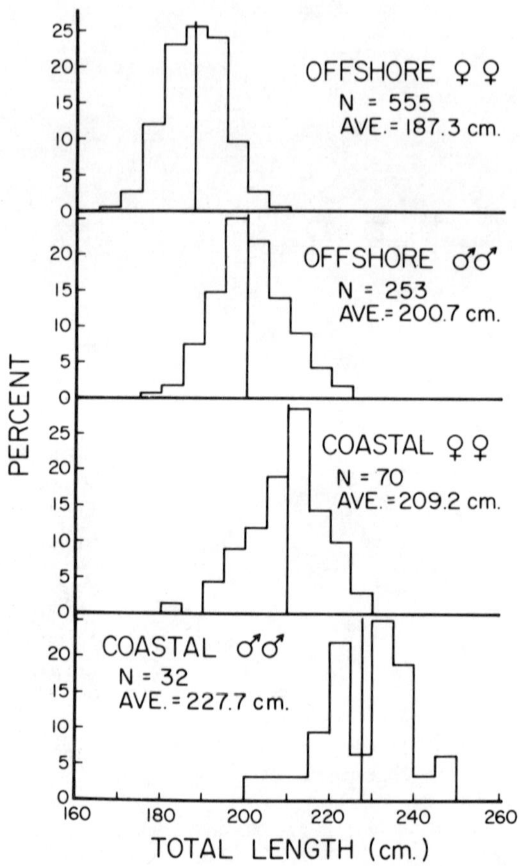

FIG. 2. Histograms of body lengths of spotted dolphins, *S. attenuata*, of two races, collected aboard tuna seiners in 1973. Specimens identified to race on the basis of locality of capture and range of lengths in sample.

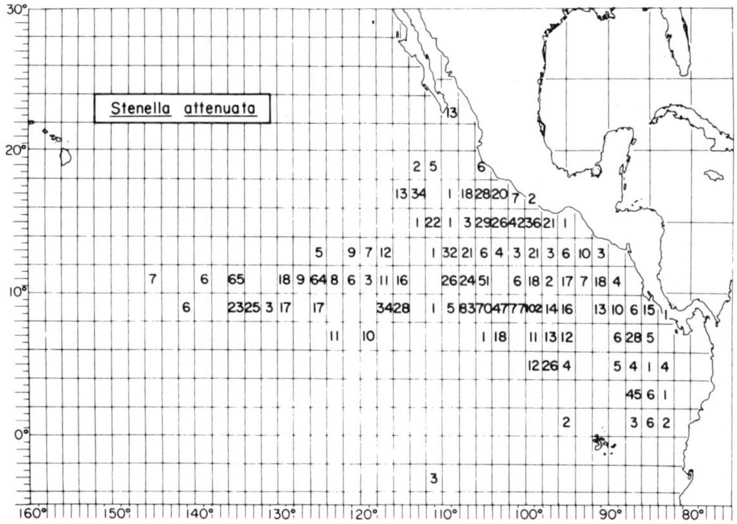

FIG. 3. Numbers of spotted dolphins, *S. attenuata*, tagged, 1969–74, by 2-degree square.

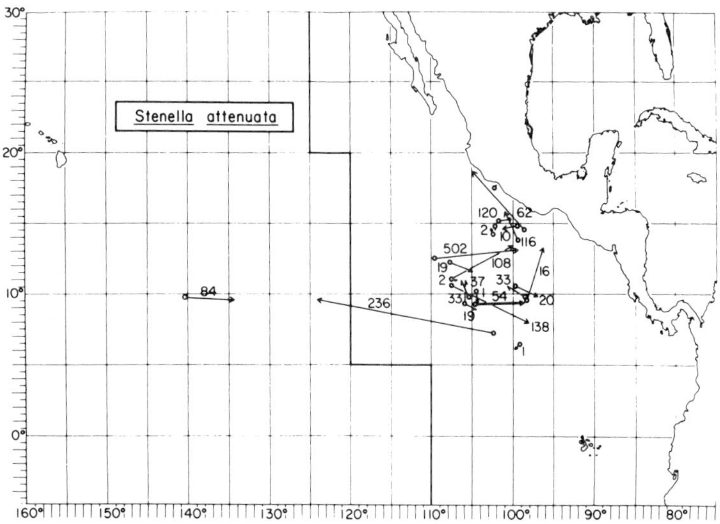

FIG. 4. Net movements of recovered tagged spotted dolphins, *S. attenuata*, with days at liberty.

Recent length data from the fishery also support the concept of two races in the eastern Pacific. Several samples of animals taken in 1973 from within 30 km of the coast of Costa Rica were clearly from a different population than samples taken from more offshore waters throughout the range (Fig. 2). The coastal adult females (those with ovarian scars) were on the average about 22 cm longer than the offshore females. The coastal adult males (those with testes weighing more than 200 g, the smallest size at which sperm has been found in the epididymis) were about 27 cm longer than the offshore males. Calculating from a length–weight relationship (unpublished data), this amounts to an average weight difference in adult males of about 35 kg (115 vs. 80 kg).

We have tagged 11 coastal and 1716 offshore

spotted dolphins (Fig. 3), and recovered 24 tagged offshore animals from known locations (Fig. 4). Time at liberty ranged from 1.7 h to 502 days, and the distance from release point to recapture point ranged from 13 to 2415 km. These results show a great deal of east–west and north–south movement, evidence for extensive mixing within the eastern half of the supposed range of the offshore race. The degree of continuity between the eastern and far-western portions of the range is still unknown because of insufficient tag returns.

The three forms clearly deserve subspecific status (Perrin 1972a). Assignment of trinomials to the offshore and Hawaiian forms will have to wait until the taxonomy of the genus at the species level has been unraveled. Many of the holotypes in *Stenella* are from unknown localities. Relationships among spotted dolphins in the Atlantic and Pacific need to be examined. The name *graffmani*, however, clearly applies to the coastal form, and I therefore provisionally call the coastal form *S. attenuata graffmani*.

Stenella longirostris — There are at least three races of spinner dolphin in the eastern Pacific (Fig. 5), differing modally in several characters (Table 3). One that occurs very near the coast of Central America, referred to below as the "Costa Rican" form, is relatively long, slender, and gray. A second race, called "eastern spinner," occurs along the coast of Mexico and seaward about 800 km and is relatively short, slender, and gray. A third, called "whitebelly spinner", occurs in far offshore waters west to about 145° W long. and is relatively short, robust, and white below.

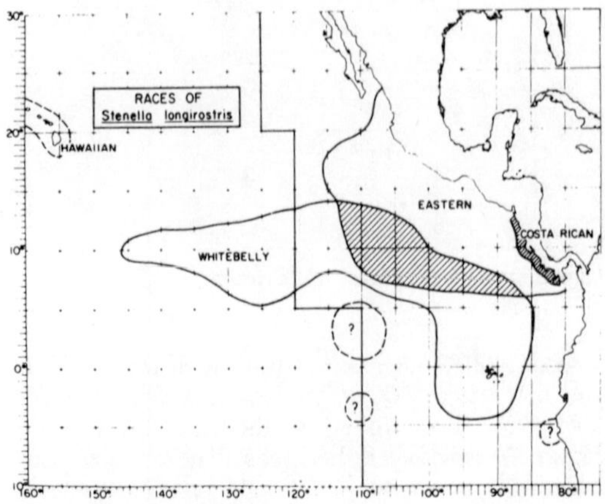

FIG. 5. Known ranges of races of *S. longirostris* in the eastern Pacific. Same data sources as Fig. 1. Boundaries are approximate only, and fluctuate seasonally and from year to year.

A fourth form occurs in Hawaiian waters and possibly to the south and west and is relatively long, robust, and white below. The ranges of the eastern and whitebelly forms overlap (Fig. 5). In the area of overlap, the two forms are occasionally captured together. A few apparent intergrades have also been collected.

The eastern and whitebelly forms are both involved in the tuna fishery (Table 1). Existence of eastern and whitebelly races was discovered through field observations of color patterns. The Costa Rican–eastern and whitebelly–Hawaiian dichotomies were discovered while examining skeletal specimens.

As with *S. attenuata*, most of the geographical variation in the skull is in features functional in feeding (Perrin 1972a). The Costa Rican spinner has a proportionately longer, more lightly built rostrum and mandible than does the eastern

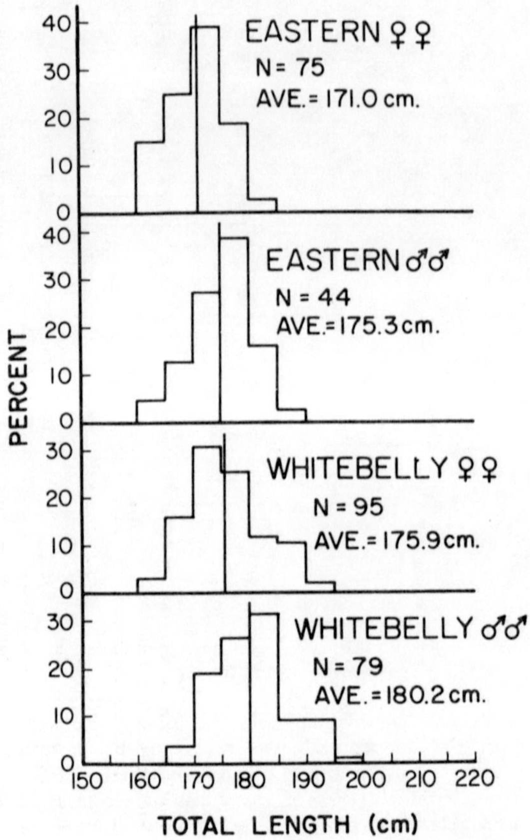

FIG. 6. Histograms of body lengths of spinner dolphins *S. longirostris*, of two races, collected aboard tuna seiners, 1968–73. Specimens identified to race on basis of locality of capture and color pattern.

TABLE 3. Races of *S. longirostris* in eastern Pacific and Hawaiian waters. Length data previously unpublished; other data from Perrin 1972a. Sample sizes in parentheses.

Race	Adult length (*cm*)		Shape of adult males		Color pattern (terminology of Perrin 1972b)	Skull	
	Males	Females	Dorsal fin	Ventral hump		Condylobasal length (*mm*)	Zygomatic width (*mm*)
Costa Rican	181–216 (2)	193–203 (3)	Canted to triangular	—	Poorly known, but like that of eastern.	416–437 (4)	149–152 (4)
Eastern	161–186 (44)	160–181 (75)	Canted to triangular	Medium to extreme	Dark dorsal field obscures cape and extends ventrad, leaving light areas in genital and axillary regions. Monotone effect.	351–407 (26)	139–153 (29)
Whitebelly	170–196 (79)	161–194 (95)	Triangular to falcate	Medium to slight	Dorsal field extends ventrad only to level of eye laterally and has irregular margin. Cape faintly visible through dorsal field. Bipartite effect.	365–419 (13)	145–158 (14)
Hawaiian	185 (1)	183–201 (2)	Falcate	Slight	Light dorsal field to level of eye, with smooth margin. Cape very obvious. Tripartite effect.	431–464 (10)	159–171 (10)

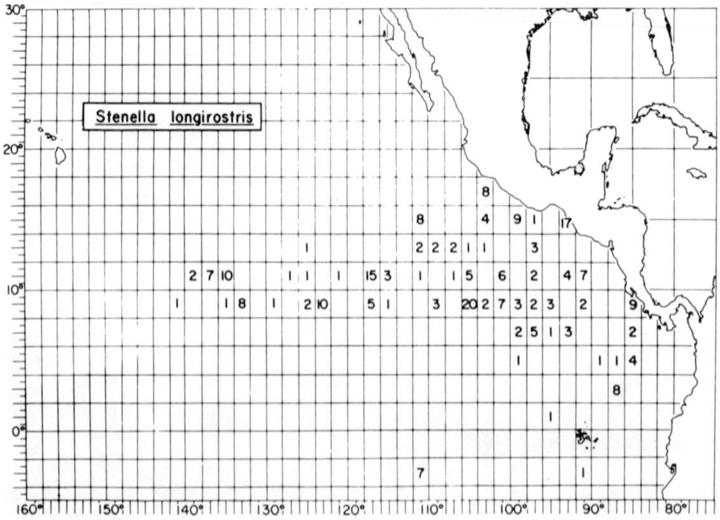

FIG. 7. Numbers of spinner dolphins, *S. longirostris*, tagged, 1969–74, by 2-degree square.

spinner, the reverse of the trend in *S. attenuata.* The skulls of the eastern and whitebelly spinners do not differ in proportion, but the whitebelly skull is on the average slightly larger. The skull of the Hawaiian spinner is longer and broader than those of the other three forms. The four series of specimens can easily be separated statistically (Perrin 1972a).

We know little about the external appearance of the Costa Rican spinner, other than it is very long and slim (a 216-cm male weighed only 57 kg) and has a color pattern very much like that of the eastern spinner. The obviously disparate color patterns of the eastern, whitebelly, and Hawaiian forms can be compared in terms of differential development of a dorsal dark field (Perrin 1972b). The dorsal field is darkest and most extensive in the eastern form and lightest and least extensive in the Hawaiian form. In the Hawaiian form, an underlying dark cape shows strongly through the light dorsal field. The result is to make the coloration of the eastern, whitebelly and Hawaiian spinners effectively uniform, bipartite, and tripartite, respectively. In some apparent intergrades from the area of overlap of the eastern and whitebelly ranges, the dark cape shows faintly through the dorsal field, but the ventral margin of the field is lower and more irregular than in the typical whitebelly spinner.

The dorsal fin of the adult male differs among the three races. In the eastern spinner the tip is forward of midlength of the fin base, giving the fin a reversed appearance. The forward cant is much more developed in some males than in others (Perrin 1972a). In the whitebelly form, the adult male fin is never "reversed" but usually triangular. In the Hawaiian form, it is always at least slightly falcate. Development of a ventral, postanal keel, of unknown function, is correlated with degree of forward cant of the dorsal fin. It is largest in eastern males and nearly absent in Hawaiian males.

The whitebelly form is on the average about 5 cm longer than the eastern form (Fig. 6) and is more robust, having a greater average girth at the axilla (Perrin 1972a). Few measurements are available for the Hawaiian spinner, but it ap-

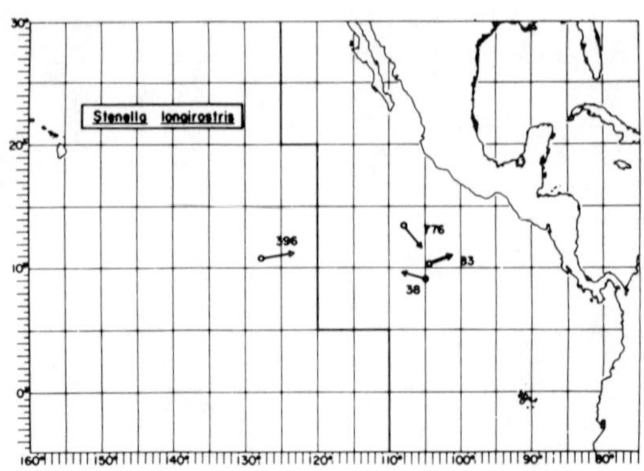

FIG. 8. Net movements of recovered tagged spinner dolphins, *S. longirostris*, with days at liberty.

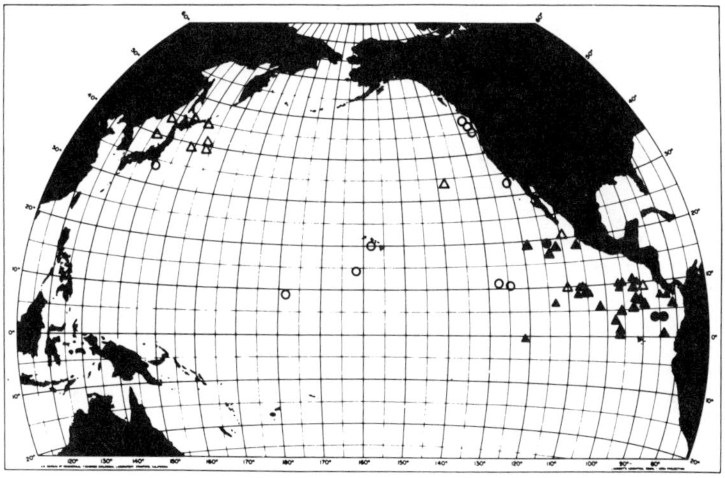

FIG. 9. *Stenella coeruleoalba* in the Pacific. Circles are specimen localities; triangles are sighting localities. Shaded symbols are new records. Open symbols are previously published records (Hubbs et al. 1973).

parently is, on the average, longer than the whitebelly spinner.

We have tagged 242 spinner dolphins (Fig. 7) and recovered five (Fig. 8). Time at liberty ranged from 38 to 776 days, and net distance traveled from 333 to 733 km. Two animals tagged together were recaptured together 83 days later. The limited results show less net movement than for *S. attenuata* and agree with the hypothesis of offshore (eastern) and far offshore (whitebelly) populations.

The four races of spinners deserve subspecific status, but, as is the case for the spotted dolphins, assignment of trinomials must wait until the question of how many species of spinner dolphin exist in the world's oceans has been settled. For the time being, they are best considered as unnamed subspecies of *S. longirostris* Gray.[2]

Stenella coeruleoalba — The striped dolphin was not known from the eastern tropical Pacific until 1970 (Hubbs et al. 1973). In addition to the published locality records (Hubbs et al. 1973) we now have 36 more records, including three for which specimens were collected (Fig. 9). The known range has been extended south to below the equator and east to the Panama Bight. There may be a break in the distribution between about 10° and 17° N lat. This would make the distribution similar to that of *Delphinus* in the eastern tropical Pacific (W. E. Evans personal communication) and may indicate that there is more than one stock. Enough specimens have now been collected from the tropical Pacific (26) so that morphological comparisons can begin to be made between this population or populations and those in the western Pacific and Japan.

Discussion

Geographical races exist for each species of small delphinid that has been closely looked at in the eastern tropical Pacific. It seems reasonable to assume that the eastern Pacific (and world) populations of the other, less well-known delphinids, e.g. *S. coeruleoalba*, *Tursiops*, and *Globicephala*, are similarly subdivided. This points out the need for emphasis on research and management at the stock level.

[2]*Note*: There is room for speculation about the identity of Gray's holotype (Perrin 1972a). The spinner skull that Schlegel in 1846 figured and redescribed as the holotype may not have been the specimen examined by Gray. Gray described his specimen as having a strongly carinated palate, a feature present in *Delphinus* but not in *Stenella*. If Schlegel was indeed mistaken about the origin of the specimen he examined, *longirostris* may be preoccupied by *Delphinus* and unavailable for *Stenella*. However, in the interest of stability of nomenclature and in the light of other known inconsistencies in Gray's work, it seems best to ignore this uncertainty now and retain the long-used *longirostris* for the spinners.

BECKETT, J. S. 1968. A harpoon adapter for tagging large free-swimming fish at the surface. J. Fish. Res. Board Can. 25: 177–179.

COE, J., AND G. SOUSA. 1972. Removing porpoise from a tuna purse seine. Mar. Fish. Rev. 34(11–12): 15–19.

EVANS, W. E., J. D. HALL, A. B. IRVINE, AND J. S. LEATHERWOOD. 1972. Methods for tagging small cetaceans. Fish. Bull. (U.S.) 70: 61–65.

HUBBS, C. L., W. F. PERRIN, AND K. C. BALCOMB. 1973. *Stenella coeruleoalba* in the eastern and central tropical Pacific. J. Mammal. 54: 549–552.

MATHER, F. J. JR. 1963. Tags and tagging techniques for large pelagic fishes. North Atlantic Fish Marking Symposium. Int. Comm. Northwest Atl. Fish. Spec. Publ. 4: 288–293.

NOAA. 1972. Report of the NOAA Tuna Porpoise Review Committee. Dep. Commerce, Washington, D.C., September 8, 1972. 63 p.

NORRIS, K. S., AND K. W. PRYOR. 1970. A tagging method for small cetaceans. J. Mammal. 51: 609–610.

PERRIN, W. F. 1968. The porpoise and the tuna. Sea Frontiers 14(3): 166–174.

1969. Using porpoise to catch tuna. World Fishing 18(6): 42–45.

1970. Color pattern of the eastern Pacific spotted porpoise *Stenella graffmani* Lönnberg (Cetacea, Delphinidae). Zoologica (N.Y.) 54(4): 135–152, pls. 1–7.

1972a. Variation and taxonomy of spotted and spinner porpoises (genus *Stenella*) of the eastern tropical Pacific and Hawaii. Ph.D. dissertation. Univ. Calif. Los Angeles, Calif. 490 p.

1972b. Color patterns of spinner porpoises (*Stenella* cf. *S. longirostris*) of the eastern Pacific and Hawaii, with comments on delphinid pigmentation. Fish. Bull. (U.S.) 70: 983–1003.

1973. Variation and taxonomy of spotted and spinner porpoises (genus *Stenella*) of the eastern tropical Pacific and Hawaii. Diss. Abstr. 33: 5069–5070-B.

PERRIN, W. F., R. W. WARNER, C. L. FISCUS, AND D. B. HOLTS. 1973. Stomach contents of porpoise, *Stenella* spp., and yellowfin tuna, *Thunnus albacares*, in mixed-species aggregations. Fish. Bull. (U.S.) 71: 1077–1092.

Reprinted from *The Wilson Bulletin*

RACIAL COMPOSITION OF MIGRANT POPULATIONS OF SANDHILL CRANES IN THE NORTHERN PLAINS STATES

Douglas H. Johnson and Robert E. Stewart

During recent years, there has been increasing pressure to expand hunting seasons on Sandhill Cranes (*Grus canadensis*) in the northern plains states. Hunters desire to take advantage of the opportunity presented by the sizable flocks of this large bird in this area. Prior to passage of the Migratory Bird Treaty Act in 1918, the Sandhill Crane was hunted and considered to have high sporting and eating qualities. In addition, complaints have been received from farmers regarding grain crop depredations by cranes. It has been suggested (Boeker et al., 1961:16) that flocks of cranes would be more widely dispersed by hunting pressure, thus alleviating local damage to crops.

The fact that three apparent subspecies of the Sandhill Crane occur in the northern plains of the United States during migration (Walkinshaw, 1965) complicates the situation since one of these, the Greater Sandhill Crane (*Grus canadensis tabida*) was until late 1971 officially designated as rare by the U.S. Bureau of Sport Fisheries and Wildlife (Anon., 1968). The other subspecies occurring in this area are the Lesser Sandhill Crane (*Grus c. canadensis*) and the Canadian Sandhill Crane (*Grus c. rowani*). We are using these subspecific designations on a tentative basis only. The validity of the *rowani* subspecies has been questioned (e.g. Stephen, 1967), and more detailed work on the breeding grounds in the future may demonstrate the existence of gradient populations of Sandhill Cranes that occur between areas occupied by typical members of the three currently recognized subspecies. These populations may be expected to differ in migrational and wintering patterns as well as in morphological characteristics. Such clines have been demonstrated for the Canada Goose (*Branta canadensis*) (Aldrich, 1946; Hanson, 1965:13), but evidence accumulated from cranes on their breeding grounds is insufficient to establish a similar pattern in this species.

Until adequate breeding-ground studies have been conducted, it is useful for the purposes of this paper to assume the existence of three populations which may be distinguished on the basis of the average size of the cranes which comprise each population. For reasons of convenience, at least, these populations can be referred to as *canadensis, rowani*, and *tabida.*

Current information concerning the breeding ranges of the three northern subspecies was derived from reports by Walkinshaw (1949, 1965), supplemented with data furnished by John W. Aldrich (pers. comm.). The Lesser Sandhill Crane, the most northern of the subspecies, breeds on the Arctic tundra of northwestern Canada, Alaska, and northeastern Siberia. Breeding populations of the Canadian Sandhill Crane favor the boreal forest and aspen parklands of west-central Canada. The breeding range of the eastern population of the Greater Sandhill Crane formerly extended throughout the glaciated prairie region in south-central Canada and north-central United States, and also included disjunct outliers of prairie or prairie-forest ecotonal communities within the Lake Forest (hardwood-pine) region to the east. At present, this population appears to be restricted to isolated groups of breeding birds in Michigan, Wisconsin, and northern Minnesota, and to groups of birds that are probably breeding in southwestern Ontario and southeastern Manitoba. Another segment of the breeding population of the Greater Sandhill Crane occurs west of the continental divide in southern British Columbia and in several western states, and therefore is not considered in this discussion.

Numerous migrating Sandhill Cranes (often many thousands) occur regularly in a few local staging areas. In North Dakota these concentrations are normally found from late August to early November on glacial outwash plains in the vicinity of wetland complexes composed of large shallow lakes that are brackish, subsaline, or saline. These include semipermanent lakes, permanent lakes, and alkali lakes (wetland habitat types IV-D, IV-E, V-E, and VI—Stewart and Kantrud, 1971). Fen pockets or man-made dugouts that occur along the margins of these lakes are utilized as roosting sites. Scattered flocks of cranes range out for several miles from these roosting sites to forage on croplands (Madsen, 1967). Similar habitat complexes are occupied in Montana but only for brief periods from late September through early November. In northwestern Minnesota roosting concentrations occur for a brief period during the latter half of September (Robert E. Farmes, pers. comm.) on large fens or alkaline bogs within the prairie-forest ecotone.

The primary purpose of this study was to determine the racial composition of Sandhill Crane populations that occur in northwestern Minnesota, North Dakota, and eastern Montana during the fall migration period. These investigations were initiated in 1970 and continued in 1971. Particular attention was given to the geographical distribution and seasonal occurrence of these populations as related to racial composition. Incidental information concerning the morphological distinctiveness of the three northern populations of Sandhill Cranes was also obtained.

METHODS

During the fall migration periods in 1970 and 1971, 180 Sandhill Cranes (153 adult and 27 immature) were collected in five major concentration areas (Fig. 1) of the northern plains states. These collections involved 11 series of specimens that were taken during early and/or late portions of the fall migration period. Attempts were made to collect the adult specimens in a random manner, whereas concurrent collections of immature

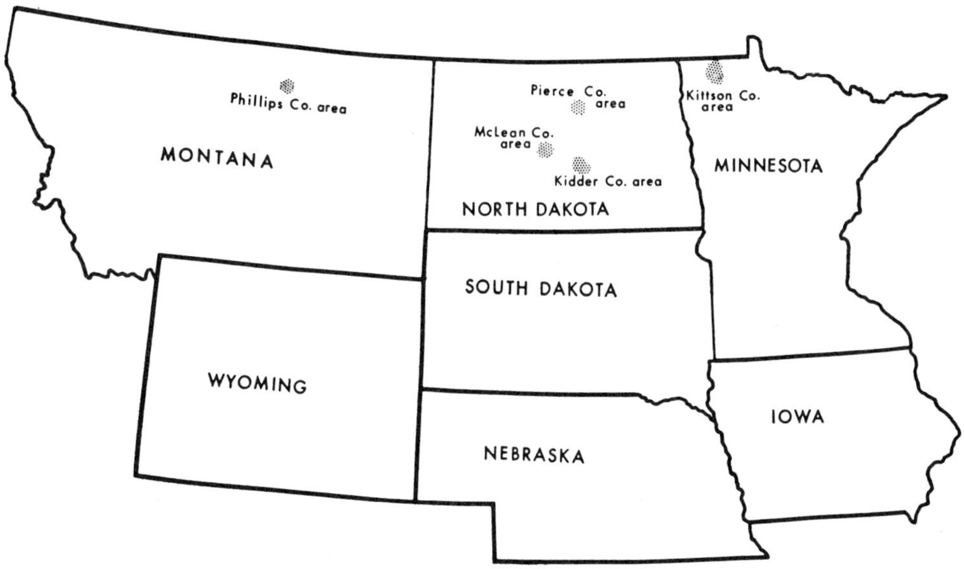

FIG. 1. Collection sites of migrant Sandhill Cranes.

birds were largely incidental. The age and sex composition of birds comprising each series of specimens and the dates of collection are indicated in Table 1.

All of the collected specimens were weighed in the field, and detailed laboratory examinations were made at the Northern Prairie Wildlife Research Center. These examinations involved sex and age determinations and the following linear measurements: wing chord, tarsus length, exposed culmen, culmen post nares (culmen from posterior edge of nostril), middle toe (without claw), and tail length.

A wing, a leg, and the head of each specimen collected in 1970, together with information on sex and age, were sent to Dr. John W. Aldrich, Division of Wildlife Research, U.S. Bureau of Sport Fisheries and Wildlife, Washington, D.C. He made independent measurements and, using those of wing chord, tarsus length, and culmen post nares, determined subspecies from them. This afforded a valuable opportunity for comparison and confirmation of our results.

MEASUREMENTS OF BREEDING SPECIMENS

The identification of the subspecies represented by an individual specimen collected during the migration period is preferably determined by comparing it with adequate series of breeding specimens that are typical of each of the relevant subspecies. Unfortunately, only minimal requirements are satisfied in this respect by the limited series of breeding Sandhill Crane specimens that are available. Walkinshaw (pers. comm.) has compiled a useful list of measurements together with data on collection dates and locations. Of these specimens and those described by Lumsden (1971), we selected individuals collected between late April and mid-July to form a sample of breeding ground specimens of known subspecies. In our analysis

TABLE 1

COLLECTIONS OF MIGRANT SANDHILL CRANES

	Adults		Immatures	
	Male	Female	Male	Female
Kittson Co., Minnesota				
22–24 September 1970	14	6	3	4
Pierce Co., North Dakota				
14 September 1971	7	5	0	1
27–28 October 1971	4	4	0	0
Kidder Co., North Dakota				
8–15 September 1970	11	14	2	1
26 October–1 November 1970	14	7	2	4
3–4 September 1971	9	6	0	0
McLean Co., North Dakota				
28 October–3 November 1970	18	9	2	2
7 September 1971	3	2	0	1
27–28 September 1971	6	2	1	0
Phillips Co., Montana				
4 October 1970	2	2	1	1
3 October 1971	5	3	1	1
Total	93	60	12	15

of the birds, we employed all available complete measurements of wing chord, tarsus length, and exposed culmen taken from adults of known sex. These measurements are summarized in Table 2.

In order to appraise the separability of the intermediate population, *rowani*, a series of statistical tests can be formulated, following Rao (1962, 1965: 490–493). Omit the presumed *rowani* specimens and assume for the moment that only two populations, *canadensis* and *tabida*, are separable. Let $\mu = (\mu_1, \mu_2, \mu_3)$ represent the vector of means of wing chord, tarsus length, and exposed culmen. Then μ_c is the mean vector for *canadensis*, and μ_t is the mean vector for *tabida*. Assume both populations have a common variance-covariance matrix, and let S be an estimate of it, based on n degrees of freedom. Then let $X = (X_1, X_2, X_3)$ be an observation of wing chord, tarsus length, and exposed culmen of any crane to be examined. The first test is whether or not X could reasonably have come from either a *canadensis* or *tabida* population or from a population intermediate to them. The form of this test is

$$F_0 = (n-1)\{(X-\mu_t)'S^{-1}(X-\mu_t) - [(\mu_c-\mu_t)'S^{-1}(X-\mu_t)]^2/(\mu_c-\mu_t)'S^{-1}(\mu_c-\mu_t)\}/2n$$

which is distributed as an F ratio with 2 and (n − 1) degrees of freedom. If

TABLE 2

MEASUREMENTS OF ADULT SANDHILL CRANES COLLECTED DURING THE BREEDING SEASON[1]

	Grus c. canadensis		*Grus c. rowani*		*Grus c. tabida*	
	Male	Female	Male	Female	Male	Female
Sample Size	36	17	10	3	11	7
Wing Chord						
Mean	469.1	445.5	503.9	472.7	541.5	538.9
Standard deviation	22.8	21.8	16.7	20.1	25.3	28.5
Extremes	418–505	420–490	480–524	456–495	502–575	503–575
Tarsus Length						
Mean	187.4	177.6	228.8	210.7	240.0	229.4
Standard deviation	12.7	11.3	7.6	5.5	9.9	5.4
Extremes	160–210	156–198	216–239	205–216	226–253	223–237
Exposed culmen						
Mean	91.4	89.7	119.4	103.0	135.2	129.3
Standard deviation	7.0	5.7	6.4	10.5	7.5	11.9
Extremes	69–102	80–101.6	109–127	93–114	122–144	113–147

[1] Linear measurements in millimeters; data from Walkinshaw (pers. comm.) and Lumsden (1971).

F_0 is not significant, it is then valid to further test the likelihood that X represents a *canadensis* member. This is tested by comparing

$$F_1 = (n-2)[(\mu_c - \mu_t)'S^{-1}(X - \mu_c)]^2/[(\mu_c - \mu_t)'S^{-1}(\mu_c - \mu_t) \times (n + 2nF_0/(n-1))]$$

to an F variate with 1 and $(n-2)$ degrees of freedom. An analogous value F_2 with μ_t replacing μ_c can be calculated to test that X belongs to the *tabida* population.

This sequence of statistical tests was applied to each of the 10 male and three female breeding specimens presumed to be *rowani*. The outcome for males is shown in Table 3; similar results were obtained for females. For no specimen was F_0 significant, even at the $P = 0.10$ level. This suggests that if the measurements are not from *canadensis* or *tabida*, they are from a population intermediate between them. This result is not intrinsically useful, but it validates further testing using F_1 and F_2. The calculated values of F_1 ranged from 11.51 to 32.63, and were all significant at the $P = 0.05$ level. It is thus highly unlikely that any of these specimens represents the *canadensis* population.

The results of testing for affinity to *tabida* were also significant, if not as conclusive. The F_2 values of eight male specimens were significant at the $P = 0.05$ level while the remaining two were nearly so. It follows then that those particular breeding specimens taken within the range of *rowani* are

TABLE 3

F VALUES FOR TESTING SEPARABILITY OF MALE BREEDING SPECIMENS PRESUMED TO BE *Grus canadensis rowani*

Specimen number	F_0	F_1	F_2
1	.37	20.43	8.91
2	.47	17.96	10.56
3	1.42	24.10	5.88
4	.90	30.15	3.71
5	1.33	11.51	15.65
6	.01	32.63	3.43
7	.22	25.28	6.26
8	.11	19.63	9.72
9	.60	14.22	13.65
10	1.70	26.10	4.76
Critical Values of F			
$P = 0.05$	3.21	4.06	4.06
$P = 0.10$	2.43	2.83	2.83

distinct from both *canadensis* and from *tabida*, although they appear somewhat closer to the latter population, at least in terms of wing chord, tarsus length, and exposed culmen measurements.

This sequence of tests provides stronger conclusions than would be obtained by merely comparing means of the populations. Finding a significant difference among means would indicate that *on the average* the presumed *rowani* group differs from the known populations. The tests above allow us to conclude that *each* of the questioned specimens differs from those taken in the ranges of typical *canadensis* and *tabida.* Such findings do not refute the possibility of a cline in morphological measurements, because the sampling (collection) of breeding specimens was not done uniformly throughout the breeding ranges and any groups intermediate between *tabida* and *rowani* or between *rowani* and *canadensis* may not be represented.

MEASUREMENTS OF ADULT MIGRANT SPECIMENS

For the purpose of making racial determinations of migrant Sandhill Cranes, certain linear measurements were found to be particularly useful. These include wing chord, tarsus length, and exposed culmen. To identify an individual migrant specimen, we compared its measurements to those taken from the series of breeding specimens that were representative of each of the three subspecies. The migrant specimen in question was assigned to a subspecies on the basis of similarity to measurements of the breeding specimens.

TABLE 4
MEASUREMENTS OF ADULT MIGRANT SANDHILL CRANES[1]

	Grus c. canadensis		*Grus c. rowani*		*Grus c. tabida*	
	Male	Female	Male	Female	Male	Female
Sample size	31	17	51	33	11	10
Wing Chord						
Mean	467.0	445.7	503.0	475.5	518.9	487.3
Standard deviation	15.5	12.0	12.9	14.3	9.4	13.8
Extremes	419–502	421–464	469–530	434–505	504–531	472–524
Tarsus Length						
Mean	187.5	179.2	230.6[2]	217.0	236.5	228.3
Standard deviation	14.4	10.8	9.5	7.6	8.3	6.3
Extremes	151–203	165–197	211–249	203–232	227–253	219–235
Exposed Culmen						
Mean	97.3[2]	92.0	119.7	114.1	131.8	120.4
Standard deviation	3.9	5.2	5.9	3.9	5.0	2.9
Extremes	91.6–105.9	83.2–99.7	99.6–130.6	107–122.5	124.4–141.1	114.9–125.3
Culmen post nares						
Mean	76.5	72.4	92.1	86.7	100.7[2]	93.5
Standard deviation	3.1	4.0	4.8	3.3	4.5	2.5
Extremes	70.3–84.0	63.7–78	74–102	82.8–94	95–107	90–98
Middle Toe						
Mean	75.4	73.4	86.4	83.3	87.2	84.5
Standard deviation	3.2	4.8	3.2	3.7	2.5	2.6
Extremes	69.3–82.2	61.5–78.8	80.6–93.5	77.2–92.7	83.3–91.0	79.4–88.5
Tail Length						
Mean	168.0	159.6	190.2	179.8	196.9	184.1
Standard deviation	6.5	6.7	7.4	6.0	4.5	6.2
Extremes	157–180	150–177	175–209	166–192	190–203	173–192
Weight						
Mean	3945.6	3459.9	4794.8[3]	4110.2[3]	4885.6	4453.7[2]
Standard deviation	299.6	250.1	385.3	251.9	370.8	427.1
Extremes	3235–4671	2866–3829	3890–5613	3676–4895	3870–5432	3645–5052

[1] Linear measurements in millimeters; weight in grams.
[2] Sample size one less than indicated.
[3] Sample size two less than indicated.

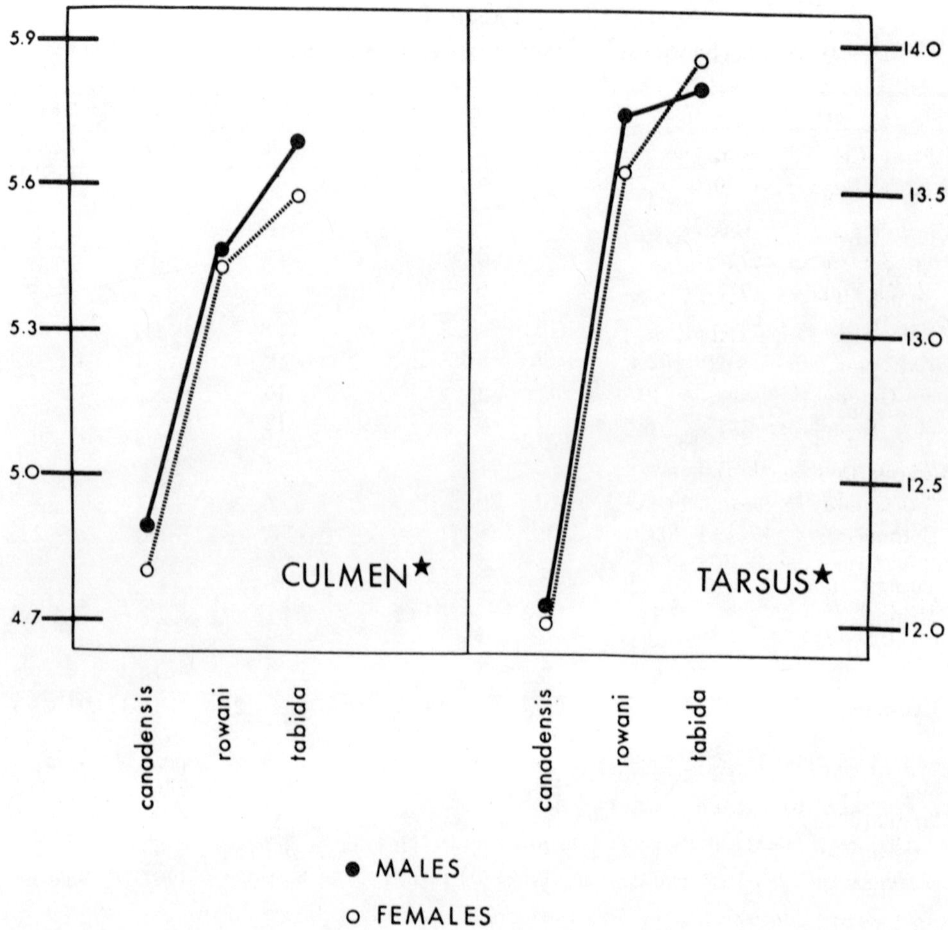

FIG. 2. Proportions of culmen post nares and tarsus measurements relative to overall body size.

Table 4 summarizes the measurements of adult migrant cranes in accordance with our racial determinations. Included are the mean, standard deviation, and extremes for wing chord, tarsus length, exposed culmen, culmen post nares, middle toe (without claw), tail length, and weight for males and females of each subspecies.

It is clear that the subspecies of *Grus canadensis* differ in body measurements; *tabida* is the largest, *canadensis* the smallest, and *rowani* intermediate. The possibility remains that the three subspecies also differ in their relative body proportions. This question was considered by investigating the ratio of linear dimensions, e.g. tarsus length, to a measure of general body size. Amadon (1943) presents a case for using the cube root of weight as a mea-

TABLE 5
GEOGRAPHIC AND SEASONAL COMPOSITION OF ADULT MIGRANT SANDHILL CRANES

	Grus c. canadensis	*Grus c. rowani*	*Grus c. tabida*
Kittson Co., Minnesota			
22–24 September 1970	0	7	13
Pierce Co., North Dakota			
14 September 1971	0	11	1
27–28 October 1971	0	8	0
Kidder Co., North Dakota			
8–15 September 1970	0	24	1
26 October–1 November 1970	2	16	3
3–4 September 1971	0	12	3
McLean Co., North Dakota			
28 October–3 November 1970	26	1	0
7 September 1971	0	5	0
27–28 September 1971	8	0	0
Phillips Co., Montana			
4 October 1970	4	0	0
3 October 1971	8	0	0

sure of overall body size. (The value of the cube root is interpreted as being comparable to a linear dimension.)

Although weights were not available for breeding specimens (nor, indeed, was any measure of general body size), all cranes collected for the present study were weighed. So by assuming that our classification of subspecies is correct, we may proceed by examining each specimen for the quantities:

$$\text{culmen}^* = \text{culmen post nares}/(\text{weight})^{1/3}$$
and
$$\text{tarsus}^* = \text{tarsus length}/(\text{weight})^{1/3}$$

Culmen* and tarsus* represent measurements relative to body size. Averages of these quantities can then be compared among subspecies. Of the remaining measurements, we considered only the corresponding function of wing chord, which did not prove useful.

The averages of culmen* and tarsus* for male and for female specimens of each race are graphed in Figure 2. One t-test was employed to test the difference between *canadensis* and *rowani*, and another for the difference between *rowani* and *tabida*. All differences were significant ($P = 0.05$) except for the difference in tarsus* between male *rowani* and *tabida*.

It would be preferable if measurements from breeding Sandhill Cranes of known origin could be used for this analysis; however, until such data become available, we may tentatively conclude that the subspecies differ not only in overall body size, but also in relative proportions. It should be noted that the *rowani* means lie closer to *tabida* than to *canadensis*, a situation analogous to one encountered earlier when testing the separability of individuals of the intermediate race. Also, the differences between *rowani* and *tabida* are more acute in culmen measurements than in tarsus length, a feature noted by Aldrich (pers. comm.).

RACIAL COMPOSITION OF MIGRANT POPULATIONS

The samples of adult migrant specimens (Table 5) exhibit considerable geographical variation in racial composition. The data from these samples indicate that three fairly distinct migrant populations are represented: a population in northwestern Minnesota that is dominated by the Greater Sandhill Crane; a population in Pierce and Kidder Counties, North Dakota, that is dominated by the intermediate Canadian Sandhill Crane; and a population in McLean County, North Dakota, and Phillips County, Montana, that is dominated by the Lesser Sandhill Crane. It is noteworthy that gross differences in racial composition exist between the migrant populations of Kidder County and those in McLean County despite a distance of less than 70 miles separating the two sites. The similarity in racial composition between the migrant populations of Kidder County and those of Pierce County, which lies about 70 miles north, suggests that these cranes may have originated from the same geographical breeding grounds. Variations in racial composition of specimens from the North Dakota collections are illustrated in Figure 3.

The Greater Sandhill Crane was found to comprise 65 percent of the late September sample from northwestern Minnesota. This relatively high proportion is of special interest since the total numbers of cranes that congregate in this area probably exceed 3000 birds. In addition, this subspecies comprised small proportions (5 to 11 percent) of the total sample specimens from the concentration areas in Kidder and Pierce Counties, North Dakota, which generally support peak fall populations of 10,000 or more birds. Apparently the present eastern population of the Greater Sandhill Crane is much larger than has been generally realized.

The Canadian Sandhill Crane comprised about 95 percent of all the sample specimens collected from Pierce County and about 85 percent of the sample specimens from Kidder County. This subspecies also is an important component of the northwestern Minnesota population since it comprised 35 percent of the sample from there.

In McLean County, North Dakota, a small population of about 250 Sandhill Cranes was present in early September 1971. The five adults collected from

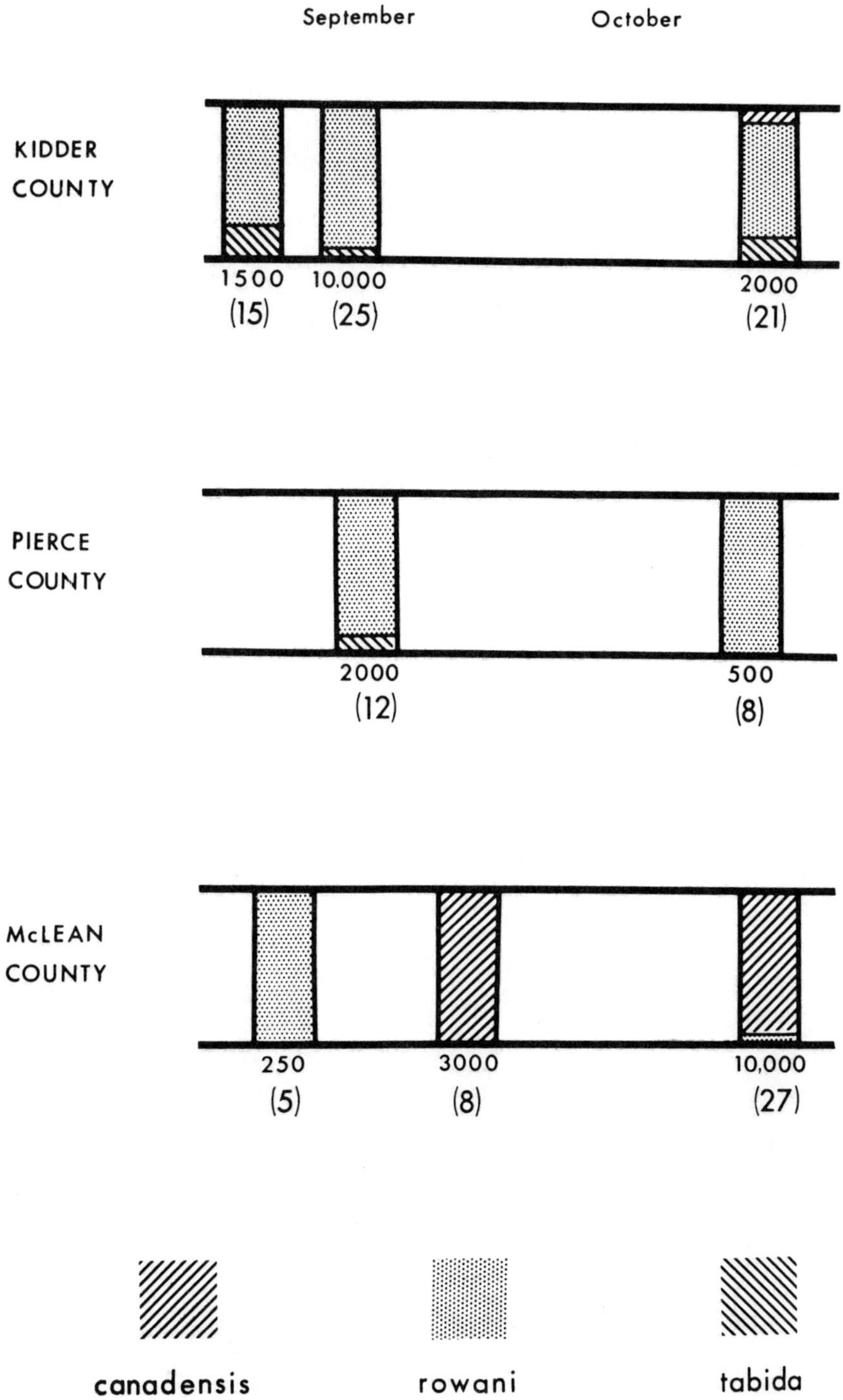

FIG. 3. Racial composition of migrant Sandhill Cranes from North Dakota, indicating seasonal variation in samples from various locations. Also shown are estimates of the population present during the collection period, and, in parentheses, the sizes of the samples.

TABLE 6
MEASUREMENTS OF IMMATURE MIGRANT SANDHILL CRANES[1]

	McLean Co., N. D. Phillips Co., Mont.		Kidder & Pierce Cos., N. D.		Kittson Co., Minn.	
	Male	Female	Male	Female	Male	Female
Sample Size	5	5	4	6	3	4
Wing Chord						
Mean	455.4	447.0	493.5	463.8	505.3	469.0
Standard deviation	9.0	16.2	9.5	9.7	8.1	26.7
Extremes	446–465	421–469	481–503	447–474	496–510	441–499
Tarsus Length						
Mean	196.8	189.2	229.5	202.5	240.7	224.5
Standard deviation	13.1	15.7	10.7	15.6	8.1	8.7
Extremes	178–212	175–215	215–240	180–226	232–248	216–232
Exposed Culmen						
Mean	85.7	85.0	109.4	98.1	115.3	103.2
Standard deviation	8.4	6.8	8.9	7.0	5.1	9.3
Extremes	73.7–95.9	77.2–95.4	96.5–116.3	86.8–105.4	111.7–121.1	89.7–109.4
Culmen post nares						
Mean	66.6	67.2	84.8	76.9	90	81.9
Standard deviation	6.2	5.0	7.5	7.3	2	8.2
Extremes	57–73	60.8–72	74–90	67.2–83	88–92	70–87.5
Middle Toe						
Mean	75.7	73.0	84.4	81.0	86.6	83.0
Standard deviation	5.6	4.1	4.4	2.9	2.8	2.3
Extremes	66.1–80.0	70.5–80.2	79.2–90.0	75.6–83.7	84.9–89.8	80.8–86.2
Tail Length						
Mean	157.0	156.0	178.2	162.0	184	170.8
Standard deviation	5.9	4.1	9.3	7.2	1	8.9
Extremes	152–167	150–161	166–188	153–171	183–185	160–179
Weight						
Mean	3475.8[2]	3157.4	4389	3715.2[3]	4310.7	3794.0
Standard deviation	428.7	287.7	738.0	505.5	295.1	535.5
Extremes	3027–4024	2880–3512	3415–5082	3270–4360	4058–4635	3022–4258

[1] Linear measurements in millimeters; weight in grams.
[2] Sample size one less than indicated.
[3] Sample size two less than indicatel.

this group were identified as the intermediate subspecies, *rowani.* Later in the season, from late September to early November, many thousands of Sandhill Cranes congregate in this area. Series of specimens that totaled 35 adults were collected from these late aggregations in 1970 and 1971. Since only one of these (3 percent) was identified as *rowani,* we hypothesize that the small numbers of this subspecies which arrive early in this staging area are subsequently greatly outnumbered when the large flocks of *canadensis* arrive.

The arrival of flocks of Lesser Sandhill Cranes on the migration staging areas is noticeably later than the arrival of the Greater and Canadian Sandhill Cranes. Ordinarily the population build-up begins in late September as compared to late August for the other two subspecies. During the period of peak numbers, the Lesser Sandhill Crane comprised about 97 percent of the sample specimens from McLean County, North Dakota, and 100 percent of the sample specimens from Phillips County, Montana. In Kidder County this subspecies represented a very small proportion (3 percent of the sample) of the population.

Buller (1967) summarized information on the racial composition of Sandhill Crane specimens that had been collected in a non-random fashion in McLean County and in the Kidder County area during the fall migration periods of previous years. A sample of five specimens collected 23–24 September 1964, in McLean County included 4 *canadensis* and 1 *rowani.* Another collection of five made one month later in the same location had an identical racial composition. The Kidder County sample of 33 specimens taken 22 September–2 November 1965, contained 29 *rowani* and four *canadensis.* Data from small samples (13 taken 27 October–3 November 1959; 12 taken 9–12 September 1960; seven taken 16–19 October 1963; four taken 8–12 September 1964; four taken 19–23 October 1964) in the Kidder County area showed that *canadensis* predominated in some of the collections made after mid-October. Additional measurements from a series of nine specimens of adult Sandhill Cranes from southeastern Wyoming (Goshen County) are also of interest since this location lies approximately 475 miles south-southeast of the staging area in Phillips County, Montana. These birds were collected by the Wyoming Game and Fish Department during 28–30 October 1971. As might be expected, all of these specimens proved to be Lesser Sandhill Cranes.

MEASUREMENT OF IMMATURE MIGRANT SPECIMENS

The collections of immature Sandhill Cranes, although largely incidental, provided some insight into the comparative size of these cranes as related to area of collection. Reliable racial determinations cannot be made for immatures because series of measurements of immature cranes of known subspecies are lacking. For this reason we grouped immatures by geographic locations within which the racial composition of adults was similar. Table 6 gives the mean, standard deviation, and extremes of measurements for males and females from (1) Kittson County, Minnesota; (2) Kidder and Pierce Counties, North Dakota; and (3) McLean County, North Dakota and Phillips County, Montana. In general, the mean linear measurements and weights indicate that immature birds from Kittson County, Minnesota, were comparatively large; those from McLean County, North Dakota, and Phillips County, Montana, were comparatively small; whereas birds from Pierce and Kidder Counties, North Dakota, were intermediate.

NEED FOR FURTHER RESEARCH

Before appropriate management procedures can be developed for migrant populations of Sandhill Cranes, additional research concerning other important aspects of their breeding biology and distribution must be undertaken. A careful delineation of the breeding range of each of the respective subspecies should be of high priority. Information thus obtained would help to clarify the separability of the various populations. The numerical status and population dynamics of each identifiable group also need to be determined, particularly if hunting becomes an appreciable mortality factor. More detailed information concerning migration patterns and locations of wintering grounds (such as obtained by Williams and Phillips, 1972) is also required.

SUMMARY

A statistical analysis employing measurements of breeding specimens representing the three migrant subspecies of Sandhill Crane (*Grus canadensis canadensis*, *Grus c. rowani*, and *Grus c. tabida*) provided supportive evidence that the intermediate population, *rowani*, is separable on the basis of wing chord, tarsus length, and exposed culmen.

A series of 180 migrant Sandhill Cranes was collected in Minnesota, North Dakota and Montana during the fall migration periods in 1970 and 1971, and racial determinations were made according to similarity with breeding specimens of known origin. On the basis of this classification, it was demonstrated that the subspecies differ in relative proportions to body size of tarsus length and culmen post nares as well as differing in overall body size. The racial composition of migrating cranes collected in different areas varied; *tabida* was predominant in Kittson County, Minnesota; *rowani* in Kidder and Pierce Counties, North Dakota; and *canadensis* in McLean County, North Dakota and Phillips County, Montana. Measurements of immature cranes were found to vary in the same general way as those of adults.

ACKNOWLEDGMENTS

These investigations could not have been undertaken without the cooperation of the North Dakota Game and Fish Department, Minnesota Department of Natural Resources, and the Montana Fish and Game Department. During the collection of specimens of

Sandhill Cranes in Minnesota and North Dakota, assistance in the field was furnished by several biologists, particularly by Jerome H. Stoudt in Minnesota and North Dakota, and by Harold A. Kantrud in North Dakota. Montana specimens were collected by Dale Witt. The examination of specimens in the laboratory was conducted with the help of Harold A. Kantrud, Dr. Charles W. Dane, and Dr. Robert E. Stewart, Jr. Dr. L. H. Walkinshaw generously provided his extensive set of measurements of Sandhill Cranes. We are especially indebted to Dr. John W. Aldrich for his help in providing data and advice concerning the determination of subspecies. In addition, George Wrakestraw made available measurements of Sandhill Cranes collected in Wyoming by the Wyoming Game and Fish Department.

LITERATURE CITED

ALDRICH, J. W. 1946. Speciation in the white-cheeked geese. Wilson Bull., 58:94–103.

AMADON, D. 1943. Bird weights as an aid in taxonomy. Wilson Bull., 55:164–177.

ANONYMOUS. 1968. Rare and endangered fish and wildlife of the United States. U.S. Bureau of Sport Fisheries and Wildlife. Resource Publ. 34. Washington, D.C.

BOEKER, E. L., J. W. ALDRICH, AND W. S. HUEY. 1961. Study of experimental Sandhill Crane hunting season in New Mexico during January 1961. U.S. Bureau of Sport Fisheries and Wildlife. Special Scientific Report—Wildlife No. 63. Washington, D.C.

BULLER, R. J. 1967. Sandhill Crane study in the central flyway. U.S. Bureau of Sport Fisheries and Wildlife. Special Scientific Report—Wildlife No. 113. Washington, D.C.

HANSON, H. C. 1965. The Giant Canada Goose. Southern Illinois Univ. Press. Carbondale.

LUMSDEN, H. G. 1971. The status of the Sandhill Crane in northern Ontario. Canadian Field-Naturalist, 85:285–293.

MADSEN, C. R. 1967. Food and habitat selection by fall migrant Sandhill Cranes in Kidder County, North Dakota. Unpubl. M. S. thesis. Michigan State Univ., East Lansing.

RAO, C. R. 1962. Use of discriminant and allied functions in multivariate analysis. Sankhya, A24:149–154.

RAO, C. R. 1965. Linear statistical inference and its applications. John Wiley & Sons, Inc. New York.

STEPHEN, W. J. D. 1967. Bionomics of the Sandhill Crane. Canadian Wildlife Service Report Series—No. 2. Ottawa.

STEWART, R. E., AND H. A. KANTRUD. 1971. Classification of natural ponds and lakes in the glaciated prairie region. U.S. Bureau of Sport Fisheries and Wildlife. Resource Publication 92. Washington, D.C.

WALKINSHAW, L. H. 1949. The Sandhill Cranes. Cranbrook Institute of Science Bulletin No. 29. Bloomfield Hills, Michigan.

WALKINSHAW, L. H. 1965. A new Sandhill Crane from central Canada. Canadian Field-Naturalist, 79:181–184.

WILLIAMS, L. E., JR., AND R. W. PHILLIPS. 1972. North Florida Sandhill Crane populations. Auk, 89:541–548.

U.S. BUREAU OF SPORT FISHERIES AND WILDLIFE, NORTHERN PRAIRIE WILDLIFE RESEARCH CENTER, JAMESTOWN, NORTH DAKOTA, 58401, 24 JULY 1972.

LEOPARD FROG SUPPLY

Joseph T. Bagnara and John S. Frost

The northern leopard frog *Rana pipiens* has been for many years the principal amphibian used for biomedical research and teaching. Unfortunately, such frogs of northern origin are no longer available in significant quantities and biological supply houses have turned to other regions, notably northwestern Mexico, as a source of frogs. The biology of these Mexican frogs is clearly different from that of the northern *R. pipiens*, a matter that was extensively discussed at a symposium on the laboratory frog (*1*), which was attended by major commercial frog suppliers. Recently it has been shown that, in the region of northern Sinaloa, Mexico, where most of these leopard frogs are collected, no *R. pipiens* are to be found. Instead, there exist two sympatric species of leopard frogs, *R. berlandieri forreri* and *R. magnaocularis* (*2*). Commercial collections made in this area from October through March contain about 80 percent *R. berlandieri forreri* and 20 percent *R. magnaocularis*. During other times of the year even fewer *R. magnaocularis* are found in these collections. Northeastern Mexico has now become an additional source of leopard frogs, and collections from this region are composed of *R. berlandieri berlandieri*.

For reasons unclear to us, commercial houses are now supplying Mexican leopard frogs designated as "*Rana pipiens berlandieri*." This is an unsatisfactory circumstance, not only because of the erroneous nomenclature, but because articles are beginning to appear about research, the meaning of which will be questionable, since the actual species on which the work was done is not known. Of more serious consequence is the fact that some investigators may think that they are working on *R. pipiens* when actually they are not. If there were no significant physiological differences among the various species of leopard frogs, the error in nomenclature would present no major problem; however, this is not the case. It has been pointed out in one demonstration of such physiological differences that melanophores of frogs of southern origin (*R. berlandieri forreri*) contain a preponderance of β-adrenergic receptors, while those from northern frogs (*R. pipiens*) contain mostly α-adrenergic receptors (*3*). Other physiological differences among leopard frog species include variations in reproductive cycles, in temperature tolerance, and in enzyme composition. The last is well manifested in the heterogeneity of isozyme patterns among leopard frog species (*4*).

We wish to emphasize that the various leopard frog species possess different biological characteristics that must be considered when they are used for experimentation.

JOSEPH T. BAGNARA
JOHN S. FROST
Department of Cellular and Developmental Biology, University of Arizona, Tucson 85721

References

1. A symposium, "The Laboratory Frog: Aquisition, Nurture, and Health," was held at the annual meeting of the American Society of Zoologists in Minneapolis, Minn., in August 1972. The papers presented were published in *Am. Zool.* **13**, 79 (1973).
2. J. S. Frost and J. T. Bagnara, *Copeia* (No. 2), 332 (1976).
3. M. E. Hadley and J. E. Goldman, *Am. J. Physiol.* **219**, 72 (1970.
4. J. E. Platz, *Copeia* (No. 4), 660 (1976).

SECTION VII.
Introduced Species

Many exotic wildlife species have been introduced into American ecosystems. Most have failed to become established or to attain a wide distribution, but some have flourished and spread with either positive or negative effect. In the 1950s and 1960s, substantial effort was directed toward the introduction of exotic game species into the United States. Some of this was trial and error, and some was based on the selection of stocks judged most likely to succeed. The rationale for the latter approach was based to a great extent on John Aldrich's early paper reprinted here. Aldrich pointed out that physiological and ecological characters of birds parallel the morphological characters on which taxonomy is based, and that museum studies can be helpful in selecting stock from appropriate climatic regimes and in predicting the outcome of transplant experiments. Much of the work of the Foreign Game Importation Program of the U.S. Fish and Wildlife Service was based on the principles set forth in this paper.

If a population of unknown origin becomes established, it is important that it be identified at the lowest possible level. This will provide an indication of at least the general area of origin and of the climatic, habitat, and other requirements of the population. Such information is particularly important if an effort to control the introduced species is contemplated. The note by Banks and Laybourne reports the identification of an accidentally introduced and established bird species; the short paper by Springer and Finucane reports the proper identification of an introduced fish. Detailed knowledge of the characters of an introduced form, based on a studiable population sample, is also necessary to evaluate any evolutionary change that may take place in the species' new home.

The transplanting of species from one ecosystem to another on the same continent is precisely the same phenomenon as introducing a species from another geographic realm. R. R. Miller's long paper cited below examines one way in which transplants may be accomplished, and examines some of the potential consequences.

George Watson's papers cited below are not management oriented. The implications are obvious, however, and clarification of the taxonomy of the Chukar group led to a change in the scientific name used for the birds widely introduced into North America. One must be cautious, when reviewing the literature, to be certain that papers on the biology of introduced (or other) forms actually relate to the taxon being studied.

Additional Readings

Aldrich, John W. 1964. The United States races of the Bob-white. *Auk* 63: 493-508.

Johnston, Richard F., and Robert K. Selander. 1971. Evolution in the House Sparrow. II. Adaptive differentiation in North American populations. *Evolution* 25: 1-28.

Lachner, Ernest A., C. Richard Robins, and Walter R. Courtenay, Jr. 1970. Exotic fishes and other aquatic organisms introduced into North America. *Smithsonian Contributions to Zoology,* no. 59.
Miller, Robert Rush. 1952. Bait fishes of the Lower Colorado River from Lake Mead, Nevada, to Yuma, Arizona. *Calif. Fish and Game* 38: 7-42.
Watson, George E. 1962. Sympatry in Palearctic *Alectoris* partridges. *Evolution* 16: 11-19.
———. 1962. Three sibling species of *Alectoris* partridge. *Ibis* 104: 353-367.

Reprinted from *Journal of Wildlife Management*

SIGNIFICANCE OF RACIAL VARIATION IN BIRDS TO WILDLIFE MANAGEMENT

John W. Aldrich
U. S. Fish and Wildlife Service, Washington, D. C.

Species

Even the most casual observer is aware of the fact that there are different kinds of birds about him. Some he recognizes by name. He may be able to distinguish the robin, blue jay, cardinal, English sparrow, or crow, as quite different from one another in appearance and habits, and in the sorts of places where they live and raise their young. The average observer would recognize them as species, distinguishable by their appearance, habits, voice, and habitats. What is more important, each bird recognizes its own kind. As a result one species does not ordinarily mate with the opposite sex of a different species; if it does, fertile offspring ordinarily do not result. This inability or disinclination of two kinds of birds to interbreed is the criterion usually accepted as distinguishing separate species.

Just as with human beings, probably no two individual birds of the same species look exactly alike. This is individual variation. Some individual variation among birds is great, as in the markedly different color phases of screech owls. One individual may be bright reddish brown and another reared in the same nest may be gray without a trace of reddishness. Red screech owls, however, recognize gray individuals as belonging to their own species and the two color phases interbreed freely. Many species of birds have rather distinct color phases, such as the red and gray phases of the ruffed grouse among game birds. In others the two sexes differ from one another more than many distinct species that live side by side without interbreeding. The brightly colored male of the ring-necked pheasant is quite unlike his brown and buff mate, yet the male and female pheasants have no difficulty in recognizing one another as of the same species. These differences, whether correlated with sex or distinct color phases independent of sex, are merely variations within the same species.

Races

Besides certain marked differences in sex and color phases, every bird is slightly different from every other because it derives from a different combination of hereditary characteristics. If a pair of birds with relatively similar hereditary features are kept apart from all others of their species they tend to produce offspring with relatively similar characters, and their offspring also tend to produce young with characters similar to the original pair. Poultry and livestock breeders are familiar with this principle, the development of "breeds" or "varieties." By selecting for breeding only individuals with certain desired characteristics they can develop strains of livestock in which most of the individuals have those characteristics.

In exactly the same way "nature" develops strains of birds particularly suited to the environments in which they live. Nature is an exacting livestock breeder. There is a fine balance between ability and inability to survive the many dangers of a natural environment. Those individuals particularly well adapted have a greater chance than others to survive long enough to breed and pass on their hereditary characters to offspring. Thus geographic regions with a uniform set of environmental features develop relatively pure strains of animals preeminently suited to live in such environments. Furthermore, Twomey (1936: 122–132) has shown that certain migratory birds winter in areas with climatic conditions similar to those in which they breed.

If a bird species becomes so numerous in a region characterized by certain environmental conditions that the area will not support as many individuals as are produced, one of two things must happen to the excess population—it will be exterminated, or it will move to another region which has a different set of environmental conditions. Some obviously will attempt the latter.

Experience with transplanted species leads to the belief that in most cases of natural population spread the pioneers will not be sufficiently well adapted to the new environment to survive; here, however, the inherent nature of all species to vary enters importantly. After many unsuccessful invasions a few individuals may chance to have a combination of hereditary characteristics that will enable them to become established in the new environment and give rise to a new population. In successive generations the offspring better suited to the environment would be selected by nature for survival and perpetuate the species by passing on favorable characteristics to their progeny; individuals less well adapted to the environment would perish. The newly invaded area thus would eventually be populated by a variety differing slightly but significantly from that in the original region whence the pioneers sprang. The differences might be entirely in physiology or in habits, suited to the new surroundings. Brooks (1940) and Kendeigh (1945) describe differences in nesting habitats chosen by certain species of warblers indicating difference in habits without racial variation; but such cases are conspicuous because of their rarity. Usually the variations in physiology and habits are accompanied by alterations in color, size, or bodily proportions.

If the two populations are enough different so that individuals may be readily separated we call them distinct races or subspecies. Thus, the eastern ruffed grouse inhabits the broad-leafed woodlands near the Atlantic coast, and farther north in the evergreen forests of southeastern Canada lives another race, the St. Lawrence ruffed grouse. The intermediate zone between any two regions occupied by different races usually will contain populations intermediate between the two races, as expected from the blending of hereditary characters through interbreeding. Furthermore, in each of the two distinct races, there may be occasional individuals which appear more like the other race. This is to be expected in populations having a common ancestry, just as there are "throw-backs" quite different from the mode among domestic breeds of animals.

In considering the above described

natural spread of populations and its role in race formation, we may fail to realize the great length of time usually consumed in occupation of new territory. Nature is an extraordinarily patient wildlife manager, not concerned with fiscal budgets or demands of impatient sportsmen for quick results. A human life span is but a moment as compared with the time required for various plants and animals to become distributed as they are today on this continent. Furthermore, for most species, there probably are places in the world environmentally suitable for successful occupation by them, but which they never reach because of formidable geographic barriers. Occasionally these barriers are passed, either accidentally or through human aid, and the species may spread very rapidly in the new area, as did the English sparrow, starling, and ring-necked pheasant when helped across the oceans to North America. Such invasion of new territory requires no evolutionary change in the species. It is preadapted to the environment. It is probably significant that the ring-necked pheasant, the racial stock of which came chiefly from eastern China, has flourished best in the grain belts of the Middle West and Northwest, areas presumably closely approximating the conditions of its highly agriculturalized ancestral home. In the Shanghai region, whence imported pheasants likely came, the climate is similar to that in the corn belt of the United States. Twomey (1936: 124) shows by climographs that the temperature and rainfall of the areas in which such aliens as the Hungarian partridge and the skylark have become successfully established in North America are similar to the regions in the Old World from which these species were imported. Any differences which may be evolved by the species in North America, portions of which were occupied rapidly after introduction, would be the result of geographic isolation from the Old World stock. This, however, should not be confused with the process discussed in this paper, of differentiation through adaptation. That geographic isolation alone can bring about differentiation is amply demonstrated by the spectacular variation of birds on many islands of the Pacific Ocean, where geographic barriers are much more marked than ecological differences. In continental areas geographic barriers are not as effective, and ecological segregation is more obvious.

If geographic isolation were the most important factor in producing variation, the most widely separated races in a chain of differentiated populations should be most different from each other. Such, however, is not always true; but is more likely to be true if the geographical separation is accompanied by continued divergence in environment. If, on the other hand, the more distant portions of the range tend environmentally to be more like the region at the opposite end than the intervening territory, the characters of the widely separated races often converge. This tendency is dramatically demonstrated in the similarity of the race of ruffed grouse of the St. Lawrence River basin to that of northern Idaho, and the resemblance of the ruffed grouse of the Appalachian Plateau to that of the west slopes of the Cascades in Oregon. In both cases the two regions are much more alike in climatic conditions than the intervening areas which support races of ruffed grouse entirely different from those on opposite sides of the continent.

In some instances the differences between two extremes of a progressive series showing racial variation are as great as would be expected in two distinct species. In fact we would not hesitate to call them distinct species were it not for the complete chain of interbreeding and intermediate populations which connect them. There is such a chain of populations in the song sparrows between the small pale reddish race of desert marshes in the southwestern United States and the large grayish birds of the virtually arctic Aleutian Islands. Almost as great a difference exists between the colorful deep chestnut ruffed grouse on the humid Olympic Peninsula in Washington, and the pale gray grouse occupying the timberline region of interior Alaska and northwestern Canada. Should the connecting populations be destroyed we would be justified in calling the remaining widely divergent races two distinct species.

There is reason to believe that most species and races of North American birds originated somewhat as follows. First, there was differentiation by gradual progressive steps in response to changing environment. Then came isolation and further differentiation until such great physical and physiological, and possibly psychological, changes were wrought as to discourage interbreeding even when the two populations again met through reinvasion of the intervening territory. The two varieties then remained distinct species, as apparently happened with the Canada and cackling geese. Thus different races of birds are in no way different from distinct species, save in the degree to which they have become differentiated from the original stocks (Dobzhansky, 1941: 155; Mayr, 1942: 362–378).

SIGNIFICANCE OF RACES TO MANAGEMENT

Of what practical significance are the above facts to wildlife management? No intelligent wildlife manager ignores the difference between species of birds in his management practices. Too often, however, differences between races have been ignored, usually with unfortunate results. Thus, repeated attempts have been made to introduce certain races of game birds into areas within the normal range of other races. Phillips (1928) and Bump (1941: 409–420) have summarized many of these. The usual result is that the birds are so poorly adapted to the new environment as to die within a year or so, leaving no trace of the introduction. This happened when Alberta ruffed grouse were introduced into Ohio, New York, Washington, Maryland, and Pennsylvania, and when prairie chickens were transplanted to the range of the eastern race (heath hen) in the Atlantic states. Many other similar failures could be enumerated. Phillips (1928: 3) has summarized the types of responses exhibited by species introduced into new environments which eventually fail to become established, as follows: (1) quick and complete disappearance; (2) lingering, but failing to breed; and (3) survival, but gradual decline over a period of years with final complete eradication by unfavorable weather or other factors. Leopold (1933: 87) elaborated on the types of failure mentioned by Phillips and added two types of success: (1) an initial vigorous spreading followed by partial decline, and (2) prompt and vigorous spreading with complete establishment. Phillips (1928) rightly points out that the actual causes of failure to survive are usually so obscure as to go undetected, and much more thorough investigation of the causes of these failures would be profitable. An example of the type of work which should throw much light on the limiting factors to survival is that described by Kendeigh (1934): careful studies of the effect of climatic factors on the physiology of the species, together with the effect of biotic interaction and competition.

If any individuals of a native race are already present and the introduced foreign stock survives long enough to interbreed with them, the progeny presumably will be less well adapted to survive than offspring of pure native stock which has developed through long selective breeding under those particular environmental conditions. A. S. Leopold (1944) has shown rather conclusively that wild turkeys in Missouri crossed with domestic stock produce offspring much less able to survive under natural conditions than the pure native stock. This indicates a heritable factor in the wild birds which fits them for life under conditions in nature; the domestic birds, derived from a different race, have been altered further by artificial selection. During this process they have lost the physiological and psychological ability to cope with conditions which are successfully met by native wild turkeys. Artificially propagated stock of any game species is likely to have lost heritable characters of "wildness," or others which fit the birds to life under natural conditions—this because of the lack of natural selection under the artificial conditions of captivity. It would seem evident, therefore, that artificially propagated stock is not as satisfactory as wild material for purposes of introduction.

Of the many attempts to transplant American birds of one race into the range of another race the bobwhite quail (*Colinus virginianus*) seems to be the only species which has attained any degree of success. Even in this bird the reports of success are not entirely conclusive. Stoddard (1932: 480) reports that of 2,516 banded Texas bobwhites (Mexican quail), liberated during his studies in Georgia, the percentage of returns in birds killed by hunters in succeeding years compared unfavorably with banded native stock taken in the same territory, particularly in the second and third years after liberation. Phillips (1928: 27) has indicated that bobwhites have been introduced into New England from so many different sources and over so long a time that the indigenous stock now probably is extinct. Restocking was particularly active there after the severe winter of 1904–05, which decimated bobwhites all over the northern states; and that period probably marked the end of the big northern birds best suited to the rigorous climatic conditions of New Hampshire, Vermont, and Maine. The opposite of this situation seems to have prevailed elsewhere. Great numbers of bobwhites from northeastern Mexico (*texanus*) and the southeastern states (*virginianus*) have been introduced into Pennsylvania and Maryland, yet large series of specimens from those states (in the U. S. National Museum) are closer to the ancestral stock (*marilandicus*)[1] than to the introduced types. Definite traces of both *texanus* and *virginianus* characters are noticeable however, in some specimens from the northeastern states, as Phillips has remarked. How long these foreign characters will persist in a population after introductions cease remains to be determined in most instances. It seems questionable that the characters of the native stock would often be entirely swamped out, as Phillips believed to be the case in New England. On the contrary, the selective nature of any characters developed by the native stock would seem to have an advantage tending to aid their perpetuation. This was pointed out by Stoddard (1932: 488), who noted a protective color advantage of native Georgia stock over introduced "Mexican quail." He also found it difficult to detect infusion of *texanus* blood into *virginianus* beyond the first cross. About 5,000 Mexican birds were released during a 5-year period over 200,000 acres in Georgia already well stocked with native birds; yet a few years later infusion of Mexican quail blood was noticeable only in the places where the imported birds had been released in considerable numbers. Stoddard now states (letter, 1945) he can detect little or no trace of the introduced birds among the several hundred examined annually.

It seems obvious that stocking with a foreign race is a waste of time and money, at least while any of the original inhabitants of a region still survive. Instead of introducing new birds of a strange race, the original occupants of the region should be given every encouragement to increase their population to the carrying capacity of the region which their hereditary constitution permits them to occupy; then all would be done that could be expected of a scientific management program. This seems to be the recommended practice of most modern wildlife managers.

If all of the original inhabitants of a certain bird species have been killed off in a region, and it is desired to restore the species, or if it is desired to stock a region in which the species never occurred, what procedure would be most likely to succeed? From the foregoing, introduction may be successful if stocks can be obtained in other regions with environmental conditions similar to or identical with the area where stocking is desired. The difficulty is to know in advance whether the two regions are similar enough. This might be determined by a thorough study of all environmental factors. It would be the more difficult but surer method and the only one possible for areas in which the species is not known to have occurred naturally.

[1] The racial divisions and nomenclature of the bobwhites referred to here differ from those of current standard publications, and are the result of recent unpublished studies by the writer. *C. v. mexicanus* is the race inhabiting the greater part of the eastern United States; *virginianus* is confined to the southern Atlantic coastal plain; *marilandicus* occupies the northern Atlantic seaboard; *floridanus* is restricted to peninsular Florida; *taylori* inhabits the Great Plains south to northern Texas; and *texanus* occurs in southern Texas and southward into northeastern Mexico.

There are innumerable failures in transplantation of birds without regard to similarity between the habitat of origin and place of introduction. The Chukar partridge in Missouri is a case recently studied (Nagel, 1945). Stock (reputedly of the race *Alectoris graeca chukar*) presumably from the southern Himalaya Mountains, was liberated in numbers in a variety of regions and habitats in Missouri. The birds wandered widely and failed to maintain their original numbers. It was concluded that the ecological conditions were not enough like the original home of the introduced stock, but that birds of some other race of this widely ranging and exceedingly plastic species possibly would be satisfactory.

There is one seemingly successful instance of bobwhite introduced into a region where the species formerly did not exist. According to Phillips (1928: 27) stock was introduced into Washington from Kansas and subsequently the species has done particularly well in the more arid plains country east of the Cascades, whence birds have been transplanted into Idaho. Recent specimens from Washington, Oregon, and Idaho seem to be predominantly of the pale *taylori* type, which evolved in the Great Plains of Nebraska, Kansas, Oklahoma, and eastern Colorado, and is the stock which would be expected to be most suitable for the similarly arid situation of the northwestern states. In this case it would seem that intelligent, or at least fortunate, choice was made.

If a species formerly occurred in a given region and specimens are available for identification, the most practical method would be to select stock of the same race as the original inhabitants of the area to be stocked. The difficulty here is that birds of two regions may appear similar enough to be considered the same race, yet may have small physiological differences (not reflected in outward appearance) in response to slightly different environmental conditions which would render them unsuited to survival in another region. In such a case, trial and error is the only method possible. The chances of success would seem greater than if stock were selected from an obviously different race which almost certainly would have physiological characteristics unsuited to the environment of another race.

This is where the museum taxonomist may be helpful in wildlife management. Specially trained ornithologists with access to the necessary reference collections can distinguish racial differences in birds, and by plotting their distribution can correlate these races with the distribution of ecological communities studied and plotted in the field. Museum specialists know that many named races of birds, and some populations not sufficiently distinct to be named, are distributed in rather close agreement with regions characterized by distinctive climatic conditions, of temperature, rainfall, and evaporation. Major climatic regions are usually recognized at a glance by the conspicuous forms of vegetation, but thorough analysis of all forms of life is necessary to distinguish their finer subdivisions. These climatic regions with their ecological subdivisions have long been known as life zones. The major characteristics and composition of life zones and their component ecological units have been worked out in a general way. More thorough and detailed description of the major biogeographical units, and particularly of their component biomes, on the basis of exhaustive ecological studies correlated with the racial variation of the animal life which inhabits them is needed. In fact such studies are essential to sound wildlife management.

LITERATURE CITED

BROOKS, M. 1940. The breeding warblers of the central Allegheny Mountain region. Wilson Bul., **52**: 249–266.

BUMP, G. 1941. The introduction and transplantation of game birds and mammals into the state of New York. N. Amer. Wildl. Conf. Trans., **5**: 409–420.

DOBZHANSKY, T. G. 1941. Genetics and the origin of species. 2d ed. New York, Columbia Univ. Press. xviii+446 pp.

KENDEIGH, S. C. 1934. The role of environment in the life of birds. Ecol. Monog., **4**: 299–417.

1945. Community selection by birds on the Helderberg Plateau of New York. Auk, **62**: 418–436.

LEOPOLD, ALDO. 1933. Game management. New York, Charles Scribner's Sons, xxi +481 pp.

LEOPOLD, A. S. 1944. The nature of heritable wildness in turkeys. Condor, **46**: 133–197.

MAYR, E. 1942. Systematics and the origin of species. New York, Columbia Univ. Press. xiv+334 pp.

NAGEL, W. O. 1945. Adaptability of the Chukar partridge to Missouri conditions. Journ. Wildl. Mgt., **9**: 207–216.

PHILLIPS, J. C. 1928. Wild birds introduced or transplanted in North America. U. S. Dept. Agric. Tech. Bul., **61**.

STODDARD, H. L. 1932. The bobwhite quail, its habits, preservation, and increase. New York, Charles Scribner's Sons. xxix+559 pp.

TWOMEY, A. C. 1936. Climographic studies of certain introduced and migratory birds. Ecology, **17**: 122–132.

Reprinted from *The Auk*

THE RED-WHISKERED BULBUL IN FLORIDA

Richard C. Banks and Roxie C. Laybourne

The establishment of a feral population of the Red-whiskered Bulbul (*Pycnonotus jocosus*) in the environs of Miami, Dade County, Florida, was reported by L. A. Stimson (*Florida Naturalist*, 35: 93, 1962). Bulbuls were present in Kendall, just south of Miami, in the spring of 1961, and apparently raised young there that summer. In the fall of that year, bulbuls were also seen in the vicinity of Princeton, approximately 35 miles southwest of Miami. Confirmation of nesting at Kendall was presented in 1962 (*Aud. Field Notes*, 16: 402, 1962). By 1964 the population of bulbuls in Kendall was estimated at 40 to 50 (*Aud. Field Notes*, 18: 349, 1964).

The 1962 report included the comment that the species "probably escaped from or was liberated by a nearby bird dealer." It was later established that the stock from which this population originated was shipped to Florida from Calcutta, India (E. J. Fisk, *Florida Naturalist*, 39: 10–11, 1966), which suggests, but does not prove, the origin of the birds.

Two juvenals were obtained as specimens in April and July, 1963, by Mary Lynn Ward and Dennis R. Paulson; these specimens are at the University of Miami. An adult male specimen taken in Kendall in June, 1963, by H. M. Stevenson, is in the Florida State University collection. Another juvenal was obtained in May, 1965, by John C. Ogden, and an adult female by Mrs. Bradley Fisk in April, 1966; both these specimens are in the U. S. National Museum.

We have examined all five of the above-mentioned specimens. The comparative material of this Asiatic species in the National Museum and in the American Museum of Natural History was inadequate for subspecific determination. Additional comparative material borrowed from the British Museum (Natural History) through the courtesy of Ian Galbraith made a positive identification possible.

The Florida birds are referable to *P. j. emeria*, a race native to the lowlands of eastern India from the State of Madras at least to Calcutta, West Bengal (H. G. Deignan, *Proc. Washington Acad. Sci.*, 38: 279–281, 1948). We are using the name *emeria* in the strict sense of Deignan (*loc. cit.*) rather than in the broader concept which includes *P. j. peguensis* of lower Burma (Deignan *in* J. L. Peters, *Birds of the world*, vol. 9, Cambridge, Massachusetts, Museum of Comparative Zoology, 1960, p. 232).

Although we are not familiar enough with the pattern of variation in *Pycnonotus jocosus* to comment definitively on the treatment of the forms, examination of the limited material at our disposal leads us to think that the two populations mentioned are separable, even though not strongly differentiated. The main difference between the three adult specimens from Burma (*P. j. peguensis*) in the USNM and the material from coastal India (*P. j. emeria*) is the darker brown on the back of *emeria* in birds of equally worn, but not seasonally similar, plumage.

Two juvenal birds from Florida taken in 1963 are dark dorsally like adults of *emeria*. The third juvenal from Florida, killed by a cat in May, 1965, has lost a large number of feathers; although this bird is paler than the others, we do not consider it comparable to them. A juvenal bird in the USNM from Burma is much paler than the adults of *emeria*, and only slightly darker than adults of *peguensis*. It is paler than the juvenals from Florida, and has a brown rather than a black cap.—RICHARD C. BANKS and ROXIE C. LAYBOURNE, *Bureau of Sport Fisheries and Wildlife, U. S. Fish and Wildlife Service, U. S. National Museum, Washington, D. C.*

Reprinted from *Transactions of the American Fisheries Society*

THE AFRICAN CICHLID, *TILAPIA HEUDELOTI* DUMERIL, IN THE COMMERCIAL FISH CATCH OF FLORIDA[1]

Victor G. Springer and John H. Finucane

Tilapia heudeloti, a native of west Africa from Senegal to Gabon, has been imported into the United States by tropical fish dealers and is well known to aquarists (Axelrod and Schultz, 1955). It was reported by these authors as *Tilapia macrocephala* (Bleeker), which is a synonym of *T. heudeloti*. Evidence has been accumulating over the past few years that this species has become well established in the Hillsborough Bay area on the west coast of Florida.

The first record of *Tilapia heudeloti* in Hillsborough Bay is based on a 204-millimeter specimen (standard length) caught in a gill net during late summer, 1959, between Mangrove Point and the Alafia River. This fish was one of two taken by mullet fishermen. The specimen is in the collection of the Florida State Board of Conservation Marine Laboratory at St. Petersburg. All subsequent records of which we have knowledge have come from points in this area (mainly Bullfrog Creek) and associated tributaries.

According to the owner of a seafood market in Tampa, tilapia were first noticed in gill-net catches during 1958. He noted that they became common in catches during early 1959 when four or five, or occasionally 10 to 15, specimens were obtained in a day's fishing. At this time he began to market the species. Catches usually occurred during the winter, and at times there were periods of 2 or 3 months when no specimens were caught.

A specimen which appeared to be a spent female was obtained from the commercial fish catch on March 27, 1960. On April 8, 1960, an attempt was made to collect tilapia along the shore and in tidal inlets in the area from which the commercial fishermen had made their catch. A 100-foot minnow seine and rotenone were used, but no specimens were obtained. On April 28, 1960, 4 of 10 specimens caught by commercial fishermen were examined. Two were females with mature or nearly mature eggs, and the other two were males. The oral cavities of both males were filled with eggs. According to Axelrod and Schultz (1955), *T. heudeloti* is known to be a mouth breeder.

Between May 1960 and October 1962, no special efforts were made to obtain information on tilapia. However, a monthly seine station was established for another study during December 1961 in a tributary of Bullfrog Creek about ½ mile above its entrance into Hillsborough Bay. The water depth at this station varies from 2 to 3 feet depending on the tide. The bottom is composed of fine mud and sand. The shore, except for one small clearing, is surrounded by mangrove thickets.

In the seine collection of October 1962 (water temperature 23.7° C.; salinity 8.22 p.p.t.; pH 7.63) from the Bullfrog Creek tributary were two small specimens of *Tilapia heudeloti* measuring 50 and 53 millimeters, standard length. The following month (water temperature 14.8° C.; salinity 1.58 p.p.t.; pH 7.80) one specimen measuring 64 millimeters was collected and another seen. The

[1] This report was initiated while the senior author was in the employ of the Florida State Board of Conservation Marine Laboratory, and it constitutes contribution number 71 of that institution.

three specimens are deposited in the United States National Museum.

In December 1962 the seafood dealer was again interviewed, and we were told that he had noticed a definite increase in catches of tilapia during the past two years. He estimated that in 1962 he had marketed between 3,500 and 4,500 pounds (presumably all *T. heudeloti*) at 25 cents a pound under the name African sunfish. Public acceptance appeared to be good, and the flavor is described as being similar to that of freshwater sunfish (*Lepomis* sp.). Apparently there is only one major supplier of tilapia in the area at the present time.

Gulf Hatcheries, a tropical fish supply house at Palmetto, Florida, reported that in the past two years (1961–62) tilapia have become firmly established at Ruskin, situated on a tributary of the Little Manatee River just south of Mangrove Point.

During December 1962, Florida was subjected to a record cold wave. The lowest water temperature for which we have a record was from Old Tampa Bay; 9.6° C. on December 13. The seafood dealer reported receiving 1,553 pounds of tilapia from nearby Bullfrog Creek between December 13–17. According to him the fish were sluggish but alive when caught. Three decomposing tilapia and numerous living adults were seen by the junior author in the Bullfrog Creek area on December 17, but there was no evidence of a massive kill. Apparently *T. heudeloti* can survive the low temperatures to which the State of Florida is occasionally subjected.

It is probable that one or more of the numerous tropical fish hatcheries in the area around Hillsborough Bay were the original source of *Tilapia heudeloti.* Some of these hatcheries maintain large outdoor ditches in which tropical fishes are reared. It is possible that individuals escaped or were released into tributaries of Hillsborough Bay. However, the possibility that the species was released by some aquarium hobbyist cannot be eliminated. The senior author recalls mention of one tropical fish dealer in the area going out of business and releasing his entire stock in local streams.

On the basis of the above information, it is evident that *Tilapia heudeloti* is now a member of the Florida fish fauna. It is not possible at the present time to predict the effect, if any, which the presence of this species will have on the indigenous fauna. Examination of stomach contents of adult specimens produced only unidentifiable vegetable material. On this basis no competition for food with the local sport or commercial fishes would be expected, as none are known to feed primarily on plants.

ACKNOWLEDGMENTS

We wish to thank Mr. Clarence Eatman for providing us with information on tilapia and Conservation Agent Rudy Guerra who first brought specimens to our attention.

LITERATURE CITED

Axelrod, H. R., and L. P. Schultz. 1955. Handbook of tropical aquarium fishes. McGraw-Hill Book Co., Inc., New York. 718 pp.

Victor G. Springer
John H. Finucane

Division of Fishes
U. S. National Museum
Washington 25, D. C.
and
U. S. Bureau of Commercial Fisheries
Biological Station
St. Petersburg Beach, Florida

SECTION VIII.
Environmental Assessment

Time and place are among the more important data associated with specimens in museum collections. Large collections, or combinations of collections, can therefore provide information on the change of fauna through time at a specific locality that has been sampled repetitively. Knowledge of the requirements of species can be used to infer, sometimes with considerable precision, the reasons for the change, in environmental terms. Extension of this logic suggests that if we intend at some future time to analyze the environmental effects of a modern action, we would do well to amass modern faunal (and floral) samples at the present time. The paper by Gray and Dauble serves as an example of this sort of activity. Welton Lee's approach involves not only assembling modern material, but using it as a central information source to further environmental studies.

Specimens originally taken for reasons far removed from environmental assessment, and housed in collections for many years, may very profitably be used to solve environmental problems. Ann Rick's paper, listed as an additional reading, cites many examples and discusses kinds of specimens available for use. Klaas et al. used zoological collections in many museums to set standards of eggshell thickness against which measurements of recent shells were compared in an evaluation of pesticide effects. Similarly, Kelly et al. used old and new fish specimens to evaluate changes in mercury residues.

Studies of this sort are vital in interpreting data on environmental contamination and in determining whether a wildlife resource may safely be used. However, the National Museum report cited below suggests that museum specimens cannot be used in this manner in all instances.

Almost any faunal survey that includes information on numbers of animals could be listed in the additional readings because of its potential utility in environmental assessment.

Additional Readings

Anderson, Daniel W., and Joseph J. Hickey. 1972. Eggshell changes in certain North American birds. *Proc. XV Internat. Ornith. Congr.*, pp. 514-540.

Gibbs, Robert H., Jr., E. Jarosewich, and Herbert L. Windom. 1974. Heavy metal concentrations in museum fish specimens: effects of preservatives and time. *Science* 184: 475-477.

Hayward, C. Lynn, Merlin L. Killpack, and Gerald L. Richards. 1963. Birds of the Nevada test site. *Brigham Young Univ. Sci. Bull. Biol. Ser.*, vol. III, no. 1.

National Museum of Natural History. 1976. The use of museum specimens for past pollutant level documentation: an evaluation. Mimeographed. 204 pp. Washington, D.C.
Rick, Anne M. 1975. Use of museum specimens in toxic chemical research. *Occ. Pap. 21, Canadian Wildlife Svc.* 21 pp.

Reprinted from *Northwest Science*

Robert H. Gray
and
Dennis D. Dauble
Freshwater Sciences Ecosystems Department
Battelle Pacific Northwest Laboratories
Richland, Washington 99352

Checklist and Relative Abundance of Fish Species from the Hanford Reach of the Columbia River

Abstract

A checklist of fish species collected in over 30 years of research is presented for the Columbia River adjoining the U.S. Energy Research and Development Administration's Hanford Reservation (km 550-629), southcentral Washington. Differences in species composition and relative abundance within the area are discussed.

Introduction

Several sources list fish species inhabiting various parts of the Columbia River system (Schultz and DeLacey, 1935; Schultz, 1936; Roebeck *et. al.,* 1954; Davis *et. al.,* 1956; Miller, 1959; McPhail, 1967; Reimers and Bond, 1967; Eddy, 1969; Patten *et. al.,* 1970). Although fisheries scientists have collected information on resident and anadromous fishes for over 30 years, little quantitative data have been published regarding the majority of species inhabiting the Columbia River adjoining the Energy Research and Development Administration's Hanford reservation in southcentral Washington.

The Columbia River adjoins the Hanford reservation beginning about 10 km (6 miles) north of Richland, Washington, between river kilometers (km) 550 and 629, and between river miles (rm) 343 and 393 (Fig. 1). The Hanford reservation is a unique area because of its closure to the general public since 1942 by federal regulations. Furthermore, it contains the last free-flowing stretch of the Columbia River remaining above Bonneville Dam.

Most early fishery investigations in the Hanford area were concerned with salmonids in association with potential thermal, chemical and radiation impacts from the operation of up to nine plutonium production reactors during World War II and until 1971. In recent years, assessment studies for future siting of nuclear power plants at Hanford have required extensive collections of all fish species. This article lists the fish fauna, relying mainly on data collected during 1973-1975 at two different Hanford locations. Less common species collected earlier by other workers from sites within the Hanford area are included to provide a composite list.

Methods and Materials

From April, 1973, to June, 1974, fish were sampled weekly between Columbia River km 605 and 613 (rm 378-383). In September, 1974, sampling began between km 557 and 566 (rm 348-354) and continued at intervals of about three weeks. After collection, species were identified by using standard taxonomic keys (Schultz, 1936;

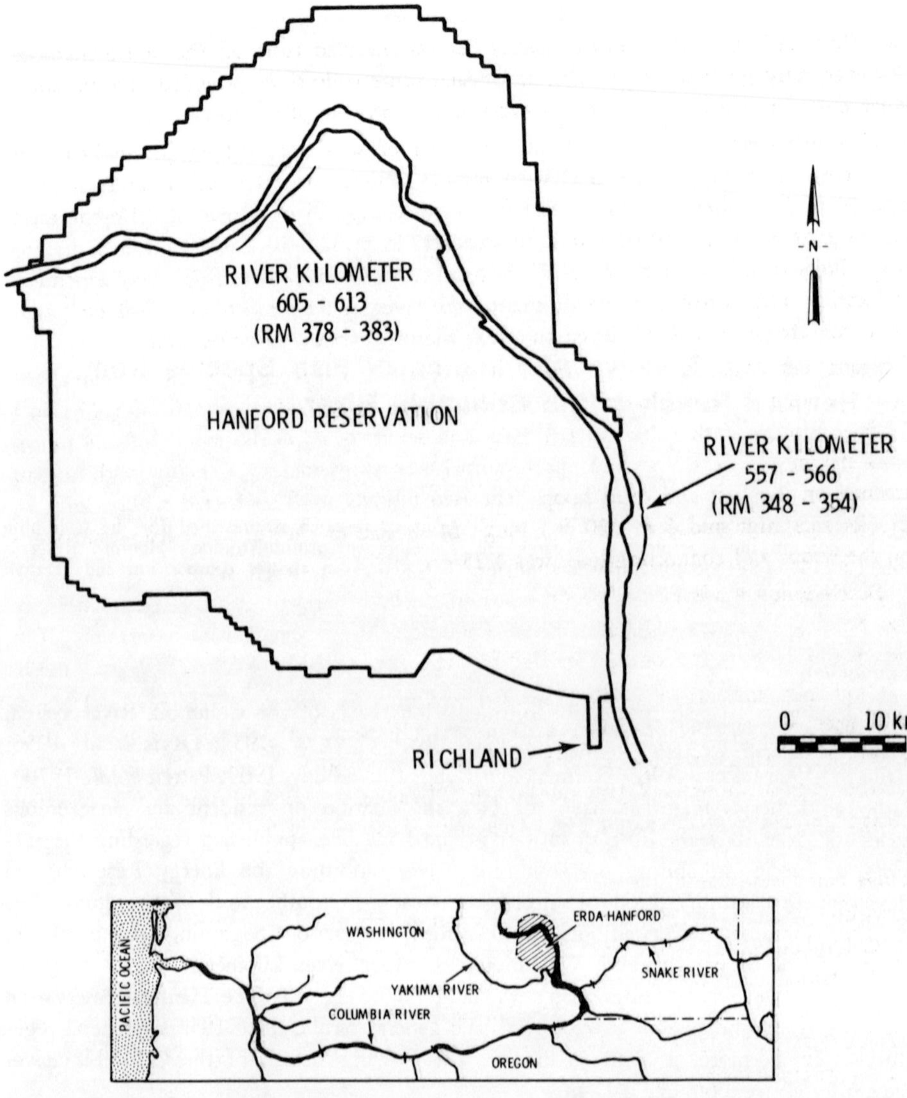

Figure 1. Location of the ERDA Hanford Reservation and fish sampling locations.

Carl *et. al.,* 1967; Eddy, 1969; Bond, 1973). Questionable identifications were confirmed by taxonomists and fisheries biologists having extensive experience with Columbia River fishes (C. D. Becker and D. G. Watson, Battelle Northwest Laboratories; C. E. Bond, Oregon State University). To insure taxonomic consistency, selected specimens were preserved in 5 percent formalin and maintained in a reference collection.

Several types of sampling gear and methods were used to suit various areas and to compensate for gear selectivity. Methods included: gill nets, trammel nets, beach seines, hoop nets, minnow traps, trotlines, electroshocker, various types of trawls, and hook and line. Some species were collected from the traveling screens of a water intake structure at Columbia rm 380 (Gray, Page, Wolf, and Schneider, 1975; Page, Gray, and Wolf, 1975). The resulting species list was derived from all the above methods. However, only gill nets, hoop nets, and beach seine collections provided relative abundance data for one complete annual cycle at the two sampling sites.

Gill nets were set from shore into the river or were set offshore perpendicular to the current in deeper waters, and were rotated among various locations at each study area. Nets were about 2 m (6-8 ft.) deep and composed of panels of different mesh size ranging in 1.25 cm (0.5 in.) increments, from 1.25-10 cm (0.5-4 in.) square-mesh. Nets varied in length from 7.5-37.5 m (25-125 ft.). Net length was dependent on location and hydrological conditions of the river at the time of set. Gill nets were set in the afternoon and retrieved the next morning about 19 hours later.

Hoop nets were usually set for 1 week periods, checked every 24-72 hours, and rotated among various locations. Hoop nets were composed of two 61 cm (24 in.) diameter tunnels, each 3 m (10 ft.) long and set in pairs so the mouth of one tunnel faced the mouth of the second. Each tunnel was supported by 6 hoops with throats attached on the first and third hoop. The two tunnels were connected with a lead, 61 cm (24 in.) high and 6 m (20 ft.) long. Mesh size of the throats, netting surrounding the hoops and connecting lead, was 1.25 cm (0.5 in.) square mesh.

Beach seining was done at four locations on both sides of the river at each study site. Seining encompassed about 30 m (100 ft.) of shoreline at each location. The seine was 1.2 m (4 ft.) deep, 9 m (30 ft.) long and consisted of 0.32 (⅛ in.) mesh.

Catch per unit effort (C/UE) of fish collected by gill net, hoop net, and beach seine was calculated as follows: Gill net C/UE was the average number of fish caught per overnight set (about 19 hours) of 33 m (100 ft.) of net. Hoop net C/UE was the average number of fish caught per 24 hours set. Beach seine C/UE was the average number of fish caught per beach seine haul.

Results and Discussion

Forty-three species of fish representing 13 families have been collected since 1943 in the Columbia River adjoining the Hanford Reservation (Table 1). These include mainly resident, but also anadromous (Clupeidae, Petromyzontidae, Salmonidae) forms.

Thirty-seven species of fish representing 12 families were collected at river km 557-566 and 605-613 in 1973-75. Thirty-one species were collected at km 605-613 and 34 species at km 557566.

Differences in species lists between the up and downriver sites were attributable to the presence or absence of various species. Mottled sculpin *(Cottus bairdi)*, reticulate sculpin *(C. perplexus)* and leopard dace *(Rhinichthys falcatus)* were only collected at km 605-613. Mountain sucker *(Catos tomus platyrhynchus)*, smallmouth bass *(Micropterus dolomieui)*, largemouth bass *(M. salmoides)*, black bullhead *(Ictalurus melas)*, yellow bullhead *(I. natalis)*, and walleye *(Stizostedion vitreum)* were only collected at km 557-566. Differences between the two sites are somewhat relative. Fish move up and downriver and all species may be present at both sites at some time during the year.

At the upriver site, resident redside shiner *(Richardsonius balteatus)*, northern squawfish *(Ptychocheilus oregonensis)*, largescale sucker *(Catostomus machrocheilus)* and the anadromous chinook salmon *(Oncorhynchus tshawytscha)* were caught in near equal numbers and collectively comprised 90 percent of the total catch.

TABLE 1. Composite list of fish species by family collected at Columbia River km 605-613 from March, 1973, through June, 1974; km 557-566 from September, 1974, through September, 1975, and elsewhere on the Hanford Reservation.

Family	Scientific Name	Common Name	Sample Location		Other Hanford Sites
			KM 605-613	KM 557-566	
Acipenseridae-Sturgeons	*Acipenser transmontanus* Richardson	White sturgeon	X	X	
Catostomidae-Suckers	*Catostomus columbianus* Eigenmann & Eigenmann	Bridgelip sucker	X	X	
	C. macrocheilus Girard	Largescale sucker	X	X	
	C. platyrhynchus Cope	Mountain sucker		X	
Centrarchidae-Bass and Sunfish	*Lepomis gibbosus* Linnaeus	Pumpkinseed	X	X	
	L. macrochirus Rafinesque	Bluegill	X	X	
	Micropterus dolomieui Lacepede	Smallmouth bass		X	
	M. salmoides Lacepede	Largemouth bass		X	
	Pomoxis annularis Rafinesque	White crappie			X[1]
	P. nigromaculatus Leseur	Black crappie	X	X	
Clupeidae-Herrings	*Alosa sapidissima* Wilson	American shad	X	X	
Cottidae-Sculpins	*Cottus asper* Richardson	Prickly sculpin	X	X	
	C. bairdi Girard	Mottled sculpin	X		
	C. beldingi Eigenmann & Eigenmann	Piute sculpin	X	X	
	C. perplexus Gilbert & Evermann	Reticulate sculpin	X		
	C. rhotheus Smith	Torrent sculpin	X	X	
Cyprinidae-Minnows and Carps	*Acrocheilus alutaceus* Agassiz & Pickering	Chiselmouth	X	X	
	Cyprinus carpio Linnaeus	Carp	X	X	
	Mylocheilus caurinus Richardson	Peamouth	X	X	
	Ptychocheilus oregonensis Richardson	Northern squawfish	X	X	
	Rhinichthys cataractae Valenciennes	Longnose dace	X	X	
	R. falcatus Eigenmann & Eigenmann	Leopard dace	X		
	R. osculus Girard	Speckled dace	X	X	
	Richardsonius balteatus Richardson	Redside shiner	X	X	
	Tinca tinca Linnaeus	Tench			X[1]

TABLE 1 (Continued)

Gadidae-Codfishes	*Lota lota* Linnaeus	Burbot			X[1,2,3]
Gasterosteidae-Sticklebacks	*Gasterosteus aculeatus* Linnaeus	Threespine stickleback	X	X	
Ictaluridae-Freshwater Catfishes	*Ictalurus melas* Rafinesque	Black bullhead		X	
	I. natalis Leseur	Yellow bullhead		X	
	I. punctatus Rafinesque	Channel catfish	X	X	
Percidae-Perches	*Perca flavescens* Mitchell	Yellow perch	X	X	
	Stizostedion vitreum Mitchell	Walleye		X	
Percopsidae-Troutperches	*Percopsis transmontana* Eigenmann & Eigenmann	Sand Roller	X	X	
Petromyzontidae-Lampreys	*Entosphenus tridentatus* Gairdner	Pacific lamprey	X	X	
	Lampetra ayresi Gunther	River lamprey			X[2]
Salmonidae-Salmon, trout, whitefish	*Coregonus clupeaformis* Mitchell	Lake Whitefish			X[1]
	Oncorhynchus kisutch Walbaum	Coho salmon	X	X	
	O. nerka Walbaum	Sockeye salmon	X	X	
	O. tshawytscha Walbaum	Chinook salmon	X	X	
	Prosopium williamsoni Girard	Mountain Whitefish	X	X	
	Salmo clarki Richardson	Cutthroat trout			X[1]
	S. gairdneri Richardson	Rainbow trout (steelhead)	X	X	
	Salvelinus malma Walbaum	Dolly Varden	X	X	

[1] D. G. Watson (personal communication).
[2] C. D. Becker (personal communication).
[3] M. P. Fujihara (personal communication).

At the downriver site, chinook salmon comprised 44 percent of the catch and resident peamouth chub *(Mylocheilus caurinus)*, largescale sucker, northern squawfish, mountain whitefish *(Prosopium williamsoni)*, bridgelip sucker *(Catostomus columbianus)*, chiselmouth *(Acrocheilus alutaceus)*, and redside shiner comprised another 49 percent. The larger numbers of chinook salmon at km 557-566 were 0-age juveniles, occurring seasonally near shore and collected by beach seine. These fish reflect the occurrence of major salmon spawning areas between km 566 and 602 (Watson, 1974), the shoreline orientation of downstream migrants (Mains and Smith, 1964), and our shore-oriented sampling methods.

To better compare species composition at the two collecting locations, species and C/UE of gill nets, hoop nets, and beach seines were evaluated for the same months of the year at each location (Table 2). Although sampling occurred in different years, these methods were used for one annual cycle at each site. For assessment purposes, comparisons were made for the periods June, 1973, through May, 1974, at km 605-613; and October, 1974, through September, 1975, at km 557-566.

A comparison of methods shows a difference in vulnerability of each species at different stages in its life cycle. However, the composite catch from gill nets, hoop nets, and beach seines provides a good representation of the species, sizes and life stages of fish present. We used this data to derive a numerical value of average relative abundance or a relative abundance index (RAI). We based the RAI on the average of three annual C/UE values for each species at each collecting site (Table 2). Species collected at the upriver location between km 605-613 are listed in descending order of abundance or rank. The rank of species collected at the downriver location between km 557-566 is shown in Table 2 parenthetically next to the RAI for that species.

Although small differences in C/UE are insignificant, analysis shows that species composition and relative abundance of some collected fish differ between the two sites. The difference may reflect our inability to sample equally all habitats, annual differences in river temperatures and flows, or habitat differences in the Columbia River adjoining the Hanford reservation.

There are several obvious differences between the two sites. The upriver sample area is a straight and deep channel with no islands, a maximum depth of 10 m and few shoreline indentations. The bottom consists mainly of boulders and large rock over 10 cm in diameter. The channel at the downriver site is shallower, wider, and contains islands; adjoining it are sloughs and backwater areas. The bottom varies from sand to cobble and boulders and is often cemented as outcroppings of upper Ringold-

TABLE 2. C/UE[a] and RAI[b] of numerically dominant species caught by gill net, hoop net and beach seine from June, 1973, through May, 1974, at km 605-613; and October, 1974, through September, 1975, at km 557-566.

Species	C/UE at km 605-613				C/UE at km 557-566				
	Gill Net	Hoop Net	Beach Seine	RAI	Gill Net	Hoop Net	Beach Seine	RAI	
Ptychocheilus oregonensis	2.79	0.48	2.98	2.08	2.13	0.28	0.78	1.06	(4)[c]
Richardsonius balteatus	0.24	0.99	4.64	1.96	0.84	0.17	0.62	0.54	(7)
Catostomus macrocheilus	0.64	0.15	2.82[d]	1.20	2.84	0.23	1.06[d]	1.37	(2)
Oncorhynchus tshawytscha	0.30	0.01	2.54	0.95	0.09	0.02	17.01	5.71	(1)
Mylocheilus caurinus	0.49	0.01	0.22	0.24	1.70	0.01	1.59	1.10	(3)
Acrocheilus alutaceus	0.49	0.01	0.01	0.17	2.23	0.03	0.01	0.76	(5)
Prosopium williamsoni	0.51		0.01	0.17	0.34		1.84	0.73	(6)
Salmo gairdneri	0.19			0.06	0.50	0.02		0.17	(8)
Acipenser transmontanus	0.19			0.06	0.29			0.10	(10)
Cyprinus carpio	0.15			0.05	0.27	0.01		0.09	(11)
Gasterosteus aculeatus		0.12	0.02	0.05					
Catostomus columbianus	0.12	0.01		0.04	2.17	0.08	0.03	0.76	(5)
Perca flavescens		0.01			0.39	0.08	0.01	0.16	(9)
Rhinicthys cataractae			0.01			0.01	0.29[e]	0.10	(10)
Ictalurus punctatus					0.20	0.01		0.07	(12)
Pomoxis nigromaculatus					0.09	0.04		0.04	(13)

[a] Catch per Unit Effort.
[b] Relative Abundance Index.
[c] Numerical Rank.
[d] Includes some juvenile *Catostomus* sp.
[e] Includes some juvenile *Rhinicthys* sp.

type material (Page, Wolf, Gray, and Schneider, 1974). Additionally, the downriver area is under the influence of the upper end of McNary Dam pool and the free-flowing Columbia River above. Both lotic and lentic habitats, as well as wider temperature ranges, are available. Thus, the downriver area has a greater habitat diversity than the upriver collecting sites.

Other species known to inhabit the Hanford Reach of the Columbia River (Table 1) are documented in fish collections at Battelle, Pacific Northwest Laboratories. These include white crappie *(Pomoxis annularis)*, tench *(Tinca tinca)*, burbot *(Lota lota)*, river lamprey *(Lampetra ayresi)*, lake whitefish *(Coregonus clupeaformis)*, and cutthroat trout *(Salmo clarki)*. Cutthroat trout were collected before construction of McNary Dam but not since (Watson, personal communication).

Several other species collected from Columbia River backwater areas south of Richland or occasionally caught by sport fishermen and known to inhabit the lower Snake and Yakima Rivers may be present. These include mosquito fish *(Gambusia affinis)*, blue catfish *(Ictalurus furcatus)*, nine spine stickleback *(Pungitius pungitius)*, and western brook lamprey *(Lampetra richardsoni)* (Becker and Watson, personal communication).

To our knowledge this is the first complete fish species list ever published for the Hanford Reach of the Columbia River, now an environmental research park, and should be of value for ichthyologists in future nuclear power plant siting considerations.

Acknowledgments

We thank C. D. Becker, C. E. Cushing, M. P. Fujihara, and D. G. Watson, who critically reviewed the manuscript and provided information on less commonly encountered fish species. This work was supported in part by the Washington Public Power Supply System, Richland, Washington.

Literature Cited

Bond, C. E. 1973. Keys to Oregon Freshwater Fishes. Tech. Bull. 58. Agric. Exp. Sta. Oregon State Univ., Corvallis.

Carl, G. C., W. A. Clemens, and C. C. Lindsey. 1967. The Freshwater Fishes of British Columbia. Handbook No. 5. British Columbia Provincial Museum, Victoria, B.C.

Davis, J. J., D. G. Watson, and C. C. Palmiter. 1956. Radiobiological Studies of the Columbia River Through 1955. USAEC Document HW 36074. Tech. Information Service, Oak Ridge, Tennessee.

Eddy, S. E. 1969. The Freshwater Fishes. Wm. C. Brown Co., Dubuque, Iowa, 2nd Edition, p. 286.

Gray, R. H., T. L. Page, E. G. Wolf, and M. J. Schneider. 1975. A Study of Screen Impingement and Fish Passage at Hanford Generating Project-A Progress Report. Battelle Northwest, Richland, Washington.

McPhail, J. D. 1967. Distribution of Freshwater Fishes in Western Washington. Northwest Science 40(1): 1-11.

Miller, R. R. 1959. Origin and Affinities of the Freshwater Fish Fauna of Western North America. *In* Zoogeography Am. Assoc. Advance Sci., Washington, D.C., Pub. 52, pp. 187-222.

Page, T. L., E. G. Wolf, R. H. Gray, and M. J. Schneider. 1974. Ecological Comparison of the Hanford No. 1 and WNP-2 Sites on the Columbia River. Battelle Northwest, Richland, Washington.

——————, R. H. Gray, and E. G. Wolf. 1975. Report on Impingement Studies Conducted at Hanford Generating Project March and April 1975, Battelle Northwest, Richland, Washington.

Patten, B. G., R. B. Thompson, and W. D. Gronlund. 1970. Distribution and Abundance of Fish in the Yakima River, Washington, April 1957 to May 1958. U.S. Fish. Wildl. Serv. Spec. Sci. Rep. Fish. No. 603.

Reimers, P. E. and C. E. Bond. 1967. Distribution of Fishes in Tributaries of the Lower Columbia River. Copeia 1967 (3): 541-550.

Robeck, G. C., C. Henderson, and R. C. Palange. 1954. Water Quality Studies on the Columbia River. U.S. Dept. Health, Education and Welfare.

Schultz, L. P. 1936. Keys to the Fishes of Washington, Oregon and closely adjoining regions. Univ. of Wash. Pub. Biol. 2(4): 103-228.

———, and A. C. DeLacey. 1935. Fishes of the American Northwest. A catalog of the fishes of Washington and Oregon with distributional records and a bibliography. Mid. Pacific Mag. Vol. X No. 4: 365-380.

Watson, D. G. 1974. Salmon Spawning Studies. *In* Annual Report for 1974 Part 2. Ecol. Sci. BNWL-1950: 88-89. Battelle Pacific Northwest Labs., Richland, Washington.

Received July 12, 1976
Accepted for publication December 17, 1976

Reprinted from ASC Newsletter

The San Francisco Bay Project: A New Approach to Using Systematics

By Welton L. Lee

In recent decades, man's activities have brought about noticeable changes in his environment. The magnitude of these changes has led to intense concern and a feeling for the need 1) to determine the nature and extent of man's impact on his own environment, and 2) to predict the future environmental impact of present technological change. This feeling has intensified in recent years and as a result, federal, state and local agencies have rushed in with a variety of environmentally oriented studies.

Of major concern is the effect of pollution in large industrial areas. The San Francisco Bay region is an example of one industrial area that has received such intensive study. By looking at what has happened here we can get a better perspective of the kinds of problems encountered and develop a background for understanding the need for, and utility of the San Francisco Bay Project.

For many years, there has been an extraordinary amount of information circulating concerning environmental change in San Francisco Bay. Apparently "everybody knows" of these changes, their extent and meaning, yet, on close inspection one discovers that most of the information is hearsay with little documentation of any consistent value. This situation led to a very thorough review by F. H. Nichols in 1973. Dr. Nichols examined the past 60 years of environmental work in San Francisco Bay, and his results were surprising and rather discouraging. In short, he found that the studies which had been undertaken were based on static measurements of a very dynamic biological system. Little useful information had been generated after the expenditure of a considerable amount of money.

Three basic reasons for this situation were identified. First, in almost no case were quantitative data taken--no standards had been set for locating stations or for replicate sampling; and there were no standards for collection management or the procedures to be employed in species identification. Nonetheless, sophisticated statistical analyses were carried out on these grossly inadequate data. A second problem identified by Dr. Nichols was our impoverished knowledge of the life cycles and details of biotic interactions of the animals collected. We had no information concerning their susceptibility to low level, chronic toxicity and the implications of such toxicity on normal life cycle functions, or to the antagonistic or synergistic effects of various compounds on these organisms.

Finally, and possibly most germaine to our present argument, Nichols pointed out that taxonomic identification of the organisms present was almost impossible given the unavailability and/or unreliability of appropriate taxonomic information. Because the available information was often of little practical use, the level of accuracy and hence, repeatability, was so low as to be almost useless. Thus, the assessment of community structure, species diversity, indices, etc., were based largely on guesswork.

Before proceeding, however, it is important to recognize a very basic and longstanding criticism relating to the identification of organisms in such studies. The question so often raised is, "Why should we bother with taxonomy?" One point of view can be seen in a comment made by Mr. Jerry Franklin as noted in a report of the proceedings of an ASC inter-agency workshop held November 25, 1974.

> "The basic problem with a species list per se is that it indicates only the presence or absence of a species in an area. This only has immediate value in identifying threatened and endangered species in the area. Beyond that, the list has relatively little value. It says nothing about relative abundance of the organisms nor does it establish the roles of the different species within the functioning ecosystem. Consequently, you cannot draw any inferences from the list--the roles of the species in terms of productivity, in nutrient cycling, in decomposition. You can't begin to answer questions about how much effect a pollutant would have on the system. You can't relate an action to the elimination of an organism and the

resulting consequence on the functioning of that ecosystem."

Although I have no quarrels with this statement in general, an impression is left that taxonomy either has no place in such studies, or is at best of little importance. Perhaps the best answer to this charge is a statement recently made by Dr. Melbourne R. Carriker [The crucial role of systematics in the identification and control of pollution. In National Estuarine Pollution and Control Assessment. Proceedings of a Conference. Superintendent of Documents. U.S. Government Printing Office. 1977.]:

> "The increasing refinement and quantitation of biological investigation requires comparable exactness in the identification of organisms used in both basic and applied research. Imprecision in identification neutralizes research rigor! If any identifications are in error, published work becomes non-replicable, and thus unscientific."

He further states:

> "To control pollution, therefore, as Turner (Pers. Comm. 1974) points out, we have to know 'what is doing what to what,' and in order to do this it is essential to know the organisms affected and to be able to recognize introduction and extinction of species."

Let us now return to the situation in San Francisco Bay and explore more fully the reasons behind the taxonomic problems encountered. These can be categorized into three general problems:

1) Most environmentally oriented studies have been prepared by inexperienced personnel.-- First, it is a general assumption that a bachelor's, master's or for that matter, even a Ph.D. degree, somehow confers with it expertise in taxonomy. Nothing could be further from the truth, especially today when so little emphasis is given to anything which doesn't utilize the latest in experimental technique or which does not "explore the frontiers of science." This country is now experiencing an appalling lack of qualified taxonomists. Both of these problems are due directly to the decreased emphasis, in many cases, scorn, with anything to do with taxonomy. This results from a general attitude that conceptually, taxonomy and systematics are uninteresting and unglamorous, and a prevailing myth that somehow we have managed to describe the existing biotic universe.

As a result our present situation is frightful. From studies in the San Francisco Bay area we have seen specimens mis-identified on a scale which staggers belief. For example, some vials were noted to contain a single species, when, in fact, representatives of no less than six families were included. Other vials, labelled as containing a variety of species of oligochaete worms, actually contain two species of polychaetes. On one occasion a specimen labelled as a polychaete worm turned out to be a clam siphon. Perhaps the worst and most flagrant example was a vial labelled as "diatoms" which contained nothing but the spines of brittle stars. It is upon this kind of "taxonomy" that many costly studies in the Bay area have been based.

2) Those who carry out such studies often show no knowledge of the extant literature and/or an inability to use the literature once it has been found.-- Certainly one prevailing problem is the inexperience of persons undertaking such studies. However, another equally important aspect relates to the nature of the literature itself. Today, taxonomic literature is published in hundreds of journals and in all languages. Furthermore, each discipline astutely maintains its own highly specialized terminology and mystique; this makes it difficult enough for an expert in the field, but almost impossible for anyone else. Furthermore, most of the keys that are available are for the most part generated by experts for the explicit use of other experts, and even at this level, the keys frequently do not work.

Another problem with the literature is that it is so compartmentalized as to be almost useless. One needs particular expertise to track down taxonomic literature, but additional expertise is also required to get at ecological information because workers in each discipline publish in separate journals. Environmental studies require the assessment of ecological, taxonomic, physiological and toxological information, each of which must be obtained by reviewing totally different and often extensive literature sources. Finally, we find that what does exist is inadequate for our needs. For example, in the San Francisco Bay area frequent use is made of the "Light's Manual," a set of keys to the more common marine intertidal organisms in central California. As fine as this manual is, it nevertheless only treats intertidal species. For the most part, benthic and pelagic organisms are not included. For this information one must either turn to individual species descriptions, found in a multitude of journals, or monographic studies covering a given group on a world-wide basis. Regional assessments are very much needed, yet few of these exist.

3) The "real" information initially obtained--the specimens themselves--are rarely if ever maintained for future use.-- This practice totally nullifies the scientific credibility of such studies. No one can repeat such work or check it for accuracy and without voucher collections we are left only with species lists which can never be verified.

The results of such problems are the same for San Francisco Bay as elsewhere: namely, we spend vast sums for studies that produce unverifiable species lists. Moreover, potentially valuable ecological, physiological or toxological information is rendered useless in the absence of any means of taxonomic verification. We must guard against accepting the results and predictions of such inadequate studies as a guide to future decisions.

In 1973, the San Francisco Bay Project was initiated as a regional program with two major objectives: 1) to involve traditionally isolated and uninvolved museums in pragmatic community problems, thereby utilizing their unique facilities and expertise, and 2) to try to solve one of the major sets of problems identified by Nichols in a manner that could be a model for other similar coastal areas--that is, to standardize and enhance the ease and reliability of species identification on a regional basis.

The program, as it now exists, consists of three phases emphasizing the concepts of a central information source, standardization of taxonomic information and making this information more readily available and more reasonable to utilize.

Phase 1 of the program is designed to produce a complete and accurately identified reference collection of invertebrates known to occur in San Francisco Bay. We are amassing a sufficiently large series for each species to be able to show age class differences, sexual dimorphism, and natural variability, as well as to represent diseased and damaged specimens. The collection is arranged systematically and is accompanied by a catalog file and all pertinent data. It is available for use by any qualified person.

As part of Phase 1 we are now offering to maintain a permanent, cataloged, and cross-referenced collection of material obtained in environmental studies in the Bay area. This includes either voucher specimens and/or total collections. These will be maintained indefinitely for the use of qualified individuals and allows, for the first time in this area, the possibility of direct "before and after" comparison.

Phase 2 includes the development of a complete, cross-referenced bibliography covering those species known to occur in San Francisco Bay. This will include material on taxonomy, physiology, development, ecology, etc. and will be cross-referenced by locality and species. We will maintain the more useful and pertinent literature at the California Academy of Sciences. For literature not available at the Academy, the nearest library where the particular paper may be found will be noted on cards. In addition, the bibliography will include hydrographic and survey work. The bibliographic file will be available to all interested individuals.

Phase 3 is perhaps the most exciting aspect of the entire program. It is concerned with the development of a series of identification manuals which will serve to guide and standardize taxonomic work for this region. The first of these is now in press and several others are in various stages of preparation. The manuals include information on the techniques used for preserving, maintaining, dissecting, and observing a particular taxonomic group along with full instructions and diagrams. For each manual there is a fully illustrated glossary that includes foreign terms and notes on equivalent terms. There is, in addition, a complete overview of the general anatomy of the group with notes on variability and the usefulness of various anatomical features. Finally, each manual includes a complete annotated bibliography.

The "keys" are really a series of distinct but complementary formats enabling the user to approach the identification problem from several different directions. We do not believe that it is necessary for keys to be so complex and esoteric that only a specialist can use them. As a result we have used a style which incorporates the complete and accurate detailed information needed by the specialist but in a manner facilitating its use by the non-specialist.

Basically we are utilizing three sets of keys. The first is a pictorial, biramous key of the type first developed and used by the Public Health Department during World War II. The second key is a much augmented dichotomous key where more than a single character is compared. The final key is tabular and, on a character by character basis, gives the user a complete detailed diagnosis at a glance.

The manuals represent a pragmatic balance between a simple guide for general use and a detailed monographic study. In those cases wherein it is deemed necessary, types are examined and the taxonomy is revised by specialists. In addition, individual species descriptions, fully illustrated, are incorporated, and the illustrations are aligned in a manner to provide direct comparison of homologous structures. The most exciting aspect of these manuals is the fact that they work well, even for inexperienced investigators.

We feel very strongly that by standardizing all taxonomic information in the manner described, by casting off the mystic and esoteric nature of much of the present-day taxonomic literature and by pulling together these various resources in a central facility that the project will provide a common base for all biological studies in the area.

We envision this program as serving as a model for similar efforts in other coastal or estuarine areas where baseline or long-term community studies are at present seriously hampered by complicated, incomplete and unusable taxonomic references and by insufficient access to supporting literature and properly identified specimens. We envision that the San Francisco Bay Project will serve as a model for other museums to utilize as a means of alleviating such problems in other coastal and estuarine areas where man has had an important impact on his environment.

Editor's Note.-- This is the second in a series of contributed papers presented at a symposium entitled "The Role of Systematics in Ecological Research." The symposium was held in May 1976 at the Fourth Annual Meeting of the Association of Systematics Collections in Blacksburg, Virginia.

The author is Chairman and Curator of the Division of Invertebrate Zoology at the California Academy of Sciences, Golden Gate Park, San Francisco, California 94118.

Reprinted from *The Wilson Bulletin*

AVIAN EGGSHELL THICKNESS: VARIABILITY AND SAMPLING

ERWIN E. KLAAS, HARRY M. OHLENDORF, AND ROBERT G. HEATH

Studies comparing recently collected eggs with those in archival collections have shown that shell thickness in certain predatory (including fish-eating) birds has decreased significantly since the introduction and widespread use of persistent chemical insecticides, particularly DDT (Ratcliffe, 1967, 1970; Hickey and Anderson, 1968; Anderson and Hickey, 1970, 1972; Peakall, 1970; Blus, 1970; Blus et al., 1972; Koeman et al., 1972). This decrease in shell thickness is correlated with reproductive failures in some of these species, with populations declining when thinning persists for a period of years. Baseline data on eggshell thickness depend on studies of archival egg collections, which have had few additions since the 1930's—when severe restrictions were placed on egg collecting as a conservation measure. For more recent data, interested researchers must often collect eggs themselves, and they are faced with many problems. These include optimal allocation of time and resources to obtain adequate samples, while minimizing the impact of egg collecting on the reproduction of a species.

We feel that a knowledge of the variability of shell thickness in species is essential to determination of the proper size and composition of samples. To illustrate this, we have measured and analyzed eggs in museum collections of the following: Black-crowned Night Heron (*Nycticorax nycticorax*), White Ibis (*Eudocimus albus*), Clapper Rail (*Rallus longirostris*), Mockingbird (*Mimus polyglottus*), and Loggerhead Shrike (*Lanius ludovicianus*). Our analyses of these data are intended (1) to probe the variability in eggshell thickness and to discuss some of the factors which contribute to this variability; (2) to show how a knowledge of variation can enable one to estimate sample sizes needed for detecting specified differences in shell thickness; and (3) to discuss the advantages and disadvantages of collecting complete clutches of eggs.

METHODS

Techniques for measuring eggshell thickness have been described by Anderson and Hickey (1970) and Ratcliffe (1970). We measured thickness using a modified Starrett Model 1010M thickness gauge, accurate to 0.01 mm. Three measurements were taken at different places around the blow-hole of each egg of the three nonpasserine species. Only those eggs with the blow-hole located near the "equator" were used. Eggs of the Mockingbird and Loggerhead Shrike are too small to measure thickness directly with the instrument described above. To estimate their thickness, the length and breadth of these eggshells were measured to the nearest 0.05 mm, using a vernier dial caliper graduated in 0.05-mm intervals, and the shells were weighed to the nearest 0.01 g on a 100-g-capacity, top-loading, pan balance. These measurements were then used to calculate Ratcliffe's (1970) thickness index, a parameter closely correlated with actual thickness. We have endeavored to use only eggs that appeared clean inside and those with small blow-holes, in order to control weight variability.

Significant variation in eggshell thickness over broad geographic areas has been demonstrated for some species (Anderson and Hickey, 1970, 1972). Thus, for each species we have restricted our analysis to eggs collected in a relatively small region on the assumption that geographic variation is reduced to a trivial level. Black-crowned Night Heron eggs were from Massachusetts, Rhode Island, Connecticut, and extreme eastern New York. White Ibis and Loggerhead Shrike eggs were from Florida, Clapper Rail eggs from Virginia, and Mockingbird eggs from South Carolina and nearby Georgia. All of these eggs were collected between 1876 and 1943.

We assume that eggs in museum collections represent random samples from the given localities. We found no indication in the collections or in the field notes of the collectors that certain eggs were selected in preference to others, except that they usually tried to take fresh *first* clutches. All collectors probably attempted to obtain complete clutches. Their original field notes verify that complete clutches were selected after repeated visits to the nest or upon flushing a bird off the nest in species where incubation generally does not begin until the last egg is laid (Lloyd Kiff, pers. comm.). Collectors also accumulated or communicated knowledge of what a full clutch of eggs should be for the species. Thus, we believe that incomplete clutches were rarely taken, and that the inadvertent inclusion of a few of these clutches in our large samples does not significantly bias our data.

We assume that shell thickness remains essentially constant after an egg is collected and dried, that is, that the materials do not sublimate or otherwise decrease.

Statistical procedures follow those presented in Sokal and Rohlf (1969).

RESULTS AND DISCUSSION

Variability—Variation in shell thickness for each of the five species has been analyzed as a Model II (random) nested analysis of variance (anova), with unequal sample sizes (Table 1), using the clutch of eggs as the primary sampling unit. For the nonpasserine species, the procedure separates variation in thickness into three hierarchical levels: among clutches, among eggshells within clutches, and among measurements (error).

Variation in thickness of shell among clutches probably depends on: differences in the stages of incubation, differences related to clutch size, genetic and physiological differences between females, differencs in diet among females within and between local populations, differences in gene pools between local populations, differences in environmental conditions between years, and other unknown factors (Kreitzer, 1972; Romanoff and Romanoff, 1949; Rothstein, 1972). Variation among eggs within clutches is probably due to day-to-day differences in calcium metabolism and eggshell deposition in individual females. The main factors which contribute to the among-measurements (error) mean square are: differences in eggshell thickness at the three

TABLE 1

NESTED ANALYSIS OF VARIANCE FOR EGGSHELL THICKNESS FOR FIVE SPECIES

Species and Source of variation	Degrees of freedom (df)	Mean squares (MS)
Black-crowned Night Heron[a]		
Among clutches	81	.00423583*
Among eggs	254	.00048310*
Among measurements (error)	672	.00003343
Total	1007	
White Ibis[a]		
Among clutches	193	.00344592*
Among eggs	407	.00081443*
Among measurements (error)	1202	.00005435
Total	1802	
Clapper Rail[a]		
Among clutches	59	.00513168*
Among eggs	466	.00033180*
Among measurements (error)	1052	.00005133
Total	1577	
Mockingbird[b]		
Among clutches	80	.00489155*
Among eggs (error)	231	.00067398
Total	311	
Loggerhead Shrike[b]		
Among clutches	72	.00383473*
Among eggs (error)	255	.0043458
Total	327	

[a] Based on three thickness measurements (mm) per egg.
[b] Based on one thickness-index measurement per egg.
* $P < .001$

points of measurement, imperfections in the micrometer within the limits specified by the manufacturer, and human error in applying and reading the instrument.

An analysis of variance was performed on White Ibis data to test differences among nine localities in Florida. These differences were found to be non-significant ($P < .01$). We conclude that most of the variation among clutches is due to individual differences in females and possibly to yearly differences in environmental conditions.

In the Mockingbird and Loggerhead Shrike, we made only one determination of the thickness index for each egg. For this reason, variation due to the intrinsic inaccuracies in measuring and to other within-egg variables could

TABLE 2

STATISTICS OF EGGSHELL THICKNESS FOR FIVE SPECIES

Items analyzed	Species Studied: Black-crowned Night Heron[a]	White Ibis[a]	Clapper Rail[a]	Mocking-bird[b]	Loggerhead Shrike[b]
Variance					
Among clutches (s^2_C)	0.0003055	0.0002832	0.0001829	0.0010949	0.0007567
Among eggs ($s^2_{E(C)}$)	0.0001499	0.0002533	0.0000935	0.0006740	0.0004346
Among measurements (s^2)	0.0000334	0.0000544	0.0000513	—	—
Percent of the variability					
Among clutches	62.50	47.93	55.81	61.90	63.52
Among eggs	30.66	42.87	28.52	38.10	36.48
Among measurements	6.84	9.20	15.66	—	—
Overall mean thickness or index	0.28348	0.34752	0.25792	0.54332	0.51506
Number of clutches	82	194	60	81	73
Number of eggs	336	601	526	312	328
Clutch size					
Mean	4.10	3.10	8.77	3.85	4.50
Range	2–7	1–5	4–12	2–5	2–6

[a] Based on direct thickness measurements (mm).
[b] Based on thickness-index measurements.

not be separated from variation among eggs within clutches. Thus, differences among eggs within clutches cannot be tested and the number of levels in the anova is reduced to two. The coefficients of variation for thickness index among eggs within clutches for these species are similar to those for shell thickness for the other three species.

Statistics derived from the anova are presented in Table 2. Variances have been calculated from expected mean squares. Variation at each level is also expressed as a percentage of the total variance. Most of the variation occurs among clutches, but a substantial proportion is found among eggs within clutches. The percentage of the variation occurring among measurements of the same egg is small.

Percentages of variation for groups and subgroups are fairly consistent from species to species. A notable exception is found in the White Ibis, in which the percentage of the total variation occurring among eggs within clutches was greater than for the other species. This finding is related, in part, to obvious differences in size and shape of eggs within clutches. Generally, eggs in clutches of other species were more uniform in these characteristics. Furthermore, because eggs of White Ibis are larger and have thicker shells than the other four species considered here, there is greater opportunity for variability.

Complete clutches vs. one egg per clutch—To get at the question of whether to collect whole clutches or some fraction of each, we computed the *relative efficiency* (RE) of one design with respect to the other. RE is a ratio, usually expressed as a percentage, of the variances resulting from the two designs being compared. The appropriate formula as given in Sokal and Rohlf (1969:289) is

$$\mathrm{RE} = \frac{s_{\bar{y}}^2 \text{ of design B}}{s_{\bar{y}}^2 \text{ of design A}} \times 100$$

where $\bar{y}$ here represents the mean of a group of clutches. The expected variance among clutch means $(s_{\bar{y}}^2)$ for each design can be derived from estimated variance components for each level in the nested analysis of variance (Table 2) according to the formula:

$$s_{\bar{y}}^2 = \frac{s^2}{mec} + \frac{s_{E \subset C}^2}{ec} + \frac{s_C^2}{c},$$

where s^2, $s_{E \subset C}^2$, and s_C^2 are the respective estimates of the variance among measurements within eggs, among eggs within clutches, and among clutches; and m is the number of measurements per egg, e the number of eggs sampled per clutch, and c the number of clutches sampled.

We computed estimated variances of $\bar{y}$ for a theoretical design A in which the entire clutch is utilized and for a second design, B, in which the sampling unit is one egg selected randomly from each clutch. In design A, the number of eggs measured per clutch (e) has been determined as the whole number nearest to the mean clutch size for the respective species (see Table 2). In design B, $e = 1$. Values of $m = 3$ and $c = 10$ were constant for all calculations of $s_{\bar{y}}^2$ for Black-crowned Night Heron, White Ibis, and Clapper Rail. These values were selected because our empirical estimates of variances were based on three measurements per egg, and ten clutches approximate the required sample sizes estimated in the next section of this paper. The value $c = 10$ was also used for Mockingbird and Loggerhead Shrike, but $m = 1$. Thus, there is no estimate of s^2 within measurements and the formula is reduced to two components,

$$s_{\bar{y}}^2 = \frac{s_{E \subset C}^2}{ec} + \frac{s_C^2}{c}.$$

Table 3 shows the gain in efficiency of design A compared to design B. The relatively large differences in thickness observed among eggs within

TABLE 3

RELATIVE EFFICIENCY (RE) OF COLLECTING WHOLE CLUTCHES (DESIGN A) VERSUS COLLECTING ONE EGG FROM A CLUTCH (DESIGN B)[1]

Species	Mean clutch size	$s^2_{\bar{y}}$A	$s^2_{\bar{y}}$B	$\frac{s^2_{\bar{y}}\text{B}}{s^2_{\bar{y}}\text{A}} \times 100$	Increase in RE (%) of Design A over Design B
Black-crowned Night Heron	4	.3457283	.4665033	134.9	35
White Ibis	3	.3736778	.5546333	148.4	48
Clapper Rail	9	.1951889	.2935000	150.4	50
Mockingbird	4	.0001263	.0001769	140.0	40
Loggerhead Shrike	5	.0000844	.0001191	141.2	41

[1] The number of measurements per egg and the number of clutches are constant in each design.

clutches (expressed as $s^2_{E \subset C}$, see Table 2) contribute substantially to the sample variance ($s^2_{\bar{y}}$).

In utilizing whole clutches, the investigator should be aware that inadvertent or intentional collection of a high proportion of incomplete clutches may bias the data, especially if sample sizes are small. Such conditions would occur if only the first one or two eggs of a clutch are collected and the shell thicknesses of successive eggs tend to differ from those of the first eggs. Thus, care should be taken to collect complete clutches. If partial clutches are collected the investigator should randomly choose the eggs.

Other designs could be compared in which the number of measurements per egg (m) or the number of clutches (c) would vary. However, variability between measurements (s^2) is small, and relative efficiency would be changed little by increasing the number of measurements per egg. If one wished to use one egg per clutch and still achieve a variance equivalent to that of design A, the percentage increase in the number of clutches to be sampled for each species would be approximately that shown in the last column in Table 2.

The relative efficiency of one design with respect to another is not very meaningful unless one considers the relative effort and time (or cost) involved in obtaining the data. Cost will vary greatly, depending on whether the species is abundant, whether colonial or solitary nester, how well it conceals its nest, and other characteristics. For most studies, the cost of removing the egg contents and measuring shell thickness is much less than that of finding sufficient nests and collecting the eggs. The investigator may also be confronted with a limited season in which eggs are available, extended travel cost and time, and other factors. Under such limitations, design A, which minimizes the number of clutches needed, will usually require the least amount of time in the field. However, the total number of eggs to be measured is greater, as entire clutches are utilized.

The investigator must weigh the efficient utilization of time and resources against the impact on a species that collecting might have. Most species, except large birds of prey and albatrosses (which have long incubation periods), will lay replacement eggs when the entire clutch is taken or destroyed (Welty, 1962:295; Thompson, 1964:242). This trait was sometimes used by early oologists to get fresh eggs and/or increase their total take. Today, it might be used to minimize the impact of egg collecting. For instance, when the nests of a small proportion of breeding pairs are actually subject to the removal of the entire clutches, less disturbance occurs than when some fraction of the clutch is taken from a greater number of nests. Such disturbance, even though only one egg is taken, may also result in nest desertion by the birds.

Sample size estimates—After the sampling design has been selected, the investigator will want to know the number of clutches needed from each of "*a*" groups of nests in order to detect a minimum true difference, δ, in mean eggshell thickness. This assumes a probability, P (power), that the difference will be detected if it exists, and a probability, α, of incorrectly claiming a difference when none exists. Sokol and Rohlf (1969:247) present a relatively simple formula for estimating the sample size if one has a priori information of the sample variance. The formula (with c substituted for Sokol's n) is as follows:

$$c \geqslant 2\left(\frac{\sigma}{\delta}\right)^2 \{t_{\alpha[\nu]} + t_{2(1-P)[\nu]}\}^2,$$

where c = number of clutches, σ = true standard deviation for clutches, δ = the smallest true difference between means that one desires to detect, ν = degrees of freedom of the sample standard deviation with a groups and c replications (clutches) per group, α = significance level, P = desired probability (power) that a difference will be found to be significant. Values of $t_{\alpha[\nu]}$ and $t_{2(1-P)[\nu]}$ are from a two-tailed t-table with ν degrees of freedom, and probability levels of α and $2(1-P)$.

The sample estimate of the coefficient of variation (C.V.) and the percent difference between means that one wishes to detect can be substituted for σ and δ, respectively, because only the ratio of σ and δ are necessary, not their actual values. We computed estimates of sample sizes (number of clutches) needed to detect thickness differences (δ) of five and 10 percent (percentages arbitrarily selected) for each of the five species at significance levels of 0.05 and powers of 0.9 (Table 4). These estimates are based on a collecting

TABLE 4

ESTIMATES OF THE NUMBER OF CLUTCHES NEEDED FROM A LOCALITY TO SHOW DIFFERENCES (δ) IN THICKNESS OF EGGS[1]

Species	Coefficient of variation	Collecting Design: $\delta = 10$ percent, Whole clutches	$\delta = 10$ percent, One-egg clutches	$\delta = 5$ percent, Whole clutches	$\delta = 5$ percent, One-egg clutches
Black-crowned Night Heron	6.55	11	15	38	51
White Ibis	5.54	8	12	27	40
Clapper Rail	5.42	8	12	26	39
Mockingbird	6.56	11	15	38	53
Loggerhead Shrike	5.67	8	11	28	40

[1] Significance level (α) of 0.05 and power (P) of 0.9.

design in which complete clutches are utilized and each egg is measured three times. The formula is very sensitive to changes in the ratio of (σ/δ). This accounts for the marked dissimilarity in numbers of sets required for detecting five and 10 percent differences.

The estimates of required sample sizes are based on estimates of the variance from relatively large samples of archival eggs. We believe these estimates are relatively precise, but they may also be conservative. For instance, we have assumed that variation has not increased in the post-DDT era. This assumption may not be valid, but we do not have sufficient samples of these five species at this time to analyze variation in such eggs. If recent eggs are sufficiently more variable in thickness, the sample sizes required would be somewhat larger.

Table 4 also presents estimated sample sizes for a one-egg-per-clutch collecting design. The number of clutches is increased by the difference in relative efficiency between the two designs.

SUMMARY

Measurements of shell thickness of the eggs of five species were subjected to nested analyses of variance. The analyses separated variation into two or three levels for which variances and percentages of the total variation were derived. The results show that differences among measurements of the same egg contribute little to the sample variance, whereas differences among eggs within clutches contribute nearly as much as differences among clutches. It is more efficient and less costly to collect entire clutches of eggs in most studies of shell thickness. Using entire clutches, sample sizes needed to detect differences of 10 percent in shell thickness (at given significance levels and power) were estimated to be eight to 11 clutches for the species studied. For differences of five percent, 26 to 38 clutches are required. Guidelines are presented which may assist other workers in evaluating the efficiency of their sampling designs, and in estimating sample sizes for detecting differences in eggshell thickness in wild birds.

ACKNOWLEDGMENTS

We thank L. J. Blus, D. R. Clark, Jr., and L. F. Stickel for their advice and critical reading of the manuscript. Data presented in this paper could not have been obtained without the generous cooperation of curators and their assistants at the following museums and universities: American Museum of Natural History, Carnegie Museum, Clemson University—Department of Zoology, Florida State Museum, Harvard University—Museum of Comparative Zoology, University of Massachusetts—Department of Zoology, Ohio State University—Museum of Zoology, Philadelphia Academy of Sciences, Smithsonian Institution—U.S. National Museum of Natural History, and Yale University—Peabody Museum of Natural History.

LITERATURE CITED

ANDERSON, D. W., AND J. J. HICKEY. 1970. Oological data on egg and breeding characteristics of Brown Pelicans. Wilson Bull., 82:14–28.

ANDERSON, D. W., AND J. J. HICKEY. 1972. Eggshell changes in certain North American birds. Proc. XV Internat. Ornith. Congress:514–540.

BLUS, L. J. 1970. Measurement of Brown Pelican eggshells from Florida and South Carolina. BioScience, 20:867–869.

BLUS, L. J., C. D. GISH, A. A. BELISLE, AND R. M. PROUTY. 1972. Logarithmic relationship of DDE residues to eggshell thinning. Nature, 235:376–377.

HICKEY, J. J., AND D. W. ANDERSON. 1968. Chlorinated hydrocarbons and eggshell changes in raptorial and fish-eating birds. Science, 162:271–273.

KOEMAN, J. H., C. F. VAN BEUSEKOM, AND J. J. M. DEGOEIJ. 1972. Eggshell and population changes in the sparrow-hawk (*Accipiter nisus*). *In*: Side-effects of persistent pesticides and other chemicals on birds and mammals in the Netherlands. Edited by J. H. Koeman. TNO-Nieuws, 27:542–550.

KREITZER, J. F. 1972. The effect of embryonic development on the thickness of the eggshells of Coturnix Quail. Poultry Sci., 51:1764–1765.

PEAKALL, D. B. 1970. Pesticides and the reproduction of birds. Scientific American, 222:73–78.

RATCLIFFE, D. A. 1967. Decrease in eggshell weight in certain birds of prey. Nature, 215:208–210.

RATCLIFFE, D. A. 1970. Changes attributable to pesticides in egg breakage frequency and eggshell thickness in some British birds. J. Appl. Ecol., 7:67–115.

ROMANOFF, A. L., AND A. J. ROMANOFF. 1949. The avian egg. John Wiley and Sons, New York.

ROTHSTEIN, S. I. 1972. Eggshell thickness and its variation in the Cedar Waxwing. Wilson Bull., 84:469–474.

SOKAL, R. R., AND R. J. ROHLF. 1969. Biometry. The principle and practice of statistics in biological research. W. H. Freeman & Co., San Francisco.

THOMPSON, A. L. 1964. A new dictionary of birds. McGraw-Hill, New York.

WELTY, J. C. 1962. The life of birds. W. B. Saunders, Philadelphia.

PATUXENT WILDLIFE RESEARCH CENTER, LAUREL, MARYLAND 20810. ACCEPTED 31 JANUARY 1974.

Reprinted from *Journal of the Fisheries Research Board of Canada*

Historical Changes in Mercury Contamination in Michigan Walleyes (*Stizostedion vitreum vitreum*)

T. M. KELLY

Great Lakes Resource Management Program, The University of Michigan, Ann Arbor, Mich. 41804, USA

J. D. JONES

Phoenix Memorial Laboratory, The University of Michigan, Ann Arbor, Mich. 41804, USA

AND G. R. SMITH

Museum of Zoology, The University of Michigan, Ann Arbor, Mich. 41804, USA

KELLY, T. M., J. D. JONES, AND G. R. SMITH. 1975. Historical changes in mercury contamination in Michigan walleyes (*Stizostedion vitreum vitreum*). J. Fish. Res. Board Can. 32: 1745–1754.

Mercury concentrations in Michigan walleyes (*Stizostedion vitreum vitreum*) are more variable among localities than between recent and historical samples within localities. Increases in mercury concentrations were noted in the recent populations of lakes Cadillac and Mitchell, Lake Gogebic, and western Lake Erie. Walleyes in four of seven localities showed no statistically significant increases in the level of mercury contamination over the past 40 yr. Mercury values over 0.5 ppm are common only on Isle Royale, in Saginaw Bay, and in western Lake Erie. Walleyes from Houghton Lake, Bear Lake, and lakes Cadillac and Mitchell show low levels of mercury contamination.

Methods for validly comparing preserved and fresh specimens were established by tests to determine the weight loss through dehydration in preservation. In 74% alcohol, muscle tissue of medium to large walleyes dehydrates to about 70% of the fresh weight in the first year.

Mercury concentration in walleye muscle from recent and old museum collections is positively correlated with age and size. Considerable variation in concentration exists not only between geographic localities, but within populations, and even within subsamples of the same tissue.

KELLY, T. M., J. D. JONES, AND G. R. SMITH. 1975. Historical changes in mercury contamination in Michigan walleyes (*Stizostedion vitreum vitreum*). J. Fish. Res. Board Can. 32: 1745–1754.

Les concentrations de mercure chez le doré (*Stizostedion vitreum vitreum*) du Michigan varient davantage entre les localités qu'entre les échantillons présents et passés au sein de mêmes localités. On a observé une augmentation des concentrations de mercure dans les populations récentes des lacs Cadillac et Mitchell, du lac Gogebic et du lac Erié occidental. Dans quatre localités sur sept, les dorés ne montrent pas d'augmentations significatives du niveau de contamination au mercure au cours des 40 dernières années. Des valeurs de mercure supérieures à 0.5 ppm ne sont communes qu'à Isle Royale, dans la baie Saginaw, et dans le lac Erié occidental. Les dorés du lac Houghton, du lac Bear et des lacs Cadillac et Mitchell ont de faibles niveaux de contamination au mercure.

A l'aide de mesures de perte de poids par déshydratation due à la préservation, nous avons mis au point des méthodes de comparaison valide entre les spécimens conservés et les spécimens frais. Dans l'alcool à 74%, le tissu musculaire de dorés de taille variant de moyenne à grande se déshydrate d'environ 70% du poids frais durant la première année.

La concentration de mercure dans le muscle de dorés provenant de collections de musée récentes et anciennes est en corrélation positive avec l'âge et la taille. Il existe des variations considérables non seulement entre les localités, mais également au sein des populations et même entre les échantillons d'un même tissu.

Received March 6, 1975
Accepted June 4, 1975

Reçu le 6 mars 1975
Accepté le 4 juin 1975

Printed in Canada (J3754)
Imprimé au Canada (J3754)

SINCE the discovery of elevated mercury concentrations in fish from the St. Clair River, Lake St. Clair, the Detroit River, and western Lake Erie, there has been interest in "natural" or preindustrial mercury concentrations. Various studies of preserved fish specimens obtained from museums have attempted to show time-related changes in mercury contamination of fish (Evans et al. 1972; Harris and Karcher 1971; Barber et al. 1972; Miller et al. 1972; and Nvorteva and Häsänen 1971). Unfortunately, the museum specimens used in past studies were not always comparable with recent specimens because of difference in size, species, locality, and probably preservation effects. In addition sample sizes were generally too small to enable conclusive interpretations.

The present study investigates: 1) effects of the preservation process, and conditions and techniques necessary for the valid use of preserved museum specimens for mercury analysis; 2) variation of mercury concentration within populations and within individuals; 3) historical changes in mercury concentration in seven populations of walleyes (*Stizostedion vitreum vitreum*) from Michigan waters.

Materials and Methods

Historical samples of walleyes had been collected between 1865 and 1936 from western Lake Erie, Saginaw Bay, and five inland lakes in Michigan (Table 1) and preserved in The University of Michigan Museum of Zoology. Comparative samples were obtained from each locality in the summer of 1971 by gillnetting. All fish were measured and several scales were removed for age determination. Changes in length due to preservation are probably insignificant (Stoba 1972).

Walleyes collected in 1971 were preserved according to a procedure that had been standard in the Museum of Zoology since about 1920. Whole fish specimens were fixed in a 10% solution of formalin for up to 1 wk, depending upon the size of specimens, washed in water for a day, and stored in 70% ethyl alcohol. Although it is possible that other methods of preservation were used in early museum collections, no evidence was found in Museum field or catalog notes to suggest this, with the exception of several Lake Erie specimens that had been frozen prior to normal preservation. As a check on possible contamination of museum specimens, stock solutions and alcohol samples from all storage containers (44 samples) were analyzed for mercury.

To expedite passage of preservation fluids into the body, the skin of recent specimens was slit at several places along the gut cavity and musculature. The fish were stored a minimum of 115 days before dissection to allow penetration of alcohol into the tissues. After the preservation period, a tissue sample weighing approximately 0.5 g was removed from the right dorsal musculature, blotted with tissue paper to remove excess surface fluid, and placed in a quartz ampoule (of known weight) which had been washed in nitric acid and heated in an oven at 200 C for 24 h. After receiving the sample, the ampoule was reweighed and sealed.

All fish specimens were analyzed for total mercury as described by Rottschafer et al. (1971). Total mercury in each sample was determined by comparing the mercury-197 in the specimen with the mercury in the analytical standard using the 77 Kev peak as the basis for comparison. Concentration in parts per billion was calculated on a live-weight basis.

Preservation Effects

To determine weight loss in fish muscle due to preservation, fillets from 11 Lake St. Clair fish (eight walleyes and three yellow perch, *Perca flavescens*) were weighed, fixed in 10% formalin, and washed in water for 1 day, again blotted and weighed, and placed in 74% alcohol. The fillets were then weighed at approximate 15-day intervals for the next 296 days. At 168 days the original alcohol was replaced by fresh 74% alcohol. This information was plotted and the proper weight-loss conversion factor was chosen from the curve, dependent upon the length of time the 1971 walleyes were stored in alcohol before mercury analysis. After preservation for 296 days, fillets were dried in an oven for 24 h at 105 C. Samples of muscle from each historical collection were also dried under the same conditions. The observed weight loss of recent fillets in alcohol for 296 days and the weight loss of recent and historical fillets due to drying were used to calculate the relative fresh weight of walleye muscle preserved for long periods in the museum. These data were used as standards to allow calculation of mercury concentration on a comparable basis.

To test for possible changes of mercury in fish when preserved and stored in alcohol, mercury analyses were performed on subsamples at several stages of preservation.

Within-Tissue Variation

The use of large numbers of valuable museum specimens precluded the removal of large sections of muscle tissue for homogenization and subsequent mercury analysis. The use of small (0.5-g) pieces of muscle for analysis allowed museum specimens to be sampled with almost no destruction of taxonomic features.

To evaluate within-tissue variability, and to determine possible information loss by the use of individual as opposed to homogenized samples, muscle tissue from two walleyes was analyzed for mercury using both procedures. Individual samples were removed at random from the dorsal muscle and analyzed for mercury. The remaining muscle tissue was completely removed from the fish and ground in a Waring Blendor to a fine fibrous mixture. Random subsamples of the fibrous mixture were then analyzed for mercury content. Sample variances were compared after both treatments.

Analysis

Least squares regressions of mercury concentrations on age and standard length were carried out on the data from each locality for each time period. Mercury concentrations were compared using one-way analysis of covariance to adjust differences in fish length or age. Mercury levels were then compared at the mean standard length for the pair of samples. The covariance model assumes equal slopes of mercury-length regression lines, but this assumption was not met in three cases (Houghton Lake, Isle Royale, and Saginaw Bay) because the length distribution of specimens from these localities was not sufficiently wide to establish one of the two regression lines needed for comparison. Relative mercury concentration was then determined by comparing the mean mercury concentration of the collection having a narrow size range with a predicted mercury concentration taken from the valid regression line. In Bear Lake neither the historical nor the recent collections had sufficient length range to establish a valid regression line.

Results

Preservation Effects

Fillets lost weight in formalin, returned to their original weight in water, and lost weight rapidly in alcohol (Fig. 1). The greatest weight loss (20% of fresh weight) occurred during the first week in alcohol, although the fillets continued to lose weight slowly until about 75% of the original weight remained after a period of 168 days in alcohol. There was a sharp decrease in weight at 174 days when the original alcohol, which was diluted to 61% was replaced by fresh 74% alcohol. A correction factor for each recent walleye collection was chosen from the weight loss curve (Fig. 1). For example, if the sample had dehydrated to 75% of its original weight, conversion of the observed mercury concentration to its fresh-weight equivalent was accomplished by multiplying the observed value by 0.75.

Inspection of Fig. 1 suggests that the muscle samples from specimens preserved in the museum

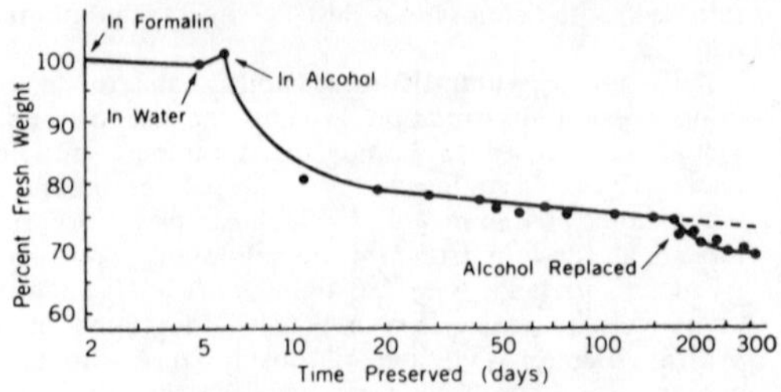

Fig. 1. Dehydration curve showing fluid weight loss of specimens in alcohol.

for many years might have dehydrated to about 70% of the original weight. Muscle samples from these museum specimens were oven dried, and yielded dry weights (D_p) of 25–35% of the preserved weight, depending on the relative volumes of fish and preservative in the containers. Muscle samples collected and preserved in 1971, which had dehydrated to 68% of the original weight, were oven dried, and yielded dry weights (D_f) of 18.8% of the fresh weight. If we assume that the dried weights of the old museum samples should also represent 18.8% of their original, fresh weight, the proportional dehydrated weight (P) of the preserved museum samples may be calculated as $P = D_f/D_p$, where D_f is the ratio of dried residue to fresh weight, and D_p is the ratio of dried residue to preserved weight. The values of D_p for each collection, and proportional dehydrated weights (P) of the preserved samples are presented in Table 1. The calculated values indicate that the old museum samples lost 22–35% of their weight during preservation, and the 1971 samples lost 25–26% during preservation. The dehydration rate of 1971 samples was maximum because fillets were cut and exposed to preservative.

Meek (1972) in a study of smaller specimens of other species suggested that old museum samples had lost several percent of their expected dry weight due to loss of fats into solution. If this is so, the estimates for dehydration in historical samples could be too high, and the correct fresh-weight equivalents for mercury concentrations would be several percent lower (depending on the mercury concentration in the fat) for all of the historical samples. Although the specimens collected in 1971 were slit to allow maximum penetration of preservatives, the rate of exchange in the experimental fillets would be a maximum estimate of the rate in the incised specimens. If a bias resulted, it would be in the direction of underestimating the mercury concentration in the 1971 specimens. However, even if the rate of fluid exchange was as extreme as 25% slower, the error would amount to only about 2%.

Mercury was not found in any of the preservation fluids used in the study either before or after use (average lower limits of sensitivity, 0.02 ppm). The average ratio of fish to preservative mass in grams was 0.23. It was estimated from these results that error due to exchange with preservatives must be less than 10%.

Within-Tissue Variation

Considerable unsuspected heterogeneity in mercury distribution within walleye fillets was discovered when samples from seperate muscle sites

TABLE 1. Sample data results of correlation and weight loss studies. Calculations are explained in the text.

Locality	Year and sample size	Weight loss experiment: Estimated preserved/fresh wt. (P)	Weight loss experiment: Dry/preserved wt., No. dried samples (D_f)	Correlation study: R Mercury vs. SL	Correlation study: R Mercury vs. age
Houghton L., 84°45′E, 44°19′N	1927(23), 36(5)	0.647	0.300 (32)	0.93	0.92
Roscommon Co.	1971(22)	0.735		0.73	0.72
L. Cadillac and Mitchell,	1927(18)	0.718	0.262 (12)	0.96	0.84
Wexford Co. 85°28′E, 44°14′N	1971(22)	0.735		0.77	0.65
Bear L., 86°9′E, 44°24′N	1932(13)	0.706	0.266 (7)	0.77	0.75
Manistee Co.	1971(16)	0.735		0.64	0.70
Saginaw Bay, 83°25′E, 44°02′N	1908(3), 17(3), 21(3), 28(2),	0.692	0.272(4)	0.96	0.87
L. Huron	29(4), 30(9), 1971(18)	0.735		0.23	0.45
W L. Erie, 83°20′E, 41°45′N	1865(2), 1920(2), 24(14),	0.709	0.261 (5)	0.48	0.49
Ohio and Michigan	26(2), 1971(24)	0.745		0.62	0.72
L. Gogebic 89°32′E, 46°28′N	1928(15), 29(4)	0.783	0.240 (16)	0.30	0.60
Ontonagon and Gogebic Co.	1971(21)	0.735		0.91	0.85
Isle Royale 88°42′E, 48°03′N					
L. Whittlesey, Dustin L.	1929(5)	0.747	0.252 (13)	0.48	0.19
Chickenbone L.	1929(8)		0.269 (6)		
L. Whittlesey	1971(26)	0.740		0.74	0.60

TABLE 2. Effects of homogenization on the variability of mercury concentrations in the muscle of two walleyes.

Collection locality	Treatment	n	$\bar{X}$	±SD	Coefficient of variation	F	Signf.
Houghton Lake	Individual	10	0.29	0.03	10.17	4.19	0.03
Houghton Lake	Homogenized	9	0.26	0.01	5.49		
Lake St. Clair	Individual	19	2.34	0.53	22.47	4.93	0.01
Lake St. Clair	Homogenized	9	2.39	0.24	9.91		

were compared with samples of a homogenate of the same fillet (Table 2). The existence of extensive within-population variation was apparent early in the study, but when repeated analyses of samples from the same fillet showed variation much in excess of that attributable to analytical error (5%), it was realized that a significant portion of the observed variation was due to heterogeneity within the muscle. The coefficient of variation (100 SD/$\bar{X}$) was 10.17 for separate sites and 5.49 for the homogenate of white muscle from a Houghton Lake walleye. The difference is slightly exaggerated in heavily contaminated fishes. In a Lake St. Clair specimen, the coefficient of variation was increased from 9.91 to 22.47 by sampling separate sites.

HISTORICAL CHANGES

Walleyes are large, predatory fish, high in the trophic system, thus there is a positive correlation between mercury concentration in the muscle and age or size (Johnels et al. 1967; Bache et al. 1971; Nvortiva and Häsänen 1971). Because of this relation, comparison of populations requires fish of similar size or age distributions for which mercury–size (or age) regressions can be compared. Owing to the vagaries of population dynamics and sampling, optimal-sized fishes were not always available for comparison. All but three of the studied populations show at least moderate positive correlation between mercury concentration and size (Tables 1, 3).

Two populations might be expected to show different mercury concentrations in one or more ways. A difference might be manifested as two regression lines of similar slope but different intercepts and different concentrations at each size. Also, populations might differ in regression slope, so that the populations with the steeper slope have significantly more mercury only in the larger size-classes. In the present analysis, populations with limited size distributions were analyzed in a special manner. Each population is discussed by locality. The data are given in Table 3 and plotted in Fig. 2 and 3.

In the analyses of three of the localities the coefficients of determination (R^2) were increased by regressing mercury concentration on age rather than standard length. Analyses of covariance of these populations, following adjustment for age, produced different results only in one case — that of Lake Gogebic (Table 3). In the other two instances, Saginaw Bay and western Lake Erie, the results and conclusions were unchanged, though in each case the tendency was to show an increase in the difference between the recent and historical sample and a decrease in the probability that the results were due to chance.

Houghton Lake — Walleyes from Houghton Lake had mercury concentrations typical for an uncontaminated lake. Specimens in the 280-mm (standard length) size-class had 0.17–0.18 ppm total mercury. Variation was not excessive. The regression slopes for mercury on length are slightly different, probably due to the inadequate recent sample. No historical differences were evident in the data (Fig. 2) and covariance analysis indicated no significant differences between mercury concentrations in the historical and recent samples (Table 3). We conclude that there is no evidence for historical changes in mercury in walleyes of Houghton Lake.

Lakes Cadillac and Mitchell — Walleyes from connected lakes Cadillac and Mitchell had rather low levels of mercury, and the mercury-length regressions are similar to those of Houghton Lake (Fig. 2, Table 3). The regression slopes are nearly equal and variation is minimal. In the 280-mm size-class, historical and recent samples had about 0.143 and 0.191 ppm mercury, respectively. Analysis of covariance indicates a small but highly significant difference between the historical and recent populations. Although the absolute increase is small, the percentage increase is between 31 and 34%.

Bear Lake — The lack of comparable age-classes from Bear Lake and the unrealistic regression lines resulting from restricted samples from each time period prevent covariance analysis and make the comparison unreliable (Table 3). The level of contamination in the historical sample was rather high, but recent values were low (not figured). There was abnormally high variation in

TABLE 3. Comparison of historical and recent concentrations of mercury in walleye populations based on analysis of covariance of mercury concentration on standard length.

Locality	Time	Pairwise adjusted mean concn	Prob. that adjusted means are equal	Coefficient of causation (R^2)	Regression slope (× 100)	Covariate slope (× 100)	Prob. that slopes are equal
Houghton L.	H	0.169	0.24	0.91	0.25	0.24	0.02
	R	0.183		0.50	0.16		
L. Cadillac,	H	0.107	0.002	0.86	0.26	0.25	0.24
Mitchell	R	0.141		0.64	0.19		
Bear L.	H	0.169	0.12	0.58	0.24	0.29	0.77
	R	0.089		0.46	0.30		
Saginaw Bay	H	0.300	0.84	0.89	0.33	0.26	0.00
	R	0.316		0.04	−0.06		
L. Erie	H	0.358	0.036	0.26	0.14	0.17	0.59
	R	0.454		0.38	0.18		
L. Gogebic	H	0.231	0.21	0.08	0.15	0.25	0.21
	R	0.292		0.89	0.28		
(Hg concn on age)	H	0.196	0.003	0.35	0.21	0.20	0.69
	R	0.339		0.87	0.19		
Isle Royale	H	0.443	0.24	0.19	0.12	0.21	0.06
	R	0.546		0.60	0.29		

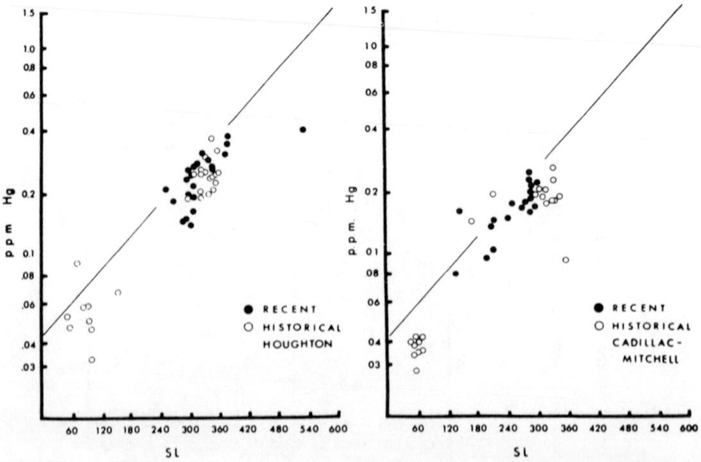

FIG. 2. Comparison of recent and historical mercury burdens in walleyes (*Stizostedion vitreum vitreum*) of Houghton Lake with lakes Cadillac and Mitchell. Concentration of mercury in walleye muscle is plotted against standard length of the fish. The line for comparison is the regression of mercury on standard length for all 283 specimens in the study.

the recent population; however, only one of the recent specimens possessed more than 0.5 ppm total mercury.

Saginaw Bay — The restricted recent sample of walleyes from Saginaw Bay did not give the size range necessary for an independent regression line (Fig. 3, Table 3). Assuming identical regression slopes, the analysis of covariance did not indicate a significant difference in total mercury between historical and recent samples. The historical regression line is slightly steeper than usual and the values may be elevated in the larger size-classes. Fishes in the 280-mm size-class had an average concentration of 0.21 ppm in the historical sample and about 0.25 ppm in the recent sample. The majority of fish over 350 mm in standard length had more than 0.5 ppm total mercury in 1971 (the unadjusted mean was 0.817 ppm).

Lake Gogebic — A peculiarity of the historical sample required a modification of the method of analysis for this locality. The correlation between mercury concentration and age was .60 (Table 1), but the correlation between mercury and standard length was low, $R = .30$, in the historical sample. These results are due to an attenuation of growth in some older individuals. Specifically, the four historical specimens with highest mercury values (Fig. 3) despite their modest size, are among the oldest fish in the sample; hence the variance is high. When the mean mercury concentration was corrected for size, the analysis of covariance did not indicate a significant difference in the two populations. However, when mercury concentration was regressed on age rather than standard length, the relation was nearly linear, and a significant difference between the recent and historical populations was indicated (Table 3).

A recent specimen 280 mm in standard length would be expected to have about 0.30 ppm total mercury in contrast to the expected historical value of 0.24 ppm. It is estimated that a 24% increase in the mercury contamination occurred in the walleye populations of Lake Gogebic.

Isle Royale — The lakes on Isle Royale in northern Lake Superior are uniquely sheltered from water-borne contamination of cultural origin, yet walleyes from several of these lakes had unusually high mercury concentrations (Fig. 3). Specimens 300 mm in standard length averaged just over 0.5 ppm mercury and large specimens commonly had more than 1.0 ppm. Unfortunately, the historical sample is variable, and lacks the mercury–length or mercury–age correlation necessary to enable a clear statistical comparison. (The high variability in the historical sample may be due to differences between lakes as well as to preservation effects.) The high mercury values in the historical and recent samples are probably due to the natural occurrence of mercury in the surrounding rocks. The surface geology of Isle Royale is unique among the localities studied.

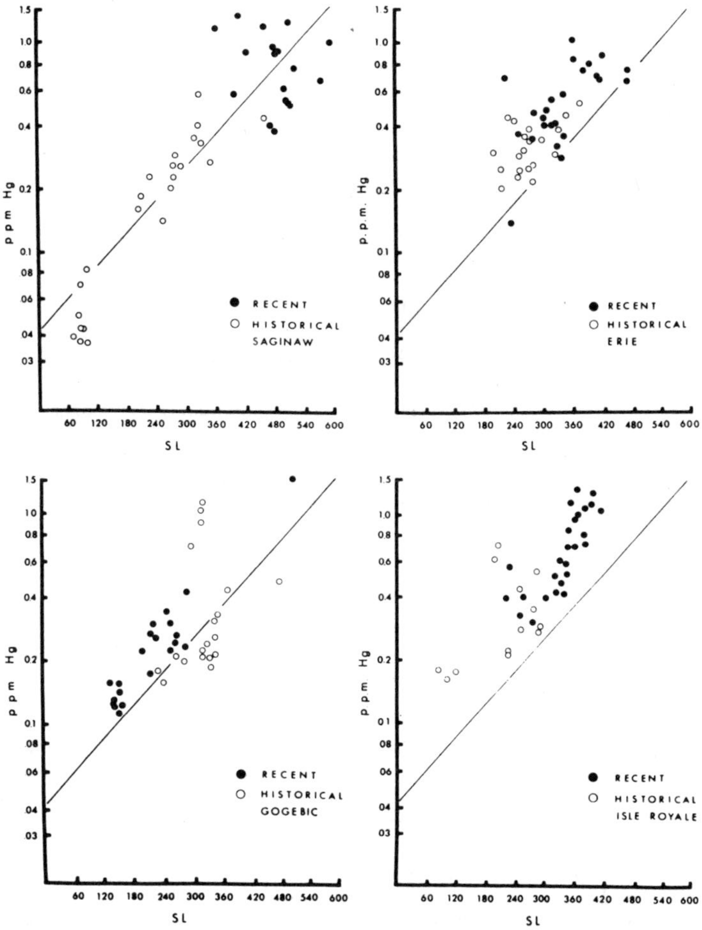

FIG. 3. Comparison of recent and historical mercury concentration in fishes of Saginaw Bay, Lake Gogebic, western Lake Erie, and lakes on Isle Royale. Concentration of mercury in walleye muscle is plotted against standard length. The line for comparison is the regression of mercury on standard length for all 283 specimens in the study.

The exposed surface of the Keweenawan basalt, conglomerate, sandstone, and associated copper deposits probably contribute more mercury than the glacial drift at the other localities.

Lake Erie, Western Basin — Analysis of covariance of the historical and recent walleyes from western Lake Erie, adjusted for size, indicated a significant increase in mercury concentration in the muscle. The covariate slopes are the same (historical = .0014, recent = .0018, $P = .59$) and adjusted means are significantly different. The mean mercury concentration, at a standard length of 307 mm is equal to 0.358 ppm for the historical sample and 0.454 ppm for the recent sample. The probability that the differences are due to chance is .036 (Table 3). The estimated increase is 27% (Fig. 3).

Discussion and Conclusions

The historical conclusions of this report depended on answers to general methodological questions concerning the effects of preservation and the nature and extent of variability in mercury concentration in fishes. The effects of preservation are considerable, but seem primarily related to changes in tissue weight due to dehydration in alcohol — we find no evidence of loss of mercury into any of the fixative or preservative fluids. Changes due to preservatives have been claimed (Gibbs et al. 1974), but are based on inadequate sample sizes and mixed tissues. The results are not distinguishable from normal variability and dehydration effects. The dehydration curve may be used as the basis of a correction factor that is crucial to any wet-weight analysis.

The loss of serum and fat molecules into the preservative, with mercury or independent of mercury, needs to be understood to allow more accurate calculation in wet-weight and dry-weight studies. An undetected weight loss (without a corresponding loss of mercury) of 2% from the historical samples would render our correction factors for those specimens too high by several percent, yielding a consistent overestimate of all historical values.

The most important methodological conclusion in this study is that there is significant variability in mercury concentration at every level in the system. The variability in the detection method is about 5%. Variability in excess of that amount exists between populations, between age- and size-classes within populations, between individuals of the same age- or size-class within populations, and between subsamples of the same fillet of an individual fish. Thus, the usual number of samples cannot be expected to give an accurate estimate of levels of contamination in a population or even in a single fish. A doubling of the standard deviation or coefficient of variation more than doubles the sample size necessary for statistical comparison. The within-tissue heterogeneity observed in this study is not necessarily general. Apparently, such variation was not observed by Westoo and Noren (1967) or Uthe (1971). Tissue differences can be corrected by using more samples or larger samples and homogenization. The increase in mercury concentration with age and size, well documented elsewhere, is the dominant component of variation in this study.

Assuming that the correction factors (Table 1) adequately compensate for the amount of dehydration in historical specimens, it can be concluded that the level of mercury contamination of walleyes increased in three of seven localities studied: lakes Cadillac and Mitchell (a connected system), Lake Gogebic, and the west end of Lake Erie. Isle Royale and Houghton Lake apparently have not undergone increases, although mercury values are, and were, unusually high in Lake Whittlesey on Isle Royale. Contamination in western Lake Erie has increased to a level approximately 27% higher than the historically high concentrations.

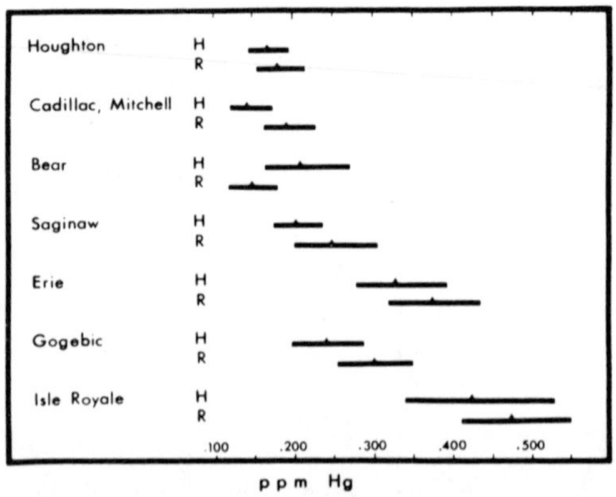

FIG. 4. Comparison of estimated means and 95% confidence limits for mercury concentration in muscle of historical and recent walleyes 280 mm in standard length, assuming a common covariate slope of .0024 for log mercury (in ppm) on standard length (millimeters).

Houghton Lake is probably representative of relatively uncontaminated conditions. Taking 280-mm specimens as an index, the level of contamination of walleyes in this lake is about 0.2 ± 0.05 ppm total mercury. Using the same index for the lakes with higher mercury concentrations, the comparable values are between 0.2 and 0.5 ppm (Fig. 4). Sample means and their confidence limits, adjusted to the average standard length (280 mm) for all 283 specimens in the study are given in Fig. 4.

The natural geographical variation in mercury levels is significant and provides an important perspective for evaluating contamination. The highest mercury levels found in this study occurred in the least "contaminated" population, on Isle Royale. Some of the lowest concentrations occurred in a system that has undergone a significant increase of more than 30% (lakes Cadillac and Mitchell), due to contamination in the past 44 yr.

Acknowledgments

We thank W. A. Willford, F. F. Hooper and C. Meek for assistance and encouragement during this project. Assistance in the field was provided by Ms. Meek and by members of the Michigan Department of Natural Resources, the National Park Service, and the Bureau of Sport Fisheries and Wildlife. Additional assistance was provided by the Fish Division, Museum of Zoology, the School of Natural Resources, and the Phoenix Memorial Laboratory, University of Michigan. The study was supported by funds from the Great Lakes Fishery Laboratory of the U.S. Bureau of Sport Fisheries and Wildlife.

BACHE, C. A., W. H. GUTENMANN, AND D. J. LISK. 1971. Residues of total mercury and methylmercuric salts in lake trout as a function of age. Science 172: 951–952.

BARBER, R. T., A. VIJAYAKUMAR, AND F. A. CROSS. 1972. Mercury concentrations in recent and ninety-year-old benthopelagic fish. Science 178: 636–639.

EVANS, R. J., J. D. BAILS, AND F. M. D'ITRI. 1972a. Total mercury content in the muscle tissues of preserved museum fish collected in the Lake St. Clair-Western Lake Erie region of the Great Lakes. Michigan St. Univ. Inst. Water Res. Tech. Rep. 22: 28 p.

1972b. Mercury levels in muscle tissues of preserved museum fish. Environ. Sci. Technol. 6(10): 901–905.

GIBBS, R. H. JR., E. JAROSEWICH, AND H. L. WINDOM. 1974. Heavy metal concentrations in museum fish specimens: effects of preservative and time. Science 184: 475–477.

HARRIS, E. J., AND R. KARCHER JR. 1971. Mercury: Its historical presence in New York State fishes. New York St. Dep. Environ. Cons. Rep. 15 p.

JOHNELS, A. G., T. WESTERMARK, W. BERG, P. I. PERSSON, AND B. SJOSTRAND. 1967. Pike (*Esox lucius* L.) and some other aquatic organisms in Sweden as indicators of mercury contamination in the environment. Oikos 18: 323–333.

MEEK, C. A. 1972. Mercury in fish populations of the Raisin and Huron River Basins, Michigan (1921–1971). M.S. Thesis. Univ. Michigan, Mich. 32 p.

MILLER, G. E., P. M. GRANT, R. KISHORE, F. J. STEINKRUGER, F. S. ROWLAND, AND V. P. GUINN. 1972. Mercury concentrations in museum specimens of tuna and swordfish. Science 175: 1121–1122.

NVORTEVA, P., AND E. HÄSÄNEN. 1971. Observations on the mercury content of *Myoxocephalus quadricornis* (L.) (Teleostei, Cottidae) in Finland. Ann. Zool. Fenn. 8: 331–335.

ROTTSCHAFER, J. M., H. B. MARK JR., AND J. D. JONES. 1971. A simple, rapid method for determining mercury in fish via neutron activation analysis. Environ. Sci. Technol. 5: 336–338.

STOBA, W. T. 1972. Effects of formalin on the length and weight of yellow perch. Trans. Am. Fish. Soc. 101: 362–364.

UTHE, J. F. 1971. A simple field technique for obtaining small samples of muscle from living fish. J. Fish. Res. Board Can. 28: 1203–1205.

WESTÖÖ, G., AND K. NOREN. 1967. Kvicksilver och metylkvicksilver i fisk. Vor Foda. 19: 135–178.

SECTION IX.
Cooperative Studies

The first two short papers included in this section could have been absorbed in the section on Identification. Whereas the papers in that section basically represent a service function of a museum, those segregated here report the cooperation of museum biologists in research conducted by other biologists in the field. In both instances, biologists on National Wildlife Refuges were carrying out local studies involving the capture and banding of birds. Their knowledge of the local avifauna of the area enabled them to recognize unusual distributional records for which they realized the value of documentation. Specimens were sent to a museum, where identifications were verified, where access to the literature in large libraries permitted further evaluation of their significance, and where the specimens could be preserved for future studies.

Although these two papers present distributional records, museums are not solely interested in the rarity or the oddity. The final paper in this section represents an increase in knowledge of birds made possible by the routine collecting and saving of specimens by field biologists not directly associated with a museum.

Many other studies are accomplished through the cooperation of field and museum biologists, where the latter takes his taxonomic techniques and expertise to the field. Several of these have been reprinted or cited elsewhere in this volume.

Many potentially valuable specimens are probably lost each year because field biologists fail to preserve them for eventual deposition in a museum. A number of studies on age determination of mammals, using dental characters, are reported in the literature with no indication that the material studied was saved or deposited in a museum. A major way in which museums cooperate with field biologists is by preserving their material for further study.

Additional Readings

Aldrich, John W., and Kenard P. Baer. 1970. Status and speciation in the Mexican Duck (*Anas diazi*). *Wilson Bull.* 82: 63-73.

Aldrich, John W., Allen J. Duvall, and Aelred D. Geis. 1958. Racial determination of origin of Mourning Doves in hunters' bags. *Jour. Wildl. Mgt.* 22: 71-75.

Avery, Michael L., and John S. Weske. 1978. Racial composition and differential migration of tower-killed Savannah Sparrows in southeastern North Dakota. *Bird-Banding* 49: 182-184.

Banks, Richard C. and Charles G. Hansen. 1970. Bird records from southern Nevada. *Condor* 72: 109-110.

Reprinted from *The Condor*

BIRD RECORDS FROM NEVADA

RICHARD C. BANKS

Bureau of Sport Fisheries and Wildlife
U. S. National Museum
Washington, D. C. 20560

AND

DONALD E. LEWIS[1]

Ruby Lake National Wildlife Refuge
Ruby Valley, Nevada 89833

Ruby Valley, a 75-mile-long hydrologically closed basin, lies in a land of high, rugged mountains and large, flat valleys in Elko County, northeastern Nevada. The southern portion of the valley is fringed by the Butte Range to the east and the majestic Ruby Mountains on the west. Nine peaks in the Rubies reach elevations of 11,000 ft or more.

Mist netting operations were carried on by Lewis in Ruby Valley during the fall migrations of 1966 and 1967 and the spring migration of 1967. The netting site was at 6000 ft elevation on Cave Creek, as it flows through the headquarters of the Ruby Lake National Wildlife Refuge. Dominant plants in this area are aspen (*Populus tremuloides*), chokecherry (*Prunus*), sumac (*Rhus*), and wild rose (*Rosa*). This island of riparian habitat at the base of the Rubies is attractive to migrant birds. Observations during periods of migration indicate a pronounced mass movement northward in the spring, whereas migration in the fall is an indefinite, casual filtering of birds to the south.

All birds netted were identified, banded, and released, with the exception of those representing important rcords for Nevada or additions to the Refuge bird list. These were frozen and sent to the Bird and Mammal Laboratories of the Bureau of Sport Fisheries and Wildlife where they were prepared as specimens. All specimens, unless otherwise noted, are in the collection of the U. S. National Museum, where they were identified by Banks. More than 40 species were added to the Refuge list, including the state and regional records listed below.

Greater Scaup. *Aythya marila nearctica.* Linsdale (1951) mentioned only a single Nevada record, from Churchill County, but had previously (1936) referred to Ridgway's (1877:343, 625) statements that the species was "abundant" as a "winter visitant at Pyramid Lake." No specimen was taken by Ridgway, and at least the portion of his record concerning abundance is probably in error. Gullion (1952) recorded a specimen taken near Riverside, Clark County, and Austin and Bradley (1966) reported a sight record in Clark County in March. On 25 November 1966 Lewis found a dead Greater Scaup near the Refuge Headquarters on Cave Creek. The bird was not sexed; plumage indicated that it was a female. The body was badly decomposed and only the head was saved. This appears to be the third specimen record of this duck in Nevada and the first in the northeastern sector. All the Nevada specimens were taken in November, during the hunting season; the recent sight record was in March. Additional observation in other months will probably reveal that this species is present in small numbers throughout the winter.

Band-tailed Pigeon. *Columba fasciata fasciata.* This species was considered a rare transient in Nevada by Linsdale (1951), who reported but a single record of the western race, *C. f. monilis*, from Churchill County. Linsdale apparently overlooked a second-hand sight record (Neff 1947:20) for the Duck Creek Range in eastern Nevada. More recently, Johnson (1965) reported that these pigeons occur as a breeding species in the Spring Range in Clark County, and tentatively assigned that population to the race *monilis* which is confined chiefly to the Pacific coastal drainage. Lewis collected a specimen of the Rocky Mountain race *C. f. fasciata* at Cave Creek in Ruby Valley on 25 October 1966. This bird was an adult male with testes measuring 6 × 3 mm and was moderately fat. The bird was accompanied by a single partner on the date of collection. One adult was observed in the same general area on 10 October 1967. This is the first record of this subspecies in the state, and the first report of the species in northeastern Nevada.

Saw-whet Owl. *Aegolius acadicus acadicus.* Linsdale (1936) reported a specimen taken in Thousand Springs Valley, Elko County, in 1868. More recent records of this species in Nevada (Linsdale 1951; Cottam 1954; Miller and Russell 1956) are from the western and southern portions of the state. Lewis collected a Saw-whet Owl at Refuge Headquarters on 28 October 1966. Two others, captured on 7 and 21 October 1966, were banded and released. The specimen is a male in immature plumage and is molting on the throat and crown. Measurements of the testes were 2 × 1.5 mm.

Yellow-shafted Flicker. *Colaptes auratus luteus.* This eastern race of flicker was first reported in Nevada by Ryser (1963) on the basis of a hawk-killed specimen from the Virginia Range near Sparks, Washoe County. Other records of Yellow-shafted Flickers are from the southern part of the state and have been referred to the race *C. a. borealis.* A female Yellow-shafted Flicker was collected on 16 October 1966 at the Refuge Headquarters and another was netted, banded, and released on that date. The specimen had only a trace of fat; the ovary measured 14 × 7 mm. It is referred to the subspecies *luteus*,

[1] Present address: Bowdoin National Wildlife Refuge, Malta, Montana 59538.

and represents the second record of this form in Nevada. As Short (1965) has advocated the merger of the race *borealis* into *luteus*, we must point out that the subspecific identification here reported applies to *luteus* even in the older, stricter sense.

Western Flycatcher. *Empidonax difficilis hellmayri.* Linsdale (1951) refers to the race *E. d. difficilis* as a "summer resident in small numbers" in middle elevations on Nevada mountain ranges; he does not mention *E. d. hellmayri*, at that time not considered to extend west of the Rocky Mountains (Brodkorb 1949). The AOU (1957) Check-list includes "eastern Nevada" in the breeding range of *hellmayri*, but without specific locality or substantiating record. As far as we can determine, Johnson (1965) was the first to document the occurrence of this subspecies in Nevada, reporting it a "fairly common summer resident between 7,600 and 9,300 feet" in the Sheep Range in the southern part of the state. A Western Flycatcher taken 1 June 1967 in Ruby Valley is referable to *E. d. hellmayri.* This bird was a female with a trace of fat and an ovary measuring 5 × 4 mm. This is, to our knowledge, the first documented report of the occurrence of this subspecies in northeastern Nevada.

Catbird. *Dumetella carolinensis ruficrissa.* There are but three reports of this species in Nevada, two based on specimens and one a sight record (Linsdale 1951; Hayward et al. 1963). A third specimen was obtained at Ruby Lake Refuge on 23 September 1966. This bird, a male, was moderately fat. The skull was incompletely ossified. The use of the trinomial follows Aldrich (1946); presumably the subspecific identification reported for this specimen applies as well to the previously recorded individuals.

Red-eyed Vireo. *Vireo olivaceus.* A female of this species was taken on 12 June 1967 at Refuge Headquarters. The bird had a trace of fat and an ovary 6 × 3 mm. This species is listed (Bureau of Sport Fisheries and Wildlife 1967) as having occurred on the Desert National Wildlife Range in southern Nevada, and Hansen (in litt. 5 January 1968) tells us that "These have been observed by me and other local ornithologists at Corn Creek. One was caught in 1966 and is now in the Malheur National Wildlife Refuge Museum as a mounted specimen." The occurrence of this species in Nevada is by no means unexpected, as it breeds in the northwestern United States and is known at least casually from all surrounding states. No trinomial has been applied to this specimen although it presumably belongs to the western population called *caniviridis* (see Burleigh 1960).

Myrtle Warbler. *Dendroica coronata.* Linsdale (1951) reported May specimens from Humboldt, Nye, and Esmeralda counties, and Hayward et al. (1963) recorded others from Nye County. Dr. Fred Ryser (pers. comm.) considers the species a rare transient in Nevada. On 27 October 1966 Lewis netted a Myrtle Warbler at the Headquarters of Ruby Lake Refuge. This specimen of undetermined sex was probably a bird of the year; it weighed 11.2 g and had little fat. This is the first report of this species in northeastern Nevada. The specimen is now in the Nevada State Museum, Carson City, where it was identified by Ryser.

Blackpoll Warbler. *Dendroica striata lurida.* On 22 May 1967 a male Blackpoll Warbler was captured on Cave Creek at Refuge Headquarters, when the northward wave of migrant warblers appeared to be at its peak. The specimen had a trace of fat; the testes measured 4 × 3 mm. This is apparently the first record of this species in Nevada. We follow Burleigh and Peters (1948) in the use of the trinomial.

Northern Waterthrush. *Seiurus noveboracensis.* An individual of this species was banded at the Refuge on 21 May 1963. On 18 September 1966 another bird was netted in Ruby Valley. This bird, saved as a specimen, was an immature female with heavy fat deposits. Although there are no references in the technical literature to prior specimens in Nevada, the species is listed (Bureau of Sport Fisheries and Wildlife 1967) as "uncommon" in spring and fall at the Desert National Wildlife Range in southern Nevada. Hansen (in litt. 5 January 1968) has informed us that "several birds of this species have been seen nearly every year in May and again in August and September and as late as the first of November" on that Range since 1960. He further states that "several have been banded at Corn Creek. There is a specimen in the Desert National Wildlife Range Museum collection that was taken by me in 1963." We follow Eaton (1957) who has suggested that the various named populations of this species are not well enough differentiated to merit recognition.

MacGillivray's Warbler. *Oporornis tolmiei tolmiei.* A male of this species was netted on 22 May 1967. Although the testes measured 7 × 5 and 6 × 4 mm, the bird was moderately fat, suggesting that it still had some distance to travel to complete its migration. Seven other birds of this species were banded and released during this study, four in the fall of 1966 and three in the spring of 1967. Linsdale (1951) considered Nevada birds of this species to be of the race *O. t. monticola*; Hayward et al. (1963) reported several specimens from southern Nevada which seemed not to be *monticola*, but expressed doubt as to the proper subspecific placement of the birds. The present record thus is the first definite one of the race *O. t. tolmiei* in the state. The trinomial is used here in the sense of the AOU (1957) Check-list, not in the strict sense of Phillips (1947).

Rose-breasted Grosbeak. *Pheucticus ludovicianus.* Sight records have been published for two occurrences of this species (three individuals) in southern Nevada (Rickard 1960). As far as we are aware, the species has not previously been collected in the state of Nevada. Lewis netted an immature male at Ruby Lake Refuge Headquarters on 15 November 1966. The bird had an incompletely ossified skull, very small gonads, and a trace of fat.

Indigo Bunting. *Passerina cyanea.* There seem to be but two previous records of this species for Nevada, both in Nye County (Linsdale 1951; Richardson 1952), but Hansen (in litt. 5 January 1968) mentioned that this species is seen almost every year in the spring and fall at Corn Creek on the Desert National Wildlife Range. Two additional specimens, the first for northeastern Nevada, were taken at the Ruby Lake Refuge Headquarters, a male (testes 8 × 5 and 8 × 6 mm, cloacal protuberance 5 × 3 mm) on 1 June 1967 and a female (ovary 3.5 × 2.5 mm) on 7 June 1967. Both specimens appear to be typical of the species and show no intergradation with the Lazuli Bunting, *P. amoena.* The earlier reported specimens were both males taken in June, and both had enlarged testes. The male reported here would certainly have been considered a member of the breeding population had it been in its proper range. Vagrant late migrants such as these (which were in company with Lazuli Buntings) may be partly responsible for the extensive hybridization between

Indigo and Lazuli Buntings (AOU 1957:552; Sibley and Short 1959).

Harris' Sparrow. *Zonotrichia querula.* A male captured at the Refuge Headquarters on 29 October 1966 had an ossified skull, testes that measured 2 × 2 mm, and no fat. This is the second record of the occurrence of this species in northeastern Nevada, and apparently the fifth record for the state. The specimen that Gullion (1957) referred to as the third record for the state was actually the fourth (see Linsdale 1936, 1951).

White-throated Sparrow. *Zonotrichia albicollis.* A female netted on 15 November 1966 at Refuge Headquarters had an unossified skull and was moderately fat. Hansen collected a male of this species (now in the U. S. National Museum) at the Corn Creek Field Station of the Desert National Wildlife Range in southern Nevada on 21 October 1960, and reported (in litt. 5 January 1968) that several individuals were present during the year. These records seem to be the third and fourth for this species in Nevada (see Gullion 1953), and the first for the northeastern part of the state.

Lincoln's Sparrow. *Melospiza lincolnii gracilis.* A specimen of this bird was taken on 12 October 1967 at the Refuge Headquarters on Cave Creek. It was an adult female with an ovary 4 × 2 mm and with moderate fat. The subspecific identification was made independently by Banks and by Roxie C. Laybourne. Although Linsdale (1951) cites records for other forms of this species, the northern Pacific coastal race *gracilis* has not previously been reported from Nevada.

ACKNOWLEDGMENTS

We are indebted to Charles G. Hansen, Wildlife Management Biologist at the Desert National Wildlife Range in southern Nevada, for providing information on the status of certain of the species in that region, and for permitting us to include his unpublished information in this report. Mrs. Roxie C. Laybourne of the Bureau of Sport Fisheries and Wildlife helped with a number of the identifications. James Creasy, Assistant Refuge Manager, aided in the mist-netting program.

LITERATURE CITED

ALDRICH, J. W. 1946. New subspecies of birds from western North America. Proc. Biol. Soc. Washington 25:129–136.

AMERICAN ORNITHOLOGISTS' UNION. 1957. Check-list of North American birds. Fifth ed. A.O.U. Baltimore.

AUSTIN, G. T., AND W. G. BRADLEY. 1966. Additional records for uncommon birds in southern Nevada. Great Basin Nat. 26:41–42.

BRODKORB, P. 1949. Variation in the North American forms of Western Flycatcher. Condor 51: 35–39.

BUREAU OF SPORT FISHERIES AND WILDLIFE. 1967. Birds of the Desert National Wildlife Range. Refuge Leaflet 132-R-3.

BURLEIGH, T. D. 1960. Three new subspecies of birds from western North America. Auk 77: 210–215.

BURLEIGH, T. D., AND H. S. PETERS. 1948. Geographic variation in Newfoundland birds. Proc. Biol. Soc. Washington 61:111–126.

COTTAM, C. 1954. Bird records for Nevada. Condor 56:223–224.

EATON, S. W. 1957. Variation in *Seiurus noveboracensis.* Auk 74:229–239.

GULLION, G. W. 1952. Recent bird records from southern Nevada. Condor 54:204.

GULLION, G. W. 1953. Additional bird records from southern Nevada. Condor 55:160.

GULLION, G. W. 1957. Miscellaneous bird records from northeastern Nevada. Condor 59:70-71.

HAYWARD, C. L., M. L. KILLPACK, AND G. L. RICHARDS. 1963. Birds of the Nevada test site. Brigham Young Univ. Sci. Bull., Biol. Ser. 3:1–27.

JOHNSON, N. K. 1965. The breeding avifaunas of the Sheep and Spring Ranges in southern Nevada. Condor 67:93–124.

LINSDALE, J. M. 1936. The birds of Nevada. Pacific Coast Avifauna no. 23.

LINSDALE, J. M. 1951. A list of the birds of Nevada. Condor 53:228–249.

MILLER, A. H., AND W. C. RUSSELL. 1956. Distributional data on the birds of the White Mountains of California and Nevada. Condor 58: 75–77.

NEFF, J. A. 1947. Habits, food and economic status of the Band-tailed Pigeon. N. Amer. Fauna no. 58.

PHILLIPS, A. R. 1947. The races of MacGillivray's Warbler. Auk 64:296–300.

RICHARDSON, F. 1952. A second record of the Indigo Bunting in Nevada. Condor 54:63.

RICKARD, W. H. 1960. An occurrence of the Rose-breasted Grosbeak in southern Nevada. Condor 62:140.

RIDGWAY, R. 1877. U. S. geological exploration of the fortieth parallel. Vol. IV pt. III. Ornithology. p. 303–669. U. S. Govt. Printing Office, Washington, D.C.

RYSER, F. A. 1963. Prothonotary Warbler and Yellow-shafted Flicker in Nevada. Condor 65: 334.

SHORT, L. L., JR. 1965. Hybridization in the flickers (*Colaptes*) of North America. Bull. Amer. Mus. Nat. Hist. 129:309–428.

SIBLEY, C. G., AND L. L. SHORT, JR. 1959. Hybridization in the buntings (*Passerina*) of the Great Plains. Auk 76:443–463.

Accepted for publication 31 July 1968.

Reprinted from *The Auk*

RECORDS, OBTAINED WHILE BANDING, OF BIRDS UNUSUAL IN SOUTHEASTERN OREGON

Eugene Kridler

Bird banding on the Malheur National Wildlife Refuge south of Burns, Harney County, in southeastern Oregon has continued to yield unusual records. Some of these were published in 1962 by me and David B. Marshall (*Condor*, 64: 162–164, 1962). Since that time, other noteworthy records have been made and are reported herein. All except that of the Red Phalarope were obtained at refuge headquarters. The last was taken at the Double-O unit of the refuge, an area located about 25 miles west of headquarters. Identification of all specimens was verified by Dr. John W. Aldrich and Mrs. Roxie C. Laybourne of the U. S. Bureau of Sport Fisheries and Wildlife at the U. S. National Museum in Washington, D.C., and the skins were deposited in the Bureau collection.

Red Phalarope (*Phalaropus fulicarius*).—On 25 June 1961 a female was taken in a mist net. Although thousands of Wilson's Phalaropes (*Steganopus tricolor*) and several hundred Northern Phalaropes (*Lobipes lobatus*) were in the vicinity, careful scrutiny with a spotting scope and binoculars of numerous phalarope flocks did not disclose any more Red Phalaropes. This species has been found previously in interior Oregon, but always in fall or winter (G. Alderson, *Condor*, 63: 97–98, 1961). Among previous records, that farthest removed from the coast was of one seen by S. G. Jewett near Summer Lake, 70 airline miles west of here (A. Einarsen, *Murrelet*, 22: 36, 1941). This is the first published summer record of the species occurring inland in Oregon, and is also the easternmost report for Oregon.

Red-eyed Vireo (*Vireo olivaceus*).—This species was first recorded in southern Oregon in 1960 (Kridler and Marshall, *op. cit.*: 163). In 1961, others were banded on 5 May, 22 August, and 14 September. The May record is the first published spring record for southern Oregon as well as the second record of the occurrence of this vireo in that part of the state. In 1962 seven more were banded, and four of these were captured in the same mist net within a few minutes of each other on 16 June. Four more were banded and two collected in the period 4 to 14 June 1963.

Tennessee Warbler (*Vermivora peregrina*).—A female (USNM no. 479637) was taken on 12 June 1963. This constitutes the first known record of the species in Oregon. Another was netted on 15 October 1963. After comparing this bird with a skin in the refuge museum, I took color photographs and the bird was banded and released. The photographs are in the refuge files.

Black-throated Blue Warbler (*Dendroica caerulescens*).—I banded and released a male on 24 October 1961. Two days later it was recaptured and released. On 2 May 1963 I closely observed a male in full breeding plumage just outside a refuge office window. I saw it again the next day, and on 4 May I pointed it out to members of Dr. Charles Quaintance's ornithology class from Eastern Oregon College. This May bird provides the first spring record of the species in Oregon. The two birds constitute the third and fourth published records for the state, and all were obtained in the same group of trees here (Marshall, *Condor*, 61: 55, 1959; Kridler and Marshall, *op. cit.*: 164).

Ovenbird (*Seiurus aurocapillus*).—A female (USNM no. 478486) taken on 4 June 1961 was identified as a member of the subspecies *S. a. aurocapillus*. On 13 June 1963 I captured, banded, and released another bird of this species. As far as I can determine these are the only Oregon records.

Northern Waterthrush (*Seiurus noveboracensis*).—I took a male (USNM no. 478475) on 9 May 1961. It was referred to the subspecies *S. n. notabilis.* On 15 September 1962 I banded and released two more. The only other published record for Oregon is that of a male collected by Reed Ferris near Beaver in western Oregon on 26 August 1931 (I. N. Gabrielson and S. G. Jewett, *Birds of Oregon*, Corvallis, Oregon State College, 1940).

American Redstart (*Setophaga ruticilla*).—The first published record for southeastern Oregon was that of a female that I collected here on 8 September 1960 (Kridler and Marshall, *op. cit.*: 164). In 1961 I banded a female on 6 June for the first summer record in this part of the state. Four were banded in that year, seven more in 1962, and one in 1963. The latest fall date among these records is 16 October 1962.

Painted Bunting (*Passerina ciris*).—On 2 June 1963 a male in full breeding plumage was frightened into one of my nets by Mr. William Bechtel of Sparks, Nevada. It was characteristic of the western race *P. c. pallidior* and now bears USNM no. 479638. This is the first known record of its occurrence not only in Oregon, but in the Pacific northwest.

Purple Finch (*Carpodacus purpureus*).—I took a male (USNM no. 478849) on 3 November 1961. Gabrielson and Jewett (*op. cit.*: 537) considered this species a permanent resident of western Oregon with a range extending to the east slope of the Cascade Mountains. They stated that one specimen was taken by L. Miller in 1899 in the Blue Mountains of Wheeler County in north-central Oregon. However, Dr. Alden H. Miller (letter) tells me that the specimen, now in the Museum of Vertebrate Zoology at the University of California in Berkeley, is a Cassin's Finch (*Carpodacus cassinii*) and not a Purple Finch as had been reported. Therefore, this 1961 specimen is the first published record for a Purple Finch in Oregon east of the Cascades.—EUGENE KRIDLER, *U. S. Bureau of Sport Fisheries and Wildlife, 835 Akumu St., Kailua, Hawaii.*

Vol. 81, pp. 485–490 30 December 1968

PROCEEDINGS
OF THE
BIOLOGICAL SOCIETY OF WASHINGTON

PENTLAND'S TINAMOU IN ARGENTINA (AVES: TINAMIDAE)

BY RICHARD C. BANKS AND WAYNE H. BOHL
Bureau of Sport Fisheries and Wildlife; U.S. National Museum, Washington, D.C., and Foreign Game Investigation Program, Utah State University, Logan, Utah

Field work in Argentina in 1965 and 1966 by personnel of the Foreign Game Investigation Program, of the Bureau of Sport Fisheries and Wildlife, had as one of its results the accumulation of a collection of nearly 300 specimens of tinamous, which have been deposited in the U.S. National Museum. Among these was a series of 15 Pentland's Tinamou, *Nothoprocta pentlandii* (Gray), from the provinces of Mendoza, San Luis, and Cordoba, at or near the southern end of the range of the species (Hellmayr and Conover, 1942; Olrog, 1963). Comparison of these southern specimens with others from the more northerly provinces of Catamarca, Tucuman, and Salta and from Bolivia revealed that two distinct southern populations were represented—a previously undescribed one in Mendoza and one in the Sierras de San Luis and Cordoba. There are thus three subspecies in Argentina, as follows.

Nothoprocta pentlandii pentlandii (Gray)

Range: From the Bolivian departments of Chuquisaca, Santa Cruz, Cochabamba, and La Paz south along the Andean foothills of western Argentina to southern Catamarca.

Comments: Comparison of this subspecies with others is made below. This is the only subspecies currently listed as occurring in Argentina by Olrog (1963), who did not recognize the next form. The type locality is in the Bolivian Andes.

Nothoprocta pentlandii doeringi Cabanis

Range: Sierras de San Luis and Cordoba, central Argentina.

Diagnosis: Similar to the northerly nominate race in size (Table 1) but grayer in the general tone of the dorsum, resulting from gray rather than brown feather edgings; more heavily spotted on the upper breast, and with a less extensive pale area centrally on the abdomen.

Comments: Nothoprocta doeringi was named by Cabanis (1878) on the basis of a specimen sent to him from Sierra Chica, near Cordoba. Sclater (1879) noted that this population was closely allied to *N. pentlandii,* and Sclater and Hudson (1889) considered the name *doeringi* synonymous with *pentlandii.* The name applied by Cabanis to the population of the Sierra Cordoba has seldom been used since that time. Examination of the series of four fully grown birds available to us from that mountain range reveals that that population is distinct from the nominate form, as indicated above. Further, five specimens from the Sierra de San Luis, in San Luis Province, are not separable from the Cordoba birds. The name bestowed by Cabanis is available and must apply to these populations, which are apparently isolated from other populations by lowlands containing unsuitable habitat.

One specimen of *N. p. pentlandii* from near La Merced, Catamarca, which of those examined is the nearest geographically to the Cordoba population, shows an approach to *N. p. doeringi* in the grayness of the dorsal coloration.

As far as we are aware, the occurrence of this species of tinamou in the Province of San Luis has not previously been reported.

Nothoprocta pentlandii mendozae new subspecies

Holotype: USNM 530855; female, Quebrada de Canota, Mendoza Province, Argentina; collected 7 May, 1966, by Wayne H. Bohl, original number 100.

Diagnosis: Differs from *N. p. pentlandii* and *N. p. doeringi* by having a much longer bill (Table 1) and additionally from *pentlandii* by a slightly longer wing, on the average. The general tone of the dorsal coloration is gray, rather than brown, this appearance resulting from the color of the lateral edges of the back feathers. In this respect, *mendozae* is similar to *doeringi* but quite distinct from *pentlandii.* The breast is more heavily spotted than in *pentlandii* and the abdomen is darker than in either *pentlandii* or *doeringi,* being suffused with pale gray or grayish-brown rather than buff.

Range: Known definitely only from the northwestern corner of the Province of Mendoza, Argentina. Other reports of *N. pentlandii* from Mendoza (Sanzin, 1917; Reed, 1919) and from northern Neuquen (Olrog, 1963) probably also apply to birds of this subspecies.

Comments: Bill size is the most distinctive character of the subspecies *N. p. mendozae.* The bill averages approximately 15 per cent longer in members of this population than in either *pentlandii* or *doeringi* (Table 1), and there is no overlap in the samples available to us. Wing length is slightly greater in *mendozae* than in *pentlandii,* but

TABLE 1. Measurements (mm.) and weight (gm.) of subspecies of *Nothoprocta pentlandii* in Argentina.

		pentlandii	*doeringi*	*mendozae*
	N	10	9	6
Wing	Range	120.2–142.8	123.0–143.2	134.5–145.7
	Mean	133.6	135.4	139.3
	N	9	9	5
Bill	Range	21.0–24.0	21.0–25.1	26.3–27.6
	Mean	22.6	23.3	26.9
Tarsus plus	N	10	9	6
Middle Toe	Range	60.2–67.8	60.2–68.4	65.2–68.6
	Mean	65.0	65.1	66.9
	N	2	10	14
Weight	Range	255–305	211–330	301–384
	Mean	280.0	265.7	339.9

doeringi is intermediate in this regard. Similarly, the length of the tarsus and middle toe combined is greater, on the average, in *mendozae* than in the other subspecies. The general impression of greater size in *mendozae* is borne out by a consideration of weight. Fourteen specimens of *mendozae* averaged 340 gm., whereas ten *doeringi* averaged only 266 gm. In both instances, equal numbers of each sex were considered; not all birds weighed were saved as specimens. Two examples of *pentlandii*, both males, weighed 255 and 305 gm.

DISCUSSION

Pentland's Tinamou is a bird of moderate to high elevations; all specimens taken in the course of field work by FGIP personnel were at elevations of 3000 feet or above. Thus the low country in the extreme northwest portion of Cordoba and the extreme southeast part of Catamarca, containing the Salinas Grandes, would be an effective barrier providing for the isolation of *N. p. doeringi* from *N. p. pentlandii*. Similarly low-lying country to the west of the Sierra de San Luis, either to the east or west of the Alto Pencosco, separates *doeringi* from *mendozae*. The Rio Bermejo and its tributaries in northern San Juan Province represent a possible barrier between *mendozae* and *pentlandii*; the country is considerably more arid to the south and west of the rios Troya, Bermejo, and Desaguadero than to the north and east.

Olrog (1963) has divided Argentina into a number of ornithological zones. The distribution of the subspecies of Pentland's Tinamou outlined in the previous paragraph fits well the zonation mapped by Olrog.

Nothoprocta p. pentlandii is distributed roughly in the "Subzona Subtropical Occidental" (p. 14), *N. p. doeringi* in the "Subzona Chaquena Occidental" and *N. p. mendozae* in the "Subzona Chacoandina" (p. 26).

The three subspecies discussed above were also compared with a population from well north of the range of *N. p. pentlandii,* called *N. p. oustaleti* Berlepsch and Stolzmann by Hellmayr and Conover (1942). The sample of this form examined averages considerably larger than any of the southern races of *N. pentlandii* in length of wing and of tarsus and middle toe. Differences in color and in feather pattern are so great that it is difficult to consider *oustaleti* conspecific with the forms of *pentlandii* seen. Indeed, *oustaleti* was considered to be a distinct species by Peters (1931).

Special comment should be made on the methods of measurement used in this study. The scutellation of the tarsus and middle toe is extremely variable in the vicinity of the juncture of these two elements. This variability made it difficult, if not impossible, to obtain measurements of either segment that would be comparable from one individual to another. However, a single scute could be used as the boundary point between tarsus and toe on a single specimen, and adding the measurements for these two segments yields a figure more nearly comparable between populations than either measurement alone. The bill chord was measured from the tip to the point of the "V" formed by the cere. Sexual dimorphism in size is slight (a maximum of 3 per cent) with a population, and measurements for both sexes are combined because of the small number of specimens available.

We wish to acknowledge gratefully the assistance in the field of Srs. Luis Arturo, Ernesto Roco, and Peter Miles. Dr. Gardiner Bump provided assistance in many ways.

Specimens Examined

N. p. pentlandii. (9)—Bolivia: Cochabamba: Vacas, 2752 and 3800 m. (3); Arani, 2752 m. (1); Pocona, 2700 m. (1). Argentina: Salta: 25 mi. W Salta, 5500 ft., (1); Tucuman: San Pedro de Colalao (2); Catamarca: near La Merced, 4000 ft. (1).

N. p. doeringi. (10)—Argentina: Cordoba: La Falda, 3000–3200 ft., (3); La Cumbre (1 plus 6 chicks); San Luis: vic. El Volcan, 3000–3400 ft. (4); 7 mi. SW San Francisco, between 2400 and 2800 ft. (1).

N. p. mendozae. (6)—Argentina: Mendoza: Quebrada de Canota, 4800–6200 ft. (5); 2 mi. SW Puesto Hoyada, 4500 ft. (1).

N. p. ? oustaleti. (6)—Peru: Lima: San Mateo (2); Huinco (1); Apurimac: Andahuaylas Hda., La Laguna, 3040 m. (1); Ecuador: Loja: Malacatos, 5000 ft. (2).

Bolivian specimens of *pentlandii* and all *oustaleti* were borrowed from the Chicago Museum of Natural History through the courtesy of Emmet R. Blake. All other specimens are in the U. S. National Museum.

Literature Cited

Cabanis, J. 1878. Ueber eine Sammlung von Vögeln der Argentinischen Republik. Jour. f. Ornith., 26: 194–199.

Hellmayr, C. E., and B. Conover. 1942. Catalogue of Birds of the Americas and the Adjacent Islands. Part I, No. 1, Field Mus. Nat. Hist., Zool. Ser., Vol. 13, pt. I, no. 1.

Olrog, C. C. 1963. Lista y Distribucion de las Aves Argentinas. Opera Lilloana IX, Instituto Miguel Lillo, Univ. Nac. de Tucuman. 377 p.

Peters, J. L. 1931. Check-list of Birds of the World. Vol. 1. Harvard Univ. Press, Cambridge. 345 pp.

Reed, C. S. 1919. Breves notas acera de nidos y huevos de algunas aves de la cordillera de Mendoza. Hornero, 1: 267–273.

Sanzin, R. 1917. Lista de aves Mendocinas. Hornero, 1: 147–152.

Sclater, P. L. 1879. [On a collection of birds.] Proc. Zool. Soc. London for 1879: 460–461.

Sclater, P. L., and W. H. Hudson. 1889. Argentine Ornithology. Vol. II. London. 251 pp.

SECTION X.
Conclusion

It seems almost mandatory that a book-length treatment of a species or group of species important in traditional management terms begin with a chapter or section devoted to taxonomy. This fact in itself is an indication of the strong ties between the disciplines of management and systematics. Such introductory chapters serve as a statement by the writers or editors that all information that follows must be placed into the geographic, physiological, behavioral, and morphological context of recognizable and named subspecies. They are warnings that the selection of management tools and techniques must be appropriate to the varying requirements of varying populations. They are suggestions that what is good for the goose is not necessarily good for the gander—particularly if it is a Giant Canada Goose and an Aleutian Canada Gander.

I have not chosen to reprint any of these introductory chapters to monographic treatments, but several are listed below. They are generally quite long, and this book has turned out to be lengthy as it is. They seldom represent original research, and frankly, they seldom give any internal indication of why they are there.

The final paper in this volume deals with birds, but its principles are far more widely applicable. It ties together the sections of this book and it states what the monograph chapters cited below imply. It is the essence of this book, and once again points out the debt that we in both disciplines owe to John Aldrich.

Additional Readings

Aldrich, John W. 1967. Taxonomy, distribution and present status. Chapter 2 in *The Wild Turkey and its management,* Oliver H. Hewitt, ed. The Wildlife Society, Washington, D.C.

Bellrose, Frank C. 1976. Classification. Chapter 2 in *Ducks, geese & swans of North America.* Stackpole Books, Harrisburg, Pa.

Collette, Bruce B., and Petru Banarescu. 1977. Systematics and Zoogeography of the fishes of the Family Percidae. *Jour. Fish. Res. Bd. Canada* 34: 1450-1463.

Cottam, Clarence, and James B. Trefethen. 1968. The family tree. Chapter II in *Whitewings, the life history, status and management of the White-winged Dove.* D. Van Nostrand Co., Princeton, N.J.

Johnsgard, Paul A. 1973. Evolution and Taxonomy. Chapter 1 in *Grouse and quails of North America.* University of Nebraska Press, Lincoln.

Tuck, Leslie M. 1972. Taxonomy and geographical distribution. Chapter 3 in *The snipes: a study of the genus* Capella. Canadian Wildlife Service, Monograph Series no. 5.

Reprinted from *The Proceedings XIIIth International Ornithological Congress*

Population Systematics in Bird Conservation

JOHN W. ALDRICH

U.S. Fish and Wildlife Service, Washington 25, D.C., U.S.A.

Population systematics as characterized by Mayr (1959) is the classification of species in terms of their component populations which tend to differ from one another in different parts of their range. This involves the determination of morphological, physiological, and behavioral differences between populations within species and the correlation of these with the isolating and selection factors in the environment that produce and maintain them.

Bird conservation is defined in the present discussion as the process of managing populations of all species of birds to obtain optimum numbers and distribution for maximum enjoyment, utilization, and conformance with economic requirements of human populations. With the rapid increase of human populations, this obviously is becoming an increasingly more complicated problem.

PURPOSE

The purpose of this paper is to call to the attention of the conservationist the potential usefulness in wildlife management of a knowledge of the ecological, physiological, and behavioral characteristics that may be associated with morphological variations of population segments within species.

Although recognizing that geographical isolation facilitates the establishment of new heritable characteristics in a population, there is much evidence that adaptation to environment through natural selection of genetically produced variations is the chief process by which biologically different population segments of species are formed (Huxley, 1938; Thorpe, 1945; Mayr, 1951; Miller, 1956). Furthermore, physiological and behavioral characteristics are frequently linked with the morphological variations (Mayr, 1951; Ross, 1962). Therefore, critical study of morphological variations within species at the subspecific level or below may give clues to the physiological and behavioral as well as morphological characteristics that are being selected by the environment. This knowledge in turn may indicate what the most critical factors in the environment are that are selecting the observed characteristics and which at the same time may be controlling the size and distribution of the population. Wildlife-conservation programs are particularly concerned with such factors and thus should utilize clues that may be discovered through the study of racial or ecotypical variants within species and apply them to the study of population management.

EVIDENCE FOR ADAPTATION TO ENVIRONMENT

The so-called Bergmann's, Allen's and Gloger's rules governing the relationship of climate to size and color variations of warm-blooded animals are

Proc. XIII Intern. Ornithol. Congr.: 819–830. 1963

well known. Mayr (1956) has recently reviewed and elaborated on these. Hamilton (1961) has further developed the concept of the adaptive significance of trends in size of birds in relation to climate.

Plant ecologists for a long time have recognized the importance of the combined climatic factors of temperature and moisture in controlling the distribution of plants of particular life form, such as coniferous forest, deciduous forest, grassland, and desert scrub. Animal ecologists are in general agreement that the distribution of animal species conforms, to a large extent, to the limits of those climax plant aggregations that characterize the areas of different combinations of climatic factors (Clements and Shelford, 1939; Pitelka, 1941; Kendeigh, 1961). In addition, the infraspecific (racial) variation of animals also tends to conform to the limits of these major ecological subdivisions.

Pitelka (1941) has commented that the species showing variations between climax areas are not necessarily those confined to the climax habitat. They may occupy the climax habitat in one area and a seral habitat in another. For example, one race of the Ovenbird (*Seiurus aurocapillus*) occupies deciduous forest of both climax and seral stages in the eastern part of North America and another race occupies only the deciduous seral stages of mountain coniferous-forest climaxes in the West. Pitelka further pointed out that geographic variation occurs chiefly in those species, such as Ruffed Grouse (*Bonasa umbellus*) Sharp-tailed Grouse (*Pedioecetes phasianellus*), Downy Woodpecker (*Dendrocopos pubescens*), Song Sparrow (*Melospiza melodia*), and Screech Owl (*Otus asio*), ranging broadly over areas with several climax types. Miller (1956) noted these same correlations in analyzing the ecological distribution of birds in California. He further commented that, although the races of vertebrates are not precisely like those ecotypes of plants that tend to occur wherever a particular ecological situation is present, they do resemble plant ecotypes in that they have evolved distinctive characters adapted to ecological conditions, but the adjustment seems to be related to a considerable range of such conditions. These conclusions are based on the relationships of named subspecies and their ecological distribution. A more critical study of morphological differences in populations correlated with areas characterized by different ecological climaxes would probably reveal greater agreement than has been demonstrated so far.

If we compare specimens representative of the different major ecological climax areas of North America, based on climax vegetation, we find that Ruffed Grouse from the dripping, mossy, dark-barked rain forest of the Olympic Peninsula in the northwestern United States are dark mahogany brown, whereas those from the dry, light-barked aspen growths of the Rocky Mountains in Wyoming are pale gray. This same general color difference, correlated with the same life areas, is repeated in a number of other species, such as the Blue Grouse (*Dendragapus obscurus*), Song Sparrow, Fox Sparrow (*Passerella iliaca*), and Red-shafted Flicker (*Colaptes cafer*), even though they have quite different habits and habitats. Grinnell (1910) noted this same difference in birds from the humid coast of Alaska and interior situations. Thus, it appears that this sort of correlation occurs in a definite geographical pattern involving species of diverse ancestry and habits, and suggests a factor or combination of factors in the environment of that area which select the same characteristics in different species for perpetuation in that population. Examination of specimens representing these same species from ecologically intermediate (not necessarily geographically intermediate) climax areas discloses intermediate color characters that are correlated with the degree of difference in the environment.

Furthermore, there is frequently a convergence of characters in populations of birds of the same species in relatively similar environments on opposite sides of the North American continent, which are separated from each other by much more distinctly different-appearing populations exposed to the quite different environments of the interior regions. A few examples are Black-capped Chickadees (*Parus atricapillus*), Ruffed Grouse, Robins (*Turdus migratorius*), Song Sparrows, Gray Jays (*Perisoreus canadensis*), Horned Larks (*Eremophila alpestris*), and Meadowlarks (*Sturnella magna* and *S. neglecta*).

One of the most convincing lines of evidence for the importance of ecological factors in guiding morphological variations is that, in geographic areas of the greatest ecological diversity, morphological variation is also greatest. This may be the case even though there are no more physiographic barriers to contribute to disruption of gene flow in that area than in others of greater ecological and morphological uniformity.

Marked morphological differences can occur in relatively short distances if sharp differences in environment occur. Examples are: differences between the appearance of Bobwhite Quail (*Colinus virginianus*) on the Wabash River bottomland and those of the surrounding uplands in Illinois, noted by Robert Ridgway (1895); the case of Song Sparrows of marsh and upland habitats in the San Francisco Bay, California, area (Marshall, 1948*a*, 1948*b*); variation in the Song Sparrow in mountain and desert associations in southwestern Utah (Marshall and Behle, 1942); color differences in Crested Larks (*Galerida cristata*) living on soils of different colors in Egypt (Moreau, 1930); apparent protective coloration of birds on desert soils in Arabia, discussed by Meinertzhagen (1950); a similar situation in Horned Larks on different-colored soils in North America, described by Behle (1942); and differences in Rock Partridges (*Alectoris graeca*) noted by Watson (1962) at different altitudes in the mountains of southern Turkey.

Friedmann (1946) has called attention to the multitude of instances of convergence in morphological, ecological, and behavioral characteristics of representatives of different taxonomic groups of birds occupying similar niches, but in different continents. Some examples cited were the resemblance of the American meadowlarks (*Sturnella*) and the African pipits

(*Macronyx*), some African species of weaver-birds (Ploceidae) and some American troupials (Icteridae), some South American tanagers (Tanagridae) and some North American finches (Fringillidae). These cases of parallelism of unrelated species that fill similar ecological niches are strong evidence of the potency of environment through natural selection in molding the morphology and behavior of birds. A study of the characteristics that these species have in common would suggest which ones are of importance to their survival.

There has been very little experimental work with birds to find out how certain population characteristics are produced through natural selection or their survival value. Avian taxonomists have been engrossed in describing the differences and naming the recognizable populations. Avian ecologists have been involved with studying the species composition and interrelationship of biotic communities, and in measuring environmental factors. The study of the relationship between these disciplines in the dynamics of speciation in birds has barely begun.

A few cases where higher vertebrates have been viewed experimentally from the ecotypical standpoint might be mentioned. Sumner (1925) and Blair (1950) found correlation between color of various species of deer mice (*Peromyscus*) and the hue of the soil on which they lived. Blair went further and found a correlation between the frequency of genes giving color to animals most closely resembling the soil color and the actual distribution of the mice. A recent study of great significance in this field is that of Bowers (1960), which demonstrated a correlation between the variation in color in Wrentits (*Chamaea fasciata*) and their environment. These little birds apparently exhibit very definitive ecotypical variations. The shade of brown in their plumage varies with the general tone of the background in their habitat from the dark-stemmed, close-growing chaparral of the more humid Pacific Coast to the light-stemmed, more open brush of the more arid interior valleys of California. However, we are still left with the unanswered question of why this correlation exists. A unique experiment to show the relative frequency of capture by owls of deer mice placed on backgrounds contrasting with the mouse's pelage color in varying degrees was performed by Dice (1947). The experiment suggested that predation may be a significant mechanism producing natural selection for color in the higher vertebrates. There is need for a great deal of controlled experimental work on this and other possible selection factors.

PROCESS OF POPULATION CHANGE

The significance of habitat in the evolution of birds has been discussed by a number of authors, including Stresemann (1943), Thorpe (1945), Mayr (1947), and Miller (1942, 1955, and 1956). The first small adaptive changes probably come about mainly through exploration and extension of ranges of species into appropriate unoccupied niches in new areas of different ecological types. Presumably this is by young individuals that have not nested before and become attached to specific sites, and which, therefore, are thought to be the pioneers in range extension. Extension of range is a complicated ecological process involving not only adaptation to the physical environment, but competition with other species occupying related ecological niches in the new environment. Stresemann (1943), Thorpe (1945), Gause (1947), Mayr (1951), and Miller (1955, 1956) believe that most distinct populations are started by individuals that are able to adjust themselves (nongenetically) to new conditions when colonizing a new region. Once such a population comes to occupy a new environment, a new set of selective factors will operate to reconstruct this population genetically.

Miller (1955) analyzed the rather considerable amount of data on the population dynamics and infraspecific variations in Song Sparrows. He concluded that random or chance fixation of genes was responsible for the initial variations noted in populations, and that in some distinct-appearing island populations this is the only obvious cause of the difference. In the continental populations, however, he noted what appeared to be ecologically controlled variations superimposed on the random variation. He favored the idea that natural selection is aided and accelerated by the availability of a local group of Song Sparrows possessing a considerable proportion of beneficial genetic traits acquired partly by chance, rather than by the occasional individual with slightly favorable characteristics which must be selected from a large population of less well-endowed individuals.

Although agreeing that genetic change after extension of range is the general situation, Mayr (1951) qualifies this by stating that in all cases of explosive range expansion, such as have occurred in the Serin Finch (*Serinus serinus*) and the Ringed Turtle Dove (*Streptopelia decaocto* [= *risoria*]), there is reason to believe that this expansion followed a genetic alteration of the peripheral populations, as postulated by Aldrich (1946). Actually, evidence is lacking as to which of these sequences is most frequently the case. However, it would seem that successful adaptation of a bird species to a new environment may be like the development of tolerance of insect species to an insecticide. A few mutants that can survive in a different environment eventually arrive there, and a new, adapted population develops from them.

Salomonsen (1955) calls attention to the different ways environment may act as a force of natural selection, depending on how populations are isolated from one another. He points out particularly that adaptive characteristics of migrant species must be in harmony with both winter and summer quarters. In some, winter conditions seem to be more important as selective forces, and in others the breeding-ground conditions. In some species, populations tend to mingle on the wintering grounds, as in the case of Savannah Sparrows (*Passerculus sandwichensis*) (Norris and Hight, 1957), and others tend to be separated. The interesting cases of leap-frog migration cited for Fox Sparrows by Swarth (1920) and for Yellow Wagtails (*Motacilla flava*) by

Stresemann (1943), in which the more northern breeding population winters farther south than the more southern breeding group, are examples of the latter.

RELATIONSHIP OF MORPHOLOGICAL, PHYSIOLOGICAL, AND BEHAVIORAL CHARACTERISTICS

Extremely pertinent to the thesis presented in this paper is the fact that physiological and behavioral differences frequently accompany morphological variation. It is not necessarily the morphological character that is adaptive but rather the underlying gene. The same gene may produce not only the visible conditions but also all sorts of invisible physiological processes which may have very pronounced selective properties (Mayr, 1951). Rigid artificial selection for long-leggedness in chickens was found in one instance to result in reduced rate of reproduction because of genetic linkage of these two characteristics (Ross, 1962). This suggests that reduced rate of reproduction in small populations of certain wild bird species may result from genetic linkage with some factor that is being subjected to intense natural selection. Innumerable cases of such character linkage have been discovered in the exhaustive genetic experiments with the insect group *Drosophila*.

There are many examples of morphologically distinct variants of bird species that also have distinct physiological or behavioral characteristics. The White-crowned Sparrows (*Zonotrichia leucophrys*) of the Pacific Coast of North America have a sedentary population (*nuttalli*) in the San Francisco Bay region and a migratory one (*pugetensis*) nesting from northern California north to southern British Columbia (American Ornithologists' Union, 1957). In Canada Geese (*Branta canadensis*), different breeding populations have differentiated spectacularly in size, color, voice, nesting sites, nest structure, and migration season. Movement of these populations has resulted in bringing two very different forms into close proximity on the Yukon Delta, Alaska, and southern Baffin Island, but apparently without interbreeding in either case (Sutton, 1932; Soper, 1946; Aldrich, 1946; Mayr, 1951.)

Ecotypical differences in populations may be entirely in physiology or in habits, without morphological manifestation. Lack (1947) shows a correlation between (a) the number of eggs laid by individual birds of the same species and (b) the climate. Larger clutches are laid in the cooler, more northern parts of the range. Brooks (1940) and Kendeigh (1945) describe differences in nesting habitat chosen by different populations of certain species of New World warblers with no detected morphological variation. However, it should be noted that the observers did not mention any study of specimens to determine whether or not morphological differences existed. Slight differences in color, size, or proportions correlated with ecological distribution can be easily overlooked. For many years this was the case in the eastern populations of the Traill's Flycatcher (*Empidonax traillii*). This species had been noted by field ornithologists to have different songs and types of nests long before the discovery that these habits, as well as prairie and boreal forest distribution, were correlated with slight morphological differences (Aldrich, 1951, 1953; Snyder, 1953; Stein, 1958).

In another instance of this sort, Bullough (1942) explained the differences in the physiological cycles of the resident British population and the migratory continental population of the European Starling (*Sturnus vulgaris*). The difference in the reproductive cycles of these two populations was considered to be the basis for all the other differences in habits and morphology. The continental Starlings reach a peak of gonad development much later than the British birds. The white tips to the feathers wear off on the underparts of British individuals sooner than the continental ones because of abrasion from entering winter roosting cavities, which are not used by continental birds. The beaks of British Starlings turn yellow in the autumn and not until February in continental examples. Thus, all differences between these two geographically distinct populations were thought to be either physiological, behavioral, or due to mechanical wear. However, even with this well-studied species which had been thought to show physiological differences in two populations without morphological manifestations, there is evidence that a slight color difference between them might have been overlooked in the earlier observations, since Meinertzhagen (1947) has mentioned a more purplish color of the heads of British Starlings.

SIGNIFICANCE TO CONSERVATION

What is the significance of all this to conservation? Knowledge of the process of natural selection of adaptive characteristics in the process of speciation is important in understanding the factors that control occurrence or abundance of a species. If a factor is critical enough to bring about evolutionary change within species through natural selection, it would seem likely that it can be critical also in controlling the distribution and abundance of that species. Having been molded by such exacting environmental forces into a special form adapted to a particular ecological situation, it is reasonable to suppose that such an adapted population segment might not thrive in a distinctly different environment as long as it possesses these modifications (Aldrich, 1946). This has been demonstrated in the case of attempted Bobwhite and Turkey (*Meleagris gallopavo*) transplanting in Georgia and Texas, respectively, and in the case of innumerable foreign bird species introduced unsuccessfully into North America. This concept should be considered in research designed to locate any species and particularly subspecies and ecotypes of game birds for transplanting into new territory or areas of former occurrence. The principle should always be to find species and their populations that are preadapted to the particular ecological niche to be filled. Whenever introductions occur, specimens should be preserved as a permanent record of the morphological characteristics of the stock that was introduced so that these may be correlated with the records of success or failure and thus will be of use in the future.

Examples of putting this principle into practice in wildlife management may be cited. The U.S. Fish and Wildlife Service is attempting to develop new breeding populations of Canada Geese on its wildlife refuges where the species does not nest at present. To do this, it is trying to use birds of the same subspecies that formerly nested in the area or those with habitat most similar to the area to be stocked. Also this Service is cooperating with the International Association of Game, Fish and Conservation Commissioners and with many states in a program of selecting geographical races of foreign game birds that are preadapted to presently unfilled ecological niches in the United States. Certain state game farms are going even farther and crossing races of these game birds in hopes of obtaining genetic types that will be better adapted than already existing stock. This is experimentation in the dynamics of speciation, attempting to find a shortcut to the long processes of evolution through natural causes. Nature works on such problems strictly by trial and error, but has the virtue of persistence. Man, through population-systematics studies, may be able to improve on nature's method by facilitating the arrival of the stock most likely to succeed at the right place for establishment.

There is much need for additional experimental breeding of different species, geographic races, and ecotypes of birds of special interest from the conservation standpoint. This should involve the study under controlled conditions of behavioral and physiological differences that are correlated with morphological differences, also of environmental factors that may be limiting to distribution or function in selection of new characters. Great strides have been made in producing desired characteristics of plumage, size, shape, egg laying, and temperament in domestic birds through artificial selection in captive breeding. A similar approach with captive wild stock offers a great challenge. Captive rearing of two races of the Sandhill Crane (*Grus canadensis*) being conducted currently by the U.S. Fish and Wildlife Service, as well as many private avicultural ventures, provide opportunities to obtain this sort of information.

The conservation of migratory birds is particularly complicated by the difficulty of determining the migratory movements of the various segments of the population within each species. Much has been learned of the migrations of some of these population units through banding representative individuals. In certain circumstances where banding is not feasible, another possible approach is the identification and recording of representatives of these populations when found away from their breeding area. Identification would be by the distinctive characteristics of size and color which we may learn to recognize as representative of a particular geographical or ecological region. Aldrich and Burleigh (unpubl. MS.) demonstrated that the abundant Lesser and rare Greater Sandhill cranes (*Grus c. canadensis* and *G. c.*

tabida) have different migration patterns, permitting hunting of the former without endangering the latter. By collecting large series of specimens, Kalmbach (pers. comm.) determined that Red-winged Blackbirds (*Agelaius phoeniceus*), causing damage to rice crops in southwestern Louisiana during the ripening period, were the resident coastal group, whereas blackbirds that caused damage to shocked grain in the fall were mostly migrants of northern groups. By identifying subspecies of Mourning Doves (*Zenaidura macroura*) in hunters' bags in southern Florida, Aldrich (1952) determined that these birds had come almost entirely from breeding areas in the United States rather than the West Indies, as had been supposed. This discovery resulted in modification of hunting regulations. Norris and Hight (1957) identified races of Savannah Sparrows caught in mist nets in ecological studies near Savannah, Georgia, as coming from several widely separated sections of North America. It has been conceived by some that such studies may be useful in geographical delineation of activity of certain viruses carried by birds.

In addition to recognizable morphological differences as indicators of distinct populations, we have evidence that identification of serological differences may be a rewarding approach to population systematics, and one that might become an important tool in wildlife management. Although not yet demonstrated with bird populations, Fujino (1962) found differences in blood types of certain populations of Antarctic whales, and Ridgway (1962) and Sindermann (1962) were able to detect serological differences in populations of fish species of commercial significance.

Preservation or creation of habitat suitable for the support of a desired species is one of the recognized methods of birds conservation. Obviously, differences in habitat requirements of different population segments of a species would be vital information to such a management program. Since variations in environment are important factors in the process of population differentiation, study of the way species vary morphologically through natural selection by the environment may give clues to the existence of physiological and behavioral differences in these population segments that will lead to a better understanding of habitat requirements. This would be of value in important programs of land acquisition for habitat preservation.

PROCEDURES

In the course of future studies of this sort we are confronted with correcting inadequate methods of indicating differences in populations which might be of significance in their management. Sibley (1954) described very well the inadequacy of trinomial nomenclature for indicating variation in infraspecific populations. This has been of growing concern to students of population systematics, and no adequate solution has been found. The general practice has been to use the formal trinomial for obviously different populations with at least 75 percent of the individuals of one distinct from

99+ percent of all others (Amadon, 1949), and merely to refer to variations within this population, including character clines of Huxley (1938), in a descriptive way. The lack of uniformity in formal description and naming of subspecies, so poignantly described by McAtee (1941), is definitely a drawback to using present nomenclature for designating the sort of meaningful adaptive differences of interest in the present paper. To be useful for management purposes, it will be necessary to work out units that group individuals with morphological differences, no matter how slight, that are correlated with ecological distribution or other factors of importance in species management. As Sibley (1954) and others have pointed out, it is undesirable to clutter up the standard trinomial nomenclature further by attempting to give each of these variants a formal name. Furthermore, it would be too cumbersome for use in management. It would be preferable to use a series of numbers or short descriptive phrases to indicate the units of interest.

In consideration of the foregoing, it seems reasonable to presume that a thorough study and classification of morphological variations through analysis of present specimen collections which can be associated with exact habitat conditions will suggest many patterns of ecotypical variation. Judicious collecting of additional specimens accompanied by precise data on environment will be necessary in most cases to establish definitely these relationships. We need more study of the correlation of variations with environment to learn which are adaptive and thus give clues as to the critical factors in the environment that control survival and distribution. When critical morphological studies are correlated with ecological, physiological, and behavioral information on the various ecotype populations, good basic biological knowledge of the species of great practical value to the wildlife manager will be at hand. Although quite different in approach from the Linnaean typological sort of systematics, neo-Darwinean population systematics still depends upon the museum study specimen as a basis for much of the necessary morphological information. Thus can research in population systematics in natural-history museums contribute in the future to basic information necessary for sound procedures in bird conservation.

SUMMARY

Population systematics classifies species in terms of population isolates, thus stressing the differences between populations within species. Adaptation to environment through natural selection of genetically produced variations is the chief process by which biologically different populations are formed. Visually distinct populations tend to be limited to the areas of distinct ecological climaxes. Marked morphological differences can occur in relatively short distances if there are sharp differences in environment. Genetic change in populations may involve extension of range of a species into new environment. Migratory populations must be adapted to both winter and summer ranges to survive. Adaptive physiological and behavioral differences frequently accompany morphological variation. Knowledge of adaptive variation in physiology, behavior, and morphology of populations aids selecting appropriate stock for introduction. Recognition of morphological and serological characteristics of populations is useful in following their movements. Population systematics needs a system of designating populations without dependence on the straitjacketed and cumbersome standard systems of trinomial scientific nomenclature. Study and classification of morphological variations of populations within species correlated with physiological and behavioral differences will give information of value to bird conservation.

LITERATURE CITED

ALDRICH, J. W. 1946. Significance of racial variation in birds to wildlife management. J. Wildl. Mgmt. 10:86–93.

———. 1951. A review of the races of the Traill's Flycatcher. Wilson Bull. 63:192–197.

———. 1952. The source of migrant Mourning Doves in southern Florida. J. Wildl. Mgmt. 16:447–456.

———. 1953. Habits and habitat differences in two races of Traill's Flycatcher. Wilson Bull. 65:8–11.

AMADON, D. 1949. The seventy-five percent rule for subspecies. Condor 51:250–258.

AMERICAN ORNITHOLOGISTS' UNION. 1957. A check-list of North American birds. 5th ed. Am. Ornithologists' Union [c/o Louisiana State Univ. Mus. Zool., Baton Rouge, La.]. xiii + 691 p.

BEHLE, W. H. 1942. Distribution and variation of the Horned Larks (*Otocoris alpestris*) of western North America. Univ. Calif. Publ. Zool. 46:205–316.

BLAIR, W. F. 1950. Ecological factors in speciation of *Peromyscus*. Evolution 4:253–275.

BOWERS, D. E. 1960. Correlation of variation in the Wrentit with environmental gradients. Condor 62:91–120.

BROOKS, M. 1940. The breeding warblers of the Central Allegheny Mountain region. Wilson Bull. 52:249–266.

BULLOUGH, W. S. 1942. On the external morphology of the British and continental races of the Starling (*Sturnus vulgaris* L.). Ibis, 14th Ser., 6:225–239.

CLEMENTS, F. E., and V. E. SHELFORD. 1939. Bio-ecology. John Wiley & Sons, Inc., New York, N.Y. 423 p.

DICE, L. R. 1947. Effectiveness of selection by owls of deer mice (*Peromyscus maniculatus*) which contrast in color with their background. Contrib. Lab. Vert. Biol., Univ. Michigan 34:1–20.

FRIEDMANN, H. 1946. Ecological counterparts in birds. Sci. Monthly 63:395–398.

FUJINO, K. 1962. Blood types of some species of antarctic whales. Am. Naturalist 96(889):205–210.

GAUSE, G. F. 1947. Problems of evolution. Trans. Conn. Acad. Sci. 37:17–68.

GRINNELL, J. 1910. Birds of the 1908 Alexander Alaska expedition. Univ. Calif. Publ. Zool. 5:361–428.

HAMILTON, T. H. 1961. The adaptive significances of intraspecific trends of variation in wing length and body size among bird species. Evolution 15:180–195.

HUXLEY, J. S. 1938. Species formation and geographical isolation. Proc. Linn. Soc., London, Session 150, 1937–38, Pt. 4:253–264.

KENDEIGH, S. C. 1945. Community selection by birds on the Helderberg Plateau of New York. Auk 62:418–436.

———. 1961. Animal ecology. Prentice-Hall, Inc., Englewood Cliffs, N.J. 468 p.

LACK, D. 1947. The significance of clutch size. Ibis 89:302–352.

MCATEE, W. L. 1941. Thoughts on subspecies. Sci. Monthly 53:368–371.

MARSHALL, J. T., JR. 1948*a*. Ecologic races of Song Sparrows in the San Francisco Bay region. Pt. I, Habitat and abundance. Condor 50:193–215.

———. 1948*b*. Ecologic races of Song Sparrows in the San Francisco Bay region. Pt. II, Geographic variation. Condor 50:233–256.

———, and W. H. BEHLE. 1942. The Song Sparrows of the Virgin River Valley, Utah. Condor 44:122–124.

MAYR, E. 1947. Ecological factors in speciation. Evolution 1:263–288.

———. 1951. Speciation in birds. Proc. X Intern. Ornithol. Congr.:91–131.

———. 1956. Geographical character gradients and climatic adaptation. Evolution 10:105–108.

———. 1959. Trends in avian systematics. Ibis 101:293–302.

MEINERTZHAGEN, R. 1947. On the Starling (British race) *Sturnus vulgaris britannicus* Bullough. Bull. Brit. Ornithol. Club 68:19–21.

———. 1950. Some problems connected with Arabian birds. Ibis 92:336–340.

MILLER, A. H. 1942. Habitat selection among higher vertebrates and its relation to intraspecific variation. Am. Naturalist 76:25–35.

———. 1955. Concepts and problems of avian systematics in relation to evolutionary processes, p. 1–22. *In* A. Wolfson [ed.], Recent studies in avian biology. Univ. of Illinois Press, Urbana. 479 p.

———. 1956. Ecologic factors that accelerate formation of races and species of terrestrial vertebrates. Evolution 10:262–277.

MOREAU, R. E. 1930. On the age of some races of birds. Ibis, 12th Ser., 6:229–239.

NORRIS, K. A., and G. L. HIGHT, JR. 1957. Subspecific variation in winter populations of Savannah Sparrows: A study in field taxonomy. Condor 59:40–52.

PITELKA, F. A. 1941. Distribution of birds in relation to major biotic communities. Am. Midland Naturalist 25:113–137.

RIDGWAY, G. J. 1962. The application of some special immunological methods to marine population problems. Am. Naturalist 96:219–224.

RIDGWAY, R. 1895. The ornithology of Illinois. Vol. 2, pt. 1. Nat. Hist. Surv. of Illinois, Springfield. 282 p. XXXII pl.

ROSS, H. H. 1962. A synthesis of evolutionary theory. Prentice-Hall, Inc., Englewood Cliffs, N.J. 387 p.

SALOMONSEN, F. 1955. Evolution and bird migration. Acta XI Congr. Intern. Ornithol. 1954:337–339.

SIBLEY, C. G. 1954. The contribution of avian taxonomy. Systematic Zool. 3:105–125.

SINDERMANN, C. J. 1962. Serology of Atlantic clupeiod fishes. Am. Naturalist 96:225–231.

SNYDER, L. L. 1953. On eastern empidonaces with particular reference to variation in *E. traillii*. Contrib. Royal Ontario Mus. Zool. and Palaeontol. 35:1–26.

SOPER, J. D. 1946. Ornithological results of the Baffin Island expedition of 1928–1929 and 1930–1931, together with more recent records. Auk 63:1–24.

STEIN, R. C. 1958. The behavioral, ecological and morphological characteristics of two populations of the Alder Flycatcher, *Empidonax traillii* (Audubon). New York State Museum and Sci. Serv. Bull. 371:1–63.

STRESEMANN, E. 1943. Oekologische Sippen-, Rassen- und Artunterschiede bei Vögeln. J. Ornithol. 91:305–328.

SUMNER, F. B. 1925. Some biological problems of our southwestern deserts. Ecology 6:352–371.

SUTTON, G. M. 1932. The exploration of Southampton Island, Hudson Bay. Mem. Carnegie Museum 12, Pt. 2, Sec. 2. p. 1–275.

SWARTH, H. S. 1920. Revision of the avian genus *Passerella*, with special reference to the distribution and migration of the races in California. Univ. Calif. Publ. Zool. 21:75–224.

THORPE, W. H. 1945. The evolutionary significance of habitat selection. J. Animal Ecol. 14:67–70.

WATSON, G. E. 1962. Three sibling species of *Alectoris* partridge. Ibis 104:353–367.

scale). When P&G replaced P&S in 1983, the suggested techniques for evaluating ecosystems used the Habitat Evaluation Procedures (HEP), which produces an estimate of effective habitat units for one or a few key animal species in a given stream segment or terrestrial area as the product of raw acres of habitat and the HSI. Reliance on spatial extent of habitat remains the rule after more than two decades. A 1996 review of monetary and non-monetary valuation of environmental investments (Feather et al., 1995) listed 15 techniques for ecological assessment, about two-thirds of which were the HEP and its variants adapted to specific types of habitats or geographical areas. Adaptations were made by several states to support their wetland mitigation programs under Section 401 of the CWA or independent state authority.

One of an array of methods used in the In-stream Flow Incremental Method for assessing the effects of minimum flow requirements to protect fish life is also a variant of the HEP (Stalnaker et al., 1995). The PHABSIM model, developed by the USFWS, is an adaptation of the HEP where the HSI values are functions of depth, cover, bottom structure, temperature and other factors affecting fish survival and propagation. The model has been widely used to establish in-stream flow requirements attached to federal and state permits for new dams, relicensing of existing dams or other actions that could alter in-stream flows.

The HEP and the PHABSIM are methods for assessing the quantity and suitability of a given set of habitat conditions for the reproduction and feeding of selected species. The given habitat conditions may be either those of existing habitats or those of predicted future habitats. For future habitat conditions, the basic hydrology, hydraulics, geomorphology and vegetation must be predicted by other means. For example, the River of Grass Evaluation Methodology (ROGEM) used by the USACE and the SFWMD to evaluate outcomes of Everglades Restoration actions is a variant of the HEP. It was used to estimate the magnitude of habitat under current conditions and under projected future conditions with the restoration project. Future habitat conditions were predicted using the SFWMD model and the Across Trophic Level System Simulation (ATLSS). It was assumed by USACE and SFWMD that optimal habitat conditions would occur if hydrologic conditions were managed to achieve natural or pre-development conditions (USACE and SFWMD, 1999).

ATLSS is a predictive model consisting of a large set of linked models for lower trophic levels; population models for functional groups of fish, macro invertebrates, amphibians and reptiles; and individual-based models for consumers (including Cape sable seaside sparrow, snail kite, wood storks, white-tailed deer and Florida panther). Developers describe the models as still evolving and identify four primary sources of uncertainty:

- Future climate and weather.
- Imperfect understanding and representation of major processes in physical and biological models.
- Important physical and biological parameters used in equations describing processes or initial conditions.
- Stochastic components.

For complex models, such as ATLSS, issues of model validation and other uncertainties are difficult to assess. It is argued that ATLSS is intended to provide relative assessments of alternative management strategies, not exact quantitative predictions. The assumption, then, is that uncertainties do not affect relative assessments (see the ATLSS attachment to USACE and SFWMD, 1999). That assumption is testable but only over a very long time period if sufficient data from continued monitoring are available.

The EPA and other federal agencies support a variety of studies of terrestrial ecosystems covering such topics as data collection on land use and land cover, regional landscape characterization, effects of chemicals and land use and related topics. Few of these efforts have advanced to the stage where they can be used in predictive models for effects of water resources management decisions.

The EPA also supports one model for freshwater ecosystems, AQUATOX (Park et al., 2004), which extends water quality models similar to those in QUAL2E and WASP6 by tracking temperature and concentrations of nutrients and organic chemicals and predicting their effects on selected species in the ecosystem, including fish, invertebrates and aquatic plants. It can represent vertically stratified lakes and streams (www.epa.gov/waterscience/models/). A sketch of interactive components in AQUATOX is shown in Figure 4.11. AQUATOX, similar mechanistic water quality models and the ATLSS ecosystem model being developed for the Everglades Project are built on a combination of mass balance equations (one for each species in the model) and biochemical kinetic expressions for rates of transformation from one species to another. Outputs are the concentrations or masses of species distributed over time and space. That approach is appropriate for programs where objectives are stated in terms of quantities of selected species of aquatic plants or animals.

Among a limited number of applications, AQUATOX has been used to track the fate and effects of PCBs in Lake Ontario and as part of the study of PCB contamination, attributed to the General Electric Company, of the Housatonic River in Massachusetts (EPA, 2004). The framework proposed to model the latter problem setting consisted of the HSPF for

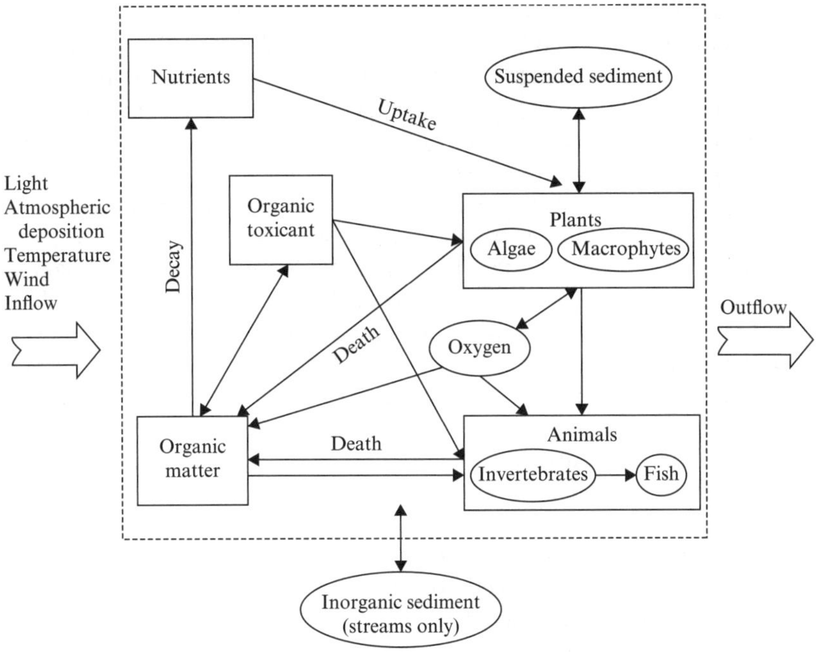

Figure 4.11 EPA's AQUATOX model for freshwater ecosystems

the watershed, a fluid dynamics model to simulate hydrodynamics and sediment transport by the stream and a food chain model to track the fate of PCBs in fish. Outputs of the proposed food chain model were compared to similar outputs of the AQUATOX to test the proposal for consistency. The model has been calibrated and explains the existing distribution of PCBs in the river very well (Weston Solutions, Inc., 2004). It remains to be seen how well the model can predict outcomes of the cleanup process.

In a series of articles Richter and his collaborators, drawing on their own work and that of others, critiqued a number of existing methods for setting stream flow targets for restoring riverine ecosystems (Richter et al., 1997; Richter et al., 2003). They argue that the flow regime of a river is now recognized as a master variable and the natural variability of a stream should be restored as nearly as possible to achieve ecologically sustainable conditions. Citations of a large number of studies support their argument that modifications in the timing, frequency and duration of floods can destroy spawning and other signals to migratory fish. Such modifications may push some food sources, young fish and eggs out of their preferred habitats, thereby reducing fish production.

Stanford (1994) provides details of such effects in a review of flow recommendations by USFWS for the Upper Colorado River. He argued that none of the methods used to estimate flows necessary to create stream conditions favorable to selected endangered species was sufficient to be used exclusive of other ecological factors. He cited effects of flow on food webs in particular. Backwater food webs provide an abundant source of food for young squawfish, one of the endangered species. However, many backwater areas in the upper Colorado have become ecologically disconnected from the river channel due to man-made revetments, sand bars or encroaching riparian vegetation. Some of the discontinuities are attributable to reductions in sediment scour that accompanied flood events prior to regulation of the stream. He also cited changes in temperature and the invasion of non-native riparian vegetation, displacing native species that flourished in an environment of more frequent and severe floods and droughts. Unstable base flow conditions also disrupt the stream ecology. For some reaches of the rivers, the most productive food web is located in the shallow area along the shoreline, but frequent flushing of these areas by hydropower operations can disrupt the web.

Stanford concludes that river ecosystems are too complex to be described by deterministic models, and that construction of an ecosystem model including all relevant processes is unreasonable. His preferred alternative is to assemble all available information about the ecosystem to derive a flow regime and quantify variables that significantly affect the extent to which endangered species may recover. His prescription, then, is a process involving multiple disciplines along the lines of an environmental impact statement.

Much of the ecological literature in the past two decades has focused on measures of ecological health other than habitant extent and has been reviewed in the context of the USACE search for improving measures of performance for environmental restoration by Stakhiv et al. (2003). Fundamental attributes of ecosystems are described in terms of physical and biological structure and the natural functions that are both controlled by and produce a particular structure at a particular time. One measure of the health of such a system is the degree of its naturalness, reflected, at least in part, by the richness of species or diversity present in the system. Thus, in contrast to the HEP and its variants, which focus on one or a few key indicator species, measures of ecosystem health tend to focus more on community. However, only a small number of operational indices of community health have reached the status of being widely accepted and widely interpretable.

One of those community health indices is the Index of Biological Integrity (IBI) developed by Karr (1981) and Karr et al. (1986) for freshwater ecosystems. The index is constructed by adding indices for a number of attributes of community structure that indicate the extent to which they vary from their natural, undisturbed condition. An equal weighting of each attribute is implicit in the rules by which IBI is constructed.

Advances in mechanistic modeling of aquatic ecosystems have resulted in our ability to include greater process (especially trophic) detail and complexity, as well as to perform dynamic simulations (Chapra, 1997; McCutcheon, 1989; Thomann and Mueller, 1987; Orlob, 1983; Schnoor, 1996). Still, these models have not advanced to the point of predicting community structure or biotic integrity.

The Modeling Process

In addition to technical advances in model development and implementation, there has been an evolution in processes by which models have been constructed and used. Some of the early environmental systems studies were undertaken with a naive view of the appropriate role and impact of models and modelers in the policy-making process. That is, policy makers were assumed to be able to foresee the need for analysis. The staff of analysts would, at their request, study the problem and, if appropriate, construct a model, identify feasible solutions and their consequences and recommend one or at most a few alternative solutions. The policy maker, after waiting patiently for these recommendations, would then make a yes or no decision. Experience to date, however, suggests the following:

- A final solution to an environmental problem rarely exists – plans and projects are dynamic. They evolve over time as facilities are added and modified to adapt to changes in system understanding, management objectives and in the demands placed on the project or policy.
- For every major decision there are many minor decisions made by different agencies or management organizations responsible for different aspects of a project.
- The time normally available to study a particular environmental problem will be shorter than the time needed, or if there is sufficient time, the objectives driving the original study will likely shift significantly before the study is completed.

This experience emphasizes some of the limitations and difficulties any water resources systems study may encounter, but more importantly, it underscores the need for constant communication among the analysts, system planners, managers, operators and policy makers. The success or failure of past water resources studies may often be traced to the level of effort expended in ensuring adequate, timely and meaningful communication. Decision makers who need information that can be derived from various models and analyses need it at particular times and in a form useful and meaningful to them. Once their window of opportunity for decision making passes, such information, no matter how well presented, will often be useless.

Over roughly the last two decades, interactions among decision makers and model builders have been enhanced by developing and implementing decision support systems (DSS) (Fedra, 1992; Georgakakos and Martin, 1996; Loucks and da Costa, 1991). These interactive modeling and display technologies can, within limits, adapt to the level of information needed and give decision makers some control over data input, model operation and data output. But will each decision maker and stakeholder trust the model output? How can they develop confidence in the models contained in a DSS? How can they modify the sub-models within a DSS to address issues the DSS developer may not have considered? An answer to these questions has been to involve the decision makers themselves, in interactive model building as well as use.

Through such model-building exercises it is just possible that those involved will reach not only a better understanding of stakeholders' concerns, but also a common or shared vision of at least how their water resources system (as represented by their model) works. Experience in stakeholder involvement in model building suggests such model-building exercises can also help multiple stakeholders reach a consensus on how their real system should be developed and managed. In the USA one of the major advocates of shared-vision modeling is the Institute for Water Resources (IWR) of USACE. They have applied their interactive general purpose model-building platform in a number of exercises where conflicts existed over the design and operation of water systems (Palmer et al., 1996; Palmer et al., 1993; Werick et al., 1996a; Werick et al., 1996b). Each of these model-building shared-vision exercises included numerous stakeholders together with experts in the use of the software. As expressed by the USACE (www.iwr/usace.army.mil/iwr/svp/svmpage.htm):

> Shared-vision models are the tools that set shared-vision planning apart from other collaborative processes. Shared-vision models provide the technical rigor necessary in a successful planning process for identifying alternatives and trade-offs in a manner where stakeholders without modeling experience can actively participate in the modeling process.
>
> Because experts and stakeholders can build these models together, including elements that interest each group, the models become trusted, consensus views of how water systems work, and how they affect stakeholders and the environment. Without adding new bureaucracies or reassigning decision-making authority, the shared-vision model and the act of developing it create a

connectedness among problems solvers that resembles the natural integration of the conditions they study.

CLOSING REMARKS

Having spent the better part of two centuries developing the use of natural resources across our continental nation, and in the process making major alterations to what we now call environmental quality, the USA has for the past 40 years embarked on an admittedly uneven effort to preserve, protect and restore both those resources and the surrounding environment. The effort has resulted in a stream of legislation, some of it quite controversial and the inevitable accompanying stream of regulations, incentives and penalties. The laws, for a variety of reasons, attack the problems in what seems to many experts a piecemeal fashion, so that the multi-dimensional and interconnected natural world seems to have been chopped into isolated bits and dealt with via very different, sometimes conflicting, policy tools, from technology specification, to ambient quality standards and from capital subsidies to tradable discharge permits. In the first half of this chapter we worked to put some flesh on that skeletal observation.

To some large extent, both past actions that caused what we now see as harmful consequences and current programs and policies that seem ad hoc and even ineffective reflect an evolving but still imperfect understanding of the natural world in all its complexity. If earlier generations had understood the consequences of damming east and west coast salmon rivers, would they have proceeded so enthusiastically and without more effort at mitigation? Would it have remained legal for so long to dispose of hazardous chemicals by stacking barrels of them on the ground if it had been understood that this amounted to waiting for rust to disperse the chemicals onto the soil and then into groundwater? Would we have gone so far down the path of creating acid precipitation by unregulated sulfur dioxide discharge from tall utility stacks if we had understood the transport and fate implications for eastern lakes and rivers? Possibly, but it would have been politically harder.

The search for understanding firm enough to allow the construction of mathematical representations of the natural world is the subject of the second major part of the chapter. One lesson we take away from that section is that much progress has been made, especially in hydrology and the simpler processes such as the breakdown of oxygen demanding organics in streams. Another, however, is that much remains to be accomplished if we are to be able to inform the process of restoration of ecological systems, such as the salmon rivers and the Florida Everglades. Complexity and interconnectedness are fine words but difficult realities, and building predictive ecosystem models that capture these realities is still in relative infancy. Advancing the state-of-the-art toward maturity will require sustained investment in serious interdisciplinary work. (As the next chapter argues, there are substantive reasons for the difficulties we have observed in trying to get such work off the ground.) There is also a continuing need for data: to enhance our understanding, to allow estimation of parameters for models and to allow validation of candidate representations of specific systems. This is a need we see being pushed far to the back of the stove top rather than being given the attention it deserves.

NOTES

1. For more background and detail on the 'Green Book' and Senate Document 97, see Chapters 2 and 3.
2. In Figure 4.9 a community outbreak refers to an outbreak in a public water system that serves year-round residents of a community, subdivision or mobile home park that has > 15 service connections or an average of > 25 residents for > 60 days per year. A non-community outbreak is one that occurs in a public water system that: (1) serves an institution, industry, camp, park, hotel or business that is used by the public for > 60 days per year but not year-round; (2) has > 15 service connections or serves an average of > 25 persons; and (3) is not a community water system (www.cdc.gov/MMWR/preview/mmwrhtml/ss5308a6.htm).
3. In this context, the term 'assessment' is used in the narrow sense, namely as the process of measuring and estimating the effects (consequences such as hydrological, biological and chemical) of a plan. By way of contrast, the term 'appraisal' is used by the P&G to refer to the process of assigning social values to effects as discussed in Chapter 3.

REFERENCES

American Society of Civil Engineers (ASCE and UNESCO) (1998), 'Sustainability criteria for water resource systems', report by the Task Committee on Sustainability Criteria and United Nations Educational, Scientific and Cultural Organization, American Society of Civil Engineers, Reston, VA.

Andreu, J.J., J. Capilla and E. Sanchis (1991), 'AQUATOOL: a computer assisted support system for water resources research management including conjunctive use', in D.P. Loucks and J.R. da Costa (eds), *Decision Support Systems*, Berlin: Springer-Verlag, pp. 333–5.

Arnold, J.G., R. Srinivasan, R.S. Muttiah and J.R. Williams (1998), 'Large area hydrologic modeling and assessment, part 1: model development', *Journal of the American Water Resources Association*, **34** (1), 73–89.

Bachmat, Y., J.D. Bredehoeft, B. Andrews, D. Holtz and S. Sebastian (1980), *Groundwater Management: The Use of Numerical Models*, water resources monograph, Washington, DC: American Geophysical Union.

Bedient, P.B. and W.C. Huber (1988), *Hydrology and Floodplain Analysis*, 1st edn, Reading, MA: Addison-Wesley Publishing Co.

Chapra, S.C. (1997), *Surface Water: Quality Modeling*, New York, NY: McGraw-Hill.

Council on Environmental Quality (CEQ) (1997), *The National Environmental Policy Act: A Study of its Effectiveness after Twenty-five Years*, Washington, DC: US Executive Office of the President.

Crawford, N.H. and R.K. Linsley (1966), 'Digital simulation in hydrology: Stanford Watershed Model IV', Stanford University Civil Engineering Department, technical report 39, Stanford, CA.

Czech, B. and P.R. Krausman (2001), *The Endangered Species Act: History, Conservation Biology and Public Policy*, Baltimore, MD: Johns Hopkins University Press.

Dahl, T.E. (1990), *Wetlands Losses in the United States 1780s to 1980s*, Washington, DC: US Fish and Wildlife Services.

Dahl, T.E. (2000), *Status and Trends of Wetlands in the Conterminous United States 1986 to 1997*, Washington, DC: US Fish and Wildlife Services.

Danish Hydraulic Institute (DHI) (1997), 'MIKE-BASIN', operating manual, Danish Hydraulic Institute, Hørsholm, Denmark.

Davies, III, J.C. (1970), *The Politics of Pollution*, New York, NY: Pegasus Books.

Delft Hydraulics (DHL) (2002), 'RIBASIM', operating manual, Delft Hydraulics, Delft, Netherlands.

Dombeck, M.P., C.A. Wood and J.E. Williams (2003), *From Conquest to Conservation – Our Public Lands Legacy*, Washington, DC: Island Press.

Fair, G.M. (1950), 'Sanitary engineering in a changing world', *Sewage and Industrial Wastes*, **22** (1), 11–16.

Feather, T.D., C.S. Russell, K. Harrington and D.T. Capan (1995), *Review of Monetary and Nonmonetary Valuation of Environmental Investments*, report 95-R-2, Alexandria, VA: US Army Corps of Engineers, Institute for Water Resources.

Federal Water Pollution Control Administration (FWPCA) (1969), *The Cost of Clean Water and its Environmental Impact*, Washington, DC: US Government Printing Office.

Fedra, K. (1992), *Advanced Computer Applications, Options*, Laxenburg, Austria: International Institute for Applied Systems Analysis.

Georgakakos, A.P. and Q.W. Martin (eds) (1996), *An International Review of Decision Support Systems in River Basin Operation, Proceedings of the Fifth Water Resources Operations Management Workshop*, Arlington, VA: American Society of Civil Engineers.

Harrison, R.W. (1961), *Alluvial Empire: A Study of State and Local Efforts Toward Land Development in the Alluvial Valley of the Lower Mississippi River*, the Delta Fund in cooperation with the Economic Research Service, Little Rock, AR: Pioneer Press.

Hays, S.P. (1959), *Conservation and the Gospel of Efficiency*, Cambridge, MA: Harvard University Press.

Hays, S.P. (1987), *Beauty, Health and Permanence: Environmental Politics in the United States, 1955–1985*, Cambridge: Cambridge University Press.

Hynning, C.J. (1939), *State Conservation of Resources*, Washington, DC: US Government Printing Office, National Resources Committee.

International Union for Conservation and Natural Resources (IUCN) (1980), *World*

Conservation Strategy: Living Resource Conservation for Sustainable Development, Gland, Switzerland: International Union for Conservation and Natural Resources, United Nations Environmental Program and World Wildlife Fund.

Karr, J.R. (1981), 'Assessment of biotic integrity using fish communities', *Fisheries*, **6** (6), 21–7.

Karr, J.R., K.D. Fausch, P.L. Angermeier, P.R. Yant and I.J. Schlosser (1986), *Assessment of Biological Integrity in Running Water: A Method and its Rationale*, Illinois Natural History Survey special publication 5, Champaign, IL: American Society of Civil Engineers.

Kempton, W., J.S. Boster and J.A. Hartley (1995), *Environmental Values in American Culture*, Cambridge, MA: MIT Press.

Kirby, J., P. O'Keefe and L. Timberlake (eds) (1995), *Sustainable Development*, London: Earthscan Publications.

Loucks, D.P. and J.R. da Costa (eds) (1991), *Decision Support Systems: Water Resources Planning*, North Atlantic Treaty Organization, Advanced Studies Institute Series, Berlin, Germany: Springer-Verlag.

Maass, A., M.M. Hufschmidt, R. Dorfman, H.A. Thomas, Jr, S.A. Marglin and G.M. Fair (1962), *Design of Water Resources Systems*, Cambridge, MA: Harvard University Press.

Martin, L.R. and E. Stakhiv (1999), *Sustainable Development: Concepts, Goals and Relevance to the Civil Works Program*, report 99-PS-1, Washington, DC: US Army Corps of Engineers, Institute for Water Resources.

McCutcheon, S.C. (1989), *Water Quality Modeling*, Boca Raton, FL: CRC Press.

McKee, J.E. and H.W. Wolf (1963), *Water Quality Criteria*, 2nd edn, Sacramento, CA: California State Water Quality Control Board.

Mehan, III, G.T. (2004), 'Better monitoring for better water management', *Water Environment Research*, **76** (1), 3–4.

Metcalf and Eddy, Inc., University of Florida and Water Resources Engineers, Inc. (1971), *Storm Water Management Model for the Environmental Protection Agency*, 11024DPC07/71-10/71, Washington, DC: US Environmental Protection Agency.

National Research Council (NRC) (1991), *Opportunities in the Hydrologic Sciences*, Washington, DC: National Academy Press.

National Resources Committee (1939), *Water Pollution in the United States*, Washington, DC: US Government Printing Office.

Office of Management and Budget (OMB) (2003), 'Informing regulatory decisions: 2003 report to Congress on the costs and benefits of Federal regulations and unfunded mandates on state, local and tribal entities', accessed September 2003 at www.whitehouse.gov/omb/inforeg/regpd-reports_congress.html.

Orlob, G.T. (ed.) (1983), *Mathematical Modeling of Water Quality: Streams, Lakes and Reservoirs*, Chichester: John Wiley and Sons, Inc.

Palmer, R.N., A.M. Keys and S. Fisher (1993), *Empowering Stakeholders Through Simulation in Water Resources Planning*, Proceedings, 20th Conference on Water Resources Planning and Management Division, New York: American Society of Civil Engineers.

Palmer, R.N., A. Hamlet, J. Trungale, S. Haddadin and D. Mekkers (1996), *A History of Shared Vision Modeling in the ACT-ACF Comprehensive Study*, September, Alexandria, VA: US Army Corps of Engineers, Institute for Water Resources.

Park, R.A., J.S. Clough and M.C. Wellman (2004), *AQUATOX (Release 2) Modeling Environmental Fate and Ecological Effects in Aquatic Ecosystems, Volumes I and II*, EPA-823-R-04-001 and EPA-823-R-002, Washington, DC: US Environmental Protection Agency, accessed 9 January, 2009 at www.epa.gov/water science/wqm/.

Platt, R.H. (1996), *Land Use and Society: Geography, Law and Public Policy*, Washington, DC: Island Press.

President's Council on Sustainable Development (PCSD) (1996), *Sustainable America: New Consensus for Prosperity, Opportunity and a Healthy Environment for the Future*, Washington, DC: US Government Printing Office.

Raskin, P., J. Sieber and A. Huber-Lee (2001), *Water Evaluation and Planning System: User Guide for WEAP21*, Boston, MA: Tellus Institute.

Richter, B.D., J.V. Bumgartner, R. Wiginton and D.P. Braun (1997), 'How much water does a river need?', *Freshwater Biology*, **37**, 231–49.

Richter, B.D., R. Mathews, D.L. Harrison and R. Wiginton (2003), 'Ecologically sustainable water management: managing river flows for ecological integrity', *Ecological Applications*, **13** (1), 226–34.

Rouse, H. and S. Ince (1957), *History of Hydraulics*, Iowa City, IA: University of Iowa Press.

Schnoor, J.L. (1996), *Environmental Modeling, Fate and Transport of Pollutants in Water, Air and Soil*, New York, NY: John Wiley and Sons.

Sierra Club (2003), 'Sierra Club history timeline', accessed 12 January, 2009 at www.sierraclub.org/history/timeline.asp

Smith, R.A., G.E. Schwarz and R.B. Alexander (1997), 'Regional interpretation of water-quality monitoring data', *Water Resources Research*, **33** (12), 2781–98.

Stakhiv, E., R. Cole, P. Scodari and L. Martin (2003), *Improving Environmental Benefits Analysis in Ecosystem Restoration Planning*, report 03-PS-3, Alexandria, VA: US Army Corps of Engineers, Institute for Water Resources.

Stalnaker, C., B.L. Lamb, J. Henriksen, K. Bovee and J. Bartholow (March 1995), *The Instream Flow Incremental Methodology: A Primer for IFIM*, monograph, Washington, DC: National Biological Service.

Stanford, J.A. (1994), 'Instream flows to assist the recovery of endangered fishes of the Upper Colorado Basin', biological report no. 24, US Fish and Wildlife Service.

Streeter, H.W. and E.B. Phelps (1925), 'A study of the pollution and natural purification of the Ohio River', public health bulletin 146, US Public Health Service.

Tarr, J., J. McCurley and T. Yosie (1980), 'The development and impact of urban wastewater technology: changing concepts of water quality control, 1850–1930', in M.V. Melosi (ed.), *Pollution and Reform in American Cities, 1870–1930*, Austin, TX: University of Texas Press, pp. 60–81.

Thomann, R.V. and J.A. Mueller (1987), *Principles of Surface Water Quality Modeling and Control*, New York, NY: Harper and Row.

US Army Corps of Engineers and South Florida Water Management District (USACE and SFWMD) (1999), *Central and Southern Florida Project: Comprehensive Review Study – Final Integrated Feasibility Report and Programmatic Environmental Impact Statement*, Jacksonville, FL: US Army Corps of Engineers.

US Army Corps of Engineers (2000), *Planning Guidance Notebook*, chapter 2 "Planning principles", engineering regulation 1105-2-100, p. 2–1.

US Bureau of the Census (2002), *Statistical Abstract of the United States*, mini historical statistics Table HS25, Washington DC: US Government Printing Office.
US Bureau of Land Management (USBLM) (January 2003), 'Disposition of the public domain', accessed 9 January, 2009 at www.blm.gov/natacq/pls02/.
US Environmental Protection Agency (EPA) (1991a), *The Watershed Approach: An Overview*, EPA 503/9-92/002, Washington, DC: US Environmental Protection Agency.
US Environmental Protection Agency (EPA) (1991b), *The Watershed Protection Approach Framework Document, Office of Wetlands, Oceans and Watersheds*, Washington, DC: US Environmental Protection Agency.
US Environmental Protection Agency (EPA) (2000), *Progress in Water Quality: An Evaluation of the National Investment in Municipal Wastewater Treatment*, EPA-832-R-00-008, Washington, DC: US Environmental Protection Agency.
US Environmental Protection Agency (EPA) (2004), 'Linking water quality and aquatic life', accessed 12 January, 2009 at www.epa.gov/waterscience/models/aquatox.
US Environmental Protection Agency (EPA) (2007), 'Superfund accomplishment figures, summary fiscal year 2003', www.epa.gov/superfund/action/process/numbers.htm, 19 March 2004.
US Fish and Wildlife Service (USFWS) (1980), *Water Resource Planning Under the Fish and Wildlife Coordination Act*, FWS/OBS-80/44, Washington, DC: US Fish and Wildlife Service.
US Fish and Wildlife Service (USFWS) (2007a), 'America's national wildlife refuge system', accessed 9 January, 2009 at www.refuges.fws.gov/centennial/pelicanisland.html.
US Fish and Wildlife Service (USFWS) (2007b), 'Migratory bird conservation commission: history', accessed 9 January, 2009 at www.realty.fws.gov/mbcc.html.
US Government Accountability Office (GAO) (1991), *Wetlands Overview: Federal and State Policies, Legislation and Programs*, Washington, DC: US Government Accountability Office.
US Interagency Committee on Water Resources (USIACWR) (1950), *Proposed Practices for Economic Analysis of River Basin Projects*, The Green Book, revised 1958, Washington, DC: US Government Printing Office.
US Senate (1962), *Policies, Standards and Procedures in the Formulation, Evaluation and Review of Plans for Use and Development of Water and Related Land Resources*, Senate Document 97, 87th Congress, 2nd Session, Washington, DC: US Government Printing Office.
US Water Resources Council (WRC) (1973), 'Establishment of principles and standards for planning water and related land resources', *Federal Register*, **38** (174), 24778.
US Water Resources Council (WRC) (1980), 'Proposed rules: principles, standards and procedures for planning water and related land resources', *Federal Register*, **45** (73), 25329–46.
US Water Resources Council (WRC) (1983), *Economic and Environmental Principles and Guidelines for Water and Related Land Resources Implementation Studies*, Washington, DC: US Government Printing Office.
Water Science and Technology Board (WSTB) (2004), *Confronting the Nation's Water Problems: The Role of Research*, Washington, DC: The National Academies Press.

Werick, W.J., R.N. Palmer, J. Trungale, A. Hamlet and W. Rowden (1996a), 'Using shared vision models in the ACT-ACF study', *Proceedings of the Hydrologic Engineering Center's Symposium on Water Resources Management*, January, Santa Barbara, CA: US Army Corps of Engineers.

Werick, W.J., W. Whipple, Jr and J. Lund (1996b), *ACT-ACF Basinwide Study*, Mobile, AL: US Army Corps of Engineers, Mobile District.

Weston Solutions, Inc. (2004), *Modeling Framework Design: Modeling Study of PCB Contamination in the Housatonic River*, report prepared for the US Army Corps of Engineers, New England District, Concord, MA and US Environmental Protection Agency, New England Region, Boston, MA, by Weston Solutions, Inc., West Chester, PA.

World Commission on Environment and Development (WCED) (1987), *Our Common Future*, Oxford: Oxford University Press.

5. On the collaboration of ecologists and economists

Clifford S. Russell and Mark Sagoff

INTRODUCTION

Chapters 3 and 4 of this book, on economic methods and ecological approaches to understanding the aquatic environment, underscore the importance of interdisciplinary collaboration in the analysis of water policies and projects. Economists believe their methods of attaching money values to prospective changes in water quantity and quality are central to making inferences about whether those changes are worth pursuing – worth at least what they will cost. But the changes themselves must be inferred by another sort of expert on the basis of the elements of the policy or the structure of the project. This is true whether what is being examined involves adding a new structure or removing an old one; imposing a new restriction on firms, farms or consumers, or relaxing an old one; causing predictable damage or attempting to mitigate earlier damage. Seen in this light, there seems to be a natural fit between economics and aquatic science and engineering (or, more broadly, economics and environmental science). This chapter deals with one part of this apparent fit, a part that does not often live up to expectations – the one between economists and ecologists.[1]

The history of attempts at collaboration between economists and ecologists is far from a happy or an encouraging one, with the efforts often ending in failure or frustration. Even when attempts have been comparatively successful, the costs in time, money and psychic energy are so large and the results apparently so dependent on contingent factors, such as the particular personalities involved, that agencies are reluctant to commission new collaborative projects. Economists and ecologists seem unable to work together even though they each have collaborated with other, related, disciplines: economists with environmental engineers and ecologists with anthropologists, sociologists and historians. Indeed, the burgeoning field of environmental history provides a testimony to the interdisciplinary affinity of ecological and historical studies.

Why, then, has the economist/ecologist divide proven so much more difficult to bridge? Two disagreements between ecologists and economists account for a large share of the problem. One bone of contention has to do with the concept of value or valuation. Ecologists tend to define the value of an object, for example, a species or an ecosystem, in terms of the uses or functions it supports, directly or indirectly. Thus, ecologists trace value to causal connections among objects embedded in systems. To answer the question, 'What is the value of *x*?' ecologists ask, 'What is the function of *x* – what other things depend upon it – in the organization of nature?'

Economists tend to think that the value of a change in an ecological system or service is most usefully defined in terms of differences in human welfare that may be associated with that change, assuming that the rest of the system continues to function. More generally, economists think in terms of trade-offs, such as are represented by the prices of marketed goods, marginal rates of substitution in production, exchange rates in international trade and answers to contingent valuation questions. To answer the question, 'What is the value of *x*?' economists generally ask, 'What will people give in exchange for *x*?' While ecologists are likely to associate valuation with an ideal or original state of an ecosystem, economists are likely to refer to an ideal or competitive market.

The second bone of contention has to do with risk and uncertainty. When is it appropriate to assume that the larger system will continue to function, more or less as before, after a proposed intervention by humanity has been undertaken? Where does the burden of proof fall to assure that changes to the natural world will not cross strategic thresholds or cause ecosystems to collapse? A clarification of these conflicts over the idea of value and over the appropriate response to uncertainty and ignorance may help to facilitate future collaborations between ecologists and economists.

Following are several brief war stories from the trenches of the history of economist/ecologist collaboration. Two of these examples draw upon the personal research experience of one of the authors (Russell). The third, to which we devote more space, is drawn from accounts of participants in, and some of the available documentation flowing from, a committee exercise, organized by the US Environmental Protection Agency (EPA) and aimed at clarifying, among other things, the value of ecological benefits of the Clean Air Act (CAA). We use these stories to suggest patterns in the dialogue between the disciplines, patterns that in turn lead to our diagnosis of the problem. We then step back to place the diagnosis in a larger perspective by describing two competing world views that leave between them a philosophical gulf, out of which arise some of the differences over valuation and uncertainty we then analyse.

Economists and ecologists may learn from each other rather than, as they often do now, confront each other, with each discipline defending its own assumptions and approaches. To the extent that ecologists and economists focus on quite different problems, they may be able to agree to disagree on the philosophical issues that divide them, but still make progress on specific cases, working together on the empirical particulars that have to be understood, rather than seeking to debunk each other at a higher level of abstraction and generality.

Each discipline may succeed better if it takes a more modest view of its own ability to solve the problems both are called upon to address. Modesty is an intellectual virtue – the opposite of self-righteousness – that stands at the foundation of every science (Rorty, 1991). In contrast, disciplinary arrogance or defensiveness may lead economists and ecologists to put more energy into chastizing and reproving each other than into learning.

We suggest modest steps aimed at fostering collaboration. Roughly speaking, we argue that continuous contact, in the context of empirical work aimed at particular cases, offers the best opportunity to skirt the fundamental philosophical differences separating the two disciplines, without denying the existence of those differences.

WAR STORIES

The following describe three efforts to be cooperative, in which ecologists and economists had great difficulty finding a common conceptual framework within which to pursue collaborative research. Indeed, in only one of the stories below was the outcome a success when judged against the original goals. The executive summary is that the representatives of the disciplines did not even engage in parallel play, trying to accommodate each other by ignoring each other. Rather, they fought, and what they fought over helps to illuminate the underlying differences.

Early Collaborative Effort (War Story 1)

In the early 1970s the Quality of the Environment program at Resources for the Future (RFF) invested in an ambitious regional modeling project. The goal was to demonstrate that the connections among pollutants, as implied by the laws of conservation of mass and energy, should be taken seriously in environmental policy formulation. The region chosen for modeling was complex; it comprised the shores and the hinterland of the Delaware River Estuary, an area stretching from Trenton, New Jersey, to the head of the Delaware Bay below Wilmington, Delaware. Both water and air quality were to be examined in the dual sense that models were to reflect both connections within discharge sources among air and water pollutants and connections between air and water in the ambient environment. The state of knowledge and skill could not support the latter effort, as it turned out, but the former effort produced large (for the time) and detailed models of petroleum refining and integrated steel production, as well as simple models of municipal wastewater treatment and fossil fuel power generation.

At the outset, the team (an engineer and an economist) represented ambient water quality and its relation to pollution by a simple Streeter-Phelps model (see Chapter 4 for more detail) of the dissolved oxygen (DO) deficit caused by the decay of biochemical oxygen-demanding organic material. The region's municipal wastewater treatment plants discharged the largest part of this matter, but some came from industry. The water quality model did not take into account toxics discharged by the industrial plants and, as important, it was clear that the dimensions of quality of direct interest to politicians and lay people generally, such as algal blooms and fish populations, while related to DO levels, were not simple one-to-one transformations. Even though the modeling of aquatic ecosystems in detail was still an emerging science, the directors of the RFF enterprise decided that the project team should include an ecologist who could introduce such a model to the regional model suite.[2]

Money was not a problem, but finding an ecologist willing to work within the structure of the project was. After the better part of a year of effort to find such a person, a newly minted ecological modeling PhD was hired. He had learned modeling heuristics and mathematical techniques and seemed interested in applying them to the Delaware River watershed. But early on it became clear that a huge gap separated his view of the purpose of the modeling exercise and of its context from that of the project framers.

That difficulty was epitomized by a remark at an early team lunch. In effect, the ecologist declared that he did not see the point of building such a complicated set of control cost models and an optimization program that could find the cheapest way of meeting any feasible ambient water quality standards. The right approach was obvious to him. Pollution discharges to the river should be minimized and then the aquatic ecosystem model would tell you what that meant for algae, DO and fish.

One can understand the basis of this viewpoint. Plainly, while an economist may regard pollution as an external cost of production to be optimized in relation to the costs and benefits of reducing it, the ecologist could regard pollution as a moral wrong, analogous to trespass or assault.

If someone tosses garbage into your backyard – or toxic substances into your liver or lungs – you may call upon the polluter to cease the nuisance. To a large extent, society may follow the example of the common law of nuisance in treating pollution as a trespass or assault to be enjoined or at least minimized (as a kind of necessary evil) rather than as a social or external cost to be optimized. The analogy between pollution and trespass or assault may explain why basic environmental statutes, such as the Clean Water Act (CWA) and the CAA, explicitly limit the use of the benefit-cost test, indeed, even the consideration of costs for reducing pollution.

On the other hand, the point of the research project was to compare the costs of different, non-minimal, discharge levels. Even if society agreed that pollution is morally wrong, it must still in fact compare the cost profiles of various strategies for controlling it. While moral principles are to be respected and may be applicable, trade-offs must also be considered. However society decides to regulate pollution – whether on legal, moral, economic, political or other grounds – information about costs at least has to be relevant.

What to the economist seemed a perfectly rational exercise – estimating minimal attainable costs as a way of clarifying a social choice – appeared to the ecologist almost worse than a waste of time; it was a bean-counter exercise in which he preferred not to engage. To say this was a rude shock to the economist on the project is an understatement. This impasse led to perhaps six months of daily debate, which constituted a frustrating background to the work of creating the integrated model. The model was being built even as its purpose was under attack by the person charged with constructing it.

In the end, the team did come together. The ecologist was persuaded (or just brow beaten) into using his skill to predict the effects of different, non-minimal levels of discharges on various ecological indicators. But, it had been a long journey from the beginning, one that the tighter funding constraints of today's world could not support.

Translating Ecological Indicators to Make Them 'Meaningful' to Policy Makers (War Story 2)

In the early 1990s the EPA funded a large-scale ecological data gathering exercise called the Environmental Monitoring and Assessment Program (EMAP). Teams set to work gathering time-series data on hundreds of dimensions of functioning in a set of representative ecological systems, ranging from the mountains to the coasts and from the deserts to the plains and forests. The idea was that sufficient data would allow both better understanding of how the systems worked and an ability to report on the status of any particular system and the direction of changes in its status.[3] Whether as part of this effort or as a parallel one aiming to get more quickly to policy-relevant outputs, a request for proposals was issued in 1993 seeking research groups interested in trying to translate the arcana of ecosystem function indicators into words and concepts more likely to excite the interest of politicians and ordinary citizens.

In response, a multidisciplinary team was assembled at Oak Ridge National Labs and Vanderbilt University. Its core group included an environmental economist, an ecologist, a psychologist with survey expertise and a sociologist. The team's proposal won support from the EPA in the form of a cooperative agreement to tackle the challenge. Several graduate students from a variety of disciplines (including religious studies) then joined the group and additional senior ecologists participated from time to time. To make the task slightly less daunting, the group chose to focus on the forest ecosystem of the Southern Appalachians, examples of which were easily accessible from Oak Ridge.

The team agreed from the beginning that the central challenge was, first, to translate the technical indicators into lay language and, second, to find a way (or several ways) to make the meaning of the resulting concepts come alive by showing what these measures, in various combinations, implied for the aesthetics and recreational uses of a forest. The ultimate goal was a large-scale survey, possibly to involve a forest video, to explore whether people could discriminate among the alternative forests implied by different values of the translated indicators. Along the way, focus groups would explore alternative wordings for the indicators and the possibilities for creating meaningful summary indicators as combinations of the original highly specific measures.

The project, the accomplishments of which were modest in relation to its goals,[4] regularly threatened to founder on the problem of giving operational meaning to the idea of discriminating among forests. The economist thought (and suggested) that a way to demonstrate that people could (or could not) do this was to ask them both preference and value questions relating to days of recreation in different forests as these were described and pictured. It was the word 'value' that caused the continuing problem. For the economist, to value meant to translate preferences into money terms, using the concept of willingness-to-pay (WTP), as in: I would be willing to pay up to $25 for a day of hiking in forest A but only $15 in forest B. If there were three or more alternative forests that could be compared two by two, a minimal test of the ability to discriminate would be to find no violation of transitivity in an individual's preference answers and consistency between the values and the preferences.

Both the ecologist and the sociologist on the team found this notion of valuation and the language used to describe it unacceptable. To equate discrimination among forests to differential WTP money to experience them struck these collaborators as at least crassly materialistic. They understood that one might distinguish automobiles or restaurants on the basis of WTP – or people at least do this – but one might not as easily apply this kind of measure to a non-use or non-consumption good, such as the integrity of a forest. To fathom our responsibilities to the forest, so to speak, we would have to come to some deliberative political consensus. According to the ecologists and the sociologist, this group deliberation would not be like making exchanges or trade-offs in a market.

More practically, the word 'value' repeatedly triggered long, often intense, sometimes acrimonious, discussions contrasting two conceptions of the basis of valuation. First, value could refer to a belief or moral principle the individual ascribes to the community as a whole. For example, the Ten Commandments represent principles of this sort. The principle Thou Shalt Not Steal represents a rule we stand for as a community. What is fundamental here is that the logical subject of choice – the valuer – is not taken to be the individual but the community as a whole. As a member of the community, the individual contributes views or reasons about common goals and thus participates in the deliberative argument in which norms and principles take shape.

Second, 'value' could refer – and in economic theory, it most often does refer – to the wants and preferences that represent what the individual believes is good for him or her. The valuer is taken to be each individual and what individuals want or prefer is to that extent supposed to increase their welfare or utility. Social valuation then becomes a matter of aggregation over individual utilities, not deliberation over views, opinions and arguments. Willingness to pay is the best measure of value in this sense. However, economists are not averse to interviewing people and conducting focus groups to determine why people prefer one kind of place or outcome to another. Economists have also devised methods, such as contingent valuation, to incorporate moral principles and aesthetic judgments in the calculus of social welfare on which policy may be based.

The research team found it difficult, if not impossible, to get beyond the threshold problem of reconciling these two conceptions of value in order to discuss the mechanical details of the how – if not via the WTP questions – one can apply judgments, preferences or principles to discriminate among forest policy outcomes. The economist was baffled by the insistence of non-economists on arguing about the meaning of a word – value or valuation – that clearly had several meanings. The others seemed outraged that such an important word had been highjacked. Possibly, if the meetings had been daily instead of roughly bi-monthly, a way around the impasse might have been found. But the project ended before that was accomplished. It is important to re-emphasize that the impasse did not arise because technical jargon interfered with understanding. Rather, the lesson seems to be that even ordinary words can become landmines when two such different views of the world try to come together to take a single research path.

A Brouhaha at the Science Advisory Board (War Story 3)

The CAA Amendments of 1990 require periodic analyses of the costs and benefits associated with various titles of the statute and mandate the creation of an Advisory Council, appointed by the Administrator of the EPA. This council met with members of other EPA committees in 2001 to discuss, among other things, the value of ecological benefits that might be attributed to policies promulgated under the Act. The meeting produced a standoff – a failure to find common ground – between ecologists representing the Health and Ecological Effects Subcommittee (HEES) of the Science Advisory Board (SAB) and economists on the Advisory Council. Members of the Advisory Council then prepared a draft Analytical Plan, later published, to summarize whatever agreement had been reached and to describe the areas where there were still differences. This draft plan provides much of the information we rely upon to analyse this impasse between ecologists and economists (EPA, 2001). To be fair, it must be noted that there was some agreement. Most importantly, according to the draft analysis, ecologists and economists agreed that, while in the past ecosystem service benefits had been measured by incremental changes in the 'market values of commercially exploited natural resources, such as forest products and commercially caught fish' (EPA, 2001, p. 26), in the future, 'benefits in terms of the non-market services of ecosystems [should] be formally acknowledged, quantified and included in the benefit-cost calculations' (EPA, 2001, p. 27). The ecologists representing HEES, however, asserted in the strongest possible terms that the benefit-cost analysis should also recognize the value of the essential life-support services ecosystems provide to the human economy. Accordingly, measures should be generated 'for the benefits of the CAA that can be attributed to ecological services protection' (ibid.). Specifically, the ecologists expressed the conviction that calculations published by Costanza et al. (1997) could be adapted to provide a placeholder estimate of ecosystem service benefits (EPA, 2001, p. 28). The Costanza team attributed a value of $33 trillion to the world's ecosystems in relation to a range of life-support services. 'However, the authors readily acknowledged that because of the nature of the uncertainties involved, this must be considered a minimum estimate'[5] (EPA, 2001, p. D-2). The economists on the Advisory Council rebuffed this suggestion in part because the assumptions and techniques behind the final number had been examined and found wanting as applications of sound economics.[6] More fundamentally, the Council asserted that even if Costanza et al. (1997) 'had actually succeeded in measuring the "true" total value of ecosystem services', this number would not be useful in measuring 'the magnitude of the decrease in ecological services that would take place without the CAA' (EPA, 2001, p. D-3).

Even if the value of ecosystems exceeded that of human health, the economists opined, 'this unfortunately tells us nothing at all about the relevant marginal measure for ecological services benefits' (EPA, 2001, p. D-6). Since ecosystems support all life, their value is immense, but that value would only be relevant here if, without the CAA, life as we experience it would disappear from the planet. Rather, 'benefit estimation is . . . predicated on knowledge of the relevant damage functions and ecological science is not yet to the point where these damage functions can easily be quantified' (EPA, 2001, p. 26). The Council economists asked the absent ecologists for 'a rigorous estimate of the *change in* ecological services . . . due to the CAA' (EPA, 2001, p. 4, emphasis added).

Economists on the Advisory Council hoped to assess ecological services in terms analogous to those used for other goods in the context of optimal resource allocation and management. Following the EPA's *A Framework for the Economic Assessment of Ecological Benefits*, economists took it as a premise 'that actions affecting the state of an ecological resource . . . will result in changes to the goods and services provided by that resource' (EPA, 2002, p. 14). A damage function for a given ecosystem would alert society to the loss of some portion of those services (for example, lower flood mitigation capacity). Absent such a connection between the CAA achievements and incremental ecological change, there would be no way to include ecosystem services in the calculation of costs and benefits for social decision making.

The ecologists may have felt they could not get a hearing for their belief that stresses on ecosystems, such as carbon dioxide loadings, re-establish, at least as a possibility, the Malthusian principle that nature places physical limits on economic production and growth. (For a recent restatement of this view, see McMichael et al., 2003.) Many ecologists believe that the value of ecosystem services must be approached from a macro rather than a micro viewpoint. That is, what matters is the totality of human effects on the larger environment. Thus, declines in ecological services, for example, the ability of biospheric systems to absorb and recycle wastes, limit the scale of the human economy. By attaching a high monetary value to ecosystem services, society would acknowledge those limits. Since the ecological system contains the economic system, on this view, it possesses a value as great as the economy itself, for which a placeholder estimate like $33 trillion may be better than no number at all.

The Concept of Valuation

From these war stories springs our basic observation: economists and ecologists understand value and valuation in quite different ways. These differences reflect both disagreements about the normative basis of environmental policy and differences over the appropriate response to ignorance and to uncertainty about the natural world. Ecologists tend to understand the value of a resource in terms of the role it plays in the functioning of a productive system. The principle might be stated as: If *a* is a necessary causal condition of *b*, then *a* is at least as valuable as *b*, especially if there is any chance that tinkering with *a* will lead to a loss of *b*. Ecologists also tend to think that changes humans make in natural systems – particularly as a result of economic activity, for example, pollution – are inherently suspect because nature knows best and the possibility of catastrophe always lurks behind human intervention.

Economists, in contrast, tend to understand value in terms of trade-offs at the margin; that is, in relation to what people are willing to give up to obtain one unit more of a commodity or service (or to receive less of a bad). The economic notion of value does not depend on the determination of general rules, laws, goals and principles that express our will as a nation or a community. Rather, economic valuation begins with the idea that we are (almost always) talking about relatively small changes in the flows of commodities and services giving rise to incremental changes in well-being. This approach seeks to value such small or incremental changes in terms of the sacrifices people, as individuals, would make in order to obtain them (would exchange for them).

In the economic approach, when commodities and services are marketed and the markets function as they should, prices are the measure of value, at least in the sense that markets function to guide resources to their most productive uses. When the goods are public, such as air quality or the functioning of an ecological system, economists still maintain that people can be thought of as holding judgments about what they would be willing to give up (pay) for more of the good, or to accept to allow less, though these judgments cannot be observed in market transactions. To the economist, then, the task of valuation involves trying to estimate what people would be willing to pay for an incremental amount of a good (or willing to accept as compensation for losing it) if faced with a market for the good in question. One can see now the gulf that separates the agendas of ecologists and economists.

TWO COMPETING WORLD SYSTEMS

Historically and philosophically, what divides many contemporary economists and ecologists is the question of whether nature represents essentially matter in flux, to which human agency imparts useful form or whether it was initially created with an organization that humans must respect. This question has been the subject of philosophical debate for at least 2000 years. On the one side, a tradition that runs from Epicurus and Lucretius to Bacon leads to contemporary political economists, such as Peter Drucker and Robert Solow, as well as to contemporary scientists and biotechnologists, who regard nature as a source of raw materials (including information) and as a sink for wastes. Human ingenuity can continually mold, manipulate and transform nature to serve human preference and desire. In this view the limits to knowledge are the only limits to economic growth.

In the period following the discovery and exploration of the New World, Europeans such as Bacon, Hobbes and Locke found it plausible to think of the value of nature as largely potential; they believed that if nature on balance benefits humanity, this is due to the application of human ingenuity and labor (for example, the axe and the plow). Thus Locke (with the New World in mind) famously wrote: 'Land which is wholly left to Nature, that hath no improvement of Pasturage, Tillage or Planting, is called, as indeed it is, waste; and we shall find the benefit of it to amount to little more than nothing' (Locke, 1963, p. 315). Karl Marx (1972, p. 73) commented, 'Nature becomes purely an object for men, something merely useful and is no longer recognized as a power working for itself.' As ecological economist Herman Daly (1996, p. 63) summarizes this view, natural resources 'are totally passive recipients of form (value) added by labor and capital' so that 'value added by labor and capital is by definition the source of all value produced and consequently of all value consumed'.

The necessary organizing principle for production, according to this tradition, is found not in nature but in the market – in the action of the invisible hand that brings together raw materials with labor and technology in response to price signals. Adam Smith wrote, 'Little else is requisite to carry a state to the highest degree of opulence from the lowest barbarianism, but peace, easy taxes and a tolerable administration of justice; all the rest being brought about by the natural course of things' (Smith, 1991 [1776]). As historian Robert Heilbroner (1961, p. 50) summarizes: 'Smith has constructed for society a giant endless chain. As regularly and as inevitably as a series of interlocked mathematical propositions, society is started on an upward march.'

On the other side, a tradition that runs from Plato and Plotinus, through the Physiocrats to ecologists and ecological economists today (such as Paul Ehrlich and Herman Daly) sees the human manipulation of nature as often counterproductive because it depletes or destroys the organization already formed in nature (thus adding to entropy) and disrupts natural systems on which all life depends. As important, the endless satisfaction of preferences, many of which must themselves be manufactured along with the goods that satisfy them, is not correlated with any conception of well-being nor trivially defined in terms of it, once basic needs are met. To rein in consumption – to minimize the flow of energy and materials from nature as resource and back to nature as sink – is both to protect essential ecosystems and to increase human well-being in a substantive and meaningful sense. Less is more, in this view.

This belief draws on the writing of classical macroeconomic theorists from Quesnay and the Physiocrats, through Malthus and Ricardo, to Jevons and Marshall. These economists argued that the fixity of natural resources (for example, the land available for the production of food) sets a physical limit on the scale of economic activity and, in particular, population. In 1890 Alfred Marshall summed up this position by emphasizing a distinction between human-made capital and 'natural agents . . . taken to include all free gifts of nature, such as mines, fisheries, etc'. Land, in which Marshall included natural resources, constitutes a 'permanent and fixed stock while appliances made by man . . . are a flow capable of being increased or diminished' (Marshall, 1920, Appendix G, p. 28).[7]

Ecologists and maverick economists associated with the discipline of ecological economics regard large-scale pollution problems, such as ozone depletion, climate change and acid rain, as the modern analogs of Malthusian population growth. They believe that global problems such as these are not susceptible to easy technical, regulatory or market solutions. In 1991 Daly indicated 'when growth pushes scale beyond the optimum, we begin to experience generalized pervasive externalities, such as the greenhouse effect, ozone layer depletion and acid rain, which are not correctable by internalization of localized external costs into a specific price' (Daly, 1991, p. 245).

The intellectual chasm that yawns between ecologists and economists, when described in terms of long-opposed philosophical traditions, may seem so deep that no attempt at collaboration can bridge it. Our position, on the contrary, is that economists and ecologists can finesse their differences on fundamental philosophical principles in order to make common recommendations in the context of specific policy proposals. For the ecologist, the problem is to reduce the impact of human activity on the natural world or at least to limit the human assault on significant ecosystems.

For the economist, the problem is to maximize the present value of the consumption of goods and services, primarily by relying on market prices to guide management and investment decisions. Both goals are consistent with our character, history and laws as a nation. We are confident that ecologists and economists can collaborate in showing political authorities how to balance these goals even if fundamental philosophical differences separate the two disciplines. In particular, the disciplines may finesse some of the philosophical differences that separate them if they agree to disagree about the concept of valuation and look for policy options that seek to maximize social wealth while restricting environmental impacts to some agreed upon – for example, politically determined – level. Policies that involve trading pollution allowances under a cap – where the cap is determined on ecological or other scientific grounds – illustrate this approach.

DISCIPLINARY ARROGANCE: A HINDRANCE TO GETTING ON WITH THE TASK

The discussions and disputes described previously did not, of course, take place between and among pure, objective intellects seeking only paths around an impasse, or at least a better understanding of that impasse. An inevitable obstacle to progress is the effect of disciplinary socialization on flesh and blood, insecure and historically situated individuals. Thus, the ecologist in war story 1 came to the table having learned from his elders and academic betters to think of mainstream economists as so involved with marginal thinking that they had lost sight of the larger picture – the sustainability of an economy within its ecological setting. The ecologist positioned economists as technocrats who could not deal with problems of scale (that is, who simply could not or would not understand that, at the scale to which economic activity had expanded, human-produced stress on the environment had become undesirable to the extent it was avoidable). In this view, economists fail to recognize that the ecosystem includes the economic system as a subsystem – not the other way around. To the ecologist, ecosystem services and other goods are not to be considered as economic beans to be counted; rather they must be respected as the basic conditions of life.

The economist, on the other hand, may see the ecologist as a mechanic who is supposed to understand how an ecosystem functions and who can relate stipulated human inputs, such as stressors, to ecological outputs. In other words, the economist expects the ecologist to present ecological relationships in economic terms, for example, to regard the ecosystem as a sort of firm, with inputs and outputs that could be entered into benefit-cost analysis. The ecologist is expected to provide a basis for a damage function and thus to help represent ecosystems properties in economic terms. Rather than respond, as it were, on the merits to the moral principles, aesthetic judgments and overall views of the integrity of ecosystems, which ecologists bring to the table, economists may try to cast these in terms of costs and benefits only as 'valuable' as there is the WTP for them. As natural places disappear the economist may feel the pain of the ecologist – and offer to price it.

More broadly, the graduate training regime, because it stresses the beauty, power and completeness of each discipline, tends to produce a similar level of self-righteousness, even arrogance, in many of its products. Rewards within each discipline go to those who master its structure and assumptions and apply them to new problems. There is hardly time, and little encouragement, to look for flaws or weak points in the conceptual structure of one's own science. Indeed, there is less time every decade, since four or five years is the norm toward which PhD programs aim, but each year the mountain of material to master grows higher and mathematical methodologies make the slopes ever more slippery. It is no great leap from gaining one's academic credentials to the notion that the discipline you enter with your hooding has grasped the world correctly in (nearly) all respects; after all, one has not spent much time discussing where the basic framework may be shaky or even flawed. For some, this acceptance leads to arrogance that is always near the surface. And nothing is more destructive of dialogue than arrogance expressed.

Therefore, it may be salutary to review briefly some of the reasons both economists and ecologists have for practicing humility in their interactions with each other and with other disciplines as well. In this enterprise we shall confine ourselves to matters directly relevant to the problems raised by normative concepts in research in valuation. In particular, we have in mind the situations, such as those described previously, involving value judgments about changes in an environmental stressor, such as pollution. The term 'pollution' itself can be seen as normative – as ecologists may, indeed, regard it – connoting corruption, trespass, assault, filth and the like. On the other hand, it could be treated, as economists are likely to do, in the neutral sense of an external cost of production, acceptable insofar as it is balanced by benefits.

Problems for Economists: Valuing Changes in Ecosystem Service Flows

In this section we shall assume with economists that the stressors in question are not obviously of catastrophic potential, so that the relevant questions are: (1) what changes in the ecosystem's service flows actually result from the proposed human action and (2) how do humans feel about (value) those changes? Consider, for example, changes in ecosystem service flows traceable to tighter control of the precursors of urban ozone formation (that is, smog). Let us say smog tends to drift downwind from the urban area over rural and natural areas, causing damage directly or indirectly to plants, animals and watercourses. We shall also assume that ecologists can produce functions relating direct and indirect physical damage to measures of exposure to ozone (extent of the cloud and concentrations within it). This is roughly what the economists on the EPA/SAB committee were seeking, for their tool kit contains the conceptual basis for valuing the decrease in damages (the benefits) attributed to CAA policy; and economists have a good deal of practical experience doing just such valuation in real situations where real money is at stake. On the question of valuation, economists would say: find out what the people concerned are willing to sacrifice (pay) in order to obtain the reduction in damage predicted by the ecological models.[8]

Behind this answer lie many assumptions. First, one must assume away the idea that smog, along with other forms of pollution, represents an evil to be minimized rather than a cost to be optimized or balanced against benefits. In other words, one must already assume that an efficiency norm, rather than, say, a property right applies in regulating pollution. (A property right entitles a person to injunctive relief against pollution in very serious cases or to a damage award more commonly.) An economist is likely to assume that pollution should be controlled according to an efficiency standard (that is, just to the point at which the cost of reducing the next unit equals the benefit). This is to accept an analysis of pollution as a social cost to be internalized into the prices of goods produced. However, it is not clear that anyone but economists – or those trained by them – accept this principle as opposed to a property-right or injunctive-relief approach, at least as a place to start the conversation.

Second, economists must assume that each person (or family) possesses a marginal willingness-to-pay (MWTP) function defined over the dimensions of change for any environmental problem.[9] It nearly goes without saying that even if MWTP constructs exist in the brains of the affected individuals, evidence of their shapes will not be seen directly because there are no transactions that would analogize to market purchases of private goods. Indeed, the prices people pay for ordinary goods result from an interplay of supply and demand and thus do not necessarily represent the maximum a person is willing to pay for the next unit of a good. The price charged may be, and often is, much lower if the good is inexpensive to produce. For example, people who finish a marathon or end a religious fast may be much thirstier or hungrier than others, yet they pay the same prices for food or drink at the supermarket or restaurant. Accordingly, their MWTPs may be represented by the minimum amount they have to pay – not the maximum price they might otherwise have paid – for the good in question.

Over the past 25 years or so, economists have developed techniques for asking people about their WTP for described changes in the provision of public goods. The majority of economists has accepted these methods, labeled as contingent valuation and contingent choice, at least in principle.[10] They have become the methods of choice in many applied environmental situations, including, just as one example, the justification of environmental project loan requests to the Inter-American Development Bank. Why and how this has happened is itself a fitting subject for analysis, but in the context of this discussion, two advantages of contingent valuation and contingent choice methods stand out:

- They are flexible, in that the analyst can directly ask about the public good of interest. With the older, so-called indirect, methods whether changes in environmental service flows could be valued in an even roughly credible way depended on whether links to actual markets could be hypothesized and whether appropriate data from those markets were accessible.[11]
- The methods in principle produce benefit estimates that reflect both use and non-use benefit accruals.

Nonetheless, there remain concerns about the robustness of the contingent methods and these seem especially serious in the contexts we are discussing here. Fundamentally, there is reason to doubt that expressed WTP measures a prospective change in welfare. Economic theory takes it as a foundational premise that preferences are to count in policy – preferences taken as they come and ranked on a WTP basis. However, economists have not explained why preferences should count – why they should be satisfied on this as-they-come basis. The connection between money (the ability to satisfy preferences) and happiness, after basic needs are met, is at best tenuous. In fact the hypothesis that people are not generally happier when they have more income and thus can obtain more of the things they have to pay for is one of the best established results of social science research.[12] Thus, the connection between WTP and welfare may be at best a tautological or stipulated one. Absent a connection between preference-satisfaction and any conception of welfare not defined in terms of it, why should society or the government seek to satisfy any and every preference, rather than just those related to need (health, poverty and so on), merit (education, science, art, environmental quality) or some socially recognized good?

Moreover, on a practical level there is both empirical evidence and a priori argument to suggest that people generally do not have any idea what they are willing to pay for changes in the natural world that are unfamiliar, possibly quite subtle and difficult to describe in easy-to-grasp ways – at least they have no idea before the process of tackling questions on a contingent valuation instrument. The a priori argument contends that people have neither time nor incentive to expend the effort to develop such self-knowledge in the abstract. The more unfamiliar and obscure the question, the less likely they are to have thought about it at all. When asked a question about environmental policy outcomes, people are more likely to think in terms of the principles relevant to policy (the right to pollute, the requirements of law, the ethical context and so on) than in terms of effects of that policy on their own welfare. Thus, people are likely to think in terms of general questions about the basis of policy to which economists have their own answers.

Technical Problems for Contingent Valuation Studies

In the absence of predetermined utility functions for the good in question, how seriously should we take the answers to contingent valuation studies? Those answers might reflect little more than the respondents' grasps at the straws offered by the survey, the content of the questions and possibly personal rules of thumb based on previous environmental involvement, such as making a contribution to an environmental advocacy group. Additionally, over the survey interaction hangs the desire, apparently widely shared, to be thought of as a good person.[13]

Artifacts of survey method and design discovered in the decades of research into contingent valuation methods tend to confirm the suspicion that respondents can do little more than grasp at straws. Examples include:

- The embedding effect, in which stated WTP does not seem to vary in reasonable ways with the scale of the problem described to the respondent (Boyle et al., 1994).
- Yea saying, in which the tendency is to agree with whatever is suggested by the question, such as a possible amount you might be willing to pay (Harris et al., 1989; Desvousges et al., 1993; and even NOAA, (1993).
- The tendency for the estimated WTP for the same hypothesized change in service flow to be higher when based on data from yes/

no (dichotomous choice) questions than from open-ended versions (Ready et al., 1996).

Generally, Peter Bohm, who did the first experiments revealing no tendency to lie about the WTP for a public good – a tendency predicted by Samuelson and naturally expected by economists – long maintained that one cannot take the answers to the purely hypothetical questions of the typical contingent value study seriously exactly because of their hypothetical character. There is no reason for respondents to rack their brains for the right answer when no consequences will likely flow from whatever answers they provide (Bohm, 1994).

Another difficulty arises when respondents are asked how much they are willing to accept in order to allow a given environmental change for the worse. The respondents may feel that they are placed in the role of extortionists selling a good they do not own. Thus, while the WTA compensation to offset the change is often conceptually the correct amount to seek and while this amount is expected to be greater than the WTP for a similar magnitude of change going in the opposite direction, efforts to obtain the WTA responses at all or to obtain responses that seem to pass a reasonableness test, have not generally been successful. Whether the reason for this failure is moral objection to even hypothetical bribes to give something up that may be seen as a birthright or that the nature of the WTA question frees the respondent from even a vague (and hypothetical) need to consider income limitations or, something else entirely, the contingent techniques do not seem to allow economists to value damage imposition, as opposed to damage mitigation (Knetsch and Sinden, 1984; Hanemann, 1991; Knetsch, 1994).

A further difficulty for the interpretation of expressions of WTP from survey methods is the apparent importance of ethical principles or moral beliefs in the answers provided by individuals. Indeed, it is remarkable how consistently researchers have found that the WTP for such effects is based on disinterested moral or cultural beliefs (Schkade and Payne, 1994; Blamey et al., 1995). In a representative study researchers found that existence value was 'almost entirely driven by ethical considerations precisely because it is a disinterested value' (Barbier et al., 1995, p. 836).

Two aspects to this observation are relevant here. One is that the ethical basis for the answers may not be a simple version of other-regarding altruism but rather something along the lines of Sen's commitment to a principle or ideal (Sen, 1977). The individual may express their view not of their own economic utility but of the principle or goal we should apply as a society. Views about policy conflict logically in the sense that they may contradict each other, not economically in the sense that resources are not adequate to satisfy them. This difference can complicate the interpretation and aggregation of the expressed WTP (Johansson-Stenman, 1999).

If people favor an outcome for principled reasons rather than for any benefit they may gain, it would be odd to say that the outcome they favor benefits them. Should a commitment to other goals or values beside personal welfare be ignored? There is widespread recognition that the WTP questions tend to produce a selfish focus in respondents, a focus that sits awkwardly at best with the inclination to consult a moral compass (Stern and Dietz, 1994; Stern et al., 1993). It may be hard for the respondent to think in terms of what is good for or benefits them in a context – or about a decision – that they would usually cast in the context of what is morally right or good for society as a whole.

Finally, a purely technical, but also not so far resolvable, challenge to the use of WTP estimates arises when the yes/no (referendum, dichotomous choice) technique is used in elicitation.[14] In essence, the challenge is rooted in the fact that with this technique no one reveals their WTP directly. Instead each respondent reveals only whether their WTP is greater or less than the bid they are presented, so that estimating either the mean or the median WTP for the sample of respondents and, thus for the population of interest, cannot be done simply by arraying or adding up stated WTP numbers. Rather, the analyst is left with a task of inference and it turns out there are at least a dozen suggested methods to perform that task. None of these may be dismissed on technical grounds and unfortunately they do not produce estimates that agree. Indeed, it is possible to find that the range of methods produces a range of answers that differ widely enough to cast doubt on any resulting benefit-cost analysis (Vaughan et al., 1999, 2000). Taken together, the previously stated problems seem to provide economists with ample grounds for humility in approaching ecological valuation problems. Even though the techniques for estimating a population's WTP for a change in the flow of environmental services have improved immensely over the past several decades, largely through the efforts of environmental economists, there are still enough philosophical, practical and technical challenges to make it clear there is much more to be done.

Problems for Ecologists: Defining What is to be Valued

Ignorance and disagreement about the structure and functions of ecosystems, how to define and delimit them – indeed, about how to reidentify a place as possessing the same ecosystem through time and change – confound attempts by ecologists to provide in any general sense the damage function economists seek. For example, the Millennium Ecosystem

Assessment defines an ecosystem as a dynamic complex of plant, animal and microorganism communities and the non-living environment interacting as a functional unit (Millennium Ecosystem Assessment, 2003, p. 15). This definition includes practically anything one can imagine (for example, a rotting carcass, a sewer system, a toxic waste lagoon, a kitchen sink, a yeast infection and a college dorm room). All of these are functional units in one sense or another. Such definitions build into the ecosystem concept the very property (for example, self-organization, autocatalysis and complex adaptiveness) that is then predicated of it. Defining the concept of an ecosystem then becomes a project of analysing terms like 'self-organization' – an effort that often depends less on empirical observation than on the derivation of the consequences of one's arbitrary mathematical assumptions. It is easiest to apply the concept of a system – including terms like 'structure' or 'function' – to objects designed for a purpose. The structure and function of a kidney, for example, can be understood in relation to the purpose of cleansing the blood. Similarly, organisms are organized by evolutionary forces to enjoy relative reproductive success. If ecosystems were organisms (that is, units of selection) or if they had a purpose (if God created them for the benefit and support of humanity), one could identify the properties of design, organization and function by which they serve an end. Absent evidence that ecosystems are organized either by natural or supernatural selection, however, it is not clear that ecosystems constitute systems rather than just contingent and idiosyncratic collections.

Whether plants and animals found at a site constitute a system of any kind, indeed, is a question much debated among ecologists. Thus, Gilbert and Owen (1990, p. 33) wrote that their observations of nature provide no evidence of 'an ontological emergence of a community level of biotic organization'. Any suggestion of pattern or structure in ecological phenomena is a 'biological epiphenomenon, a statistical abstraction, a descriptive convention without true emergent properties but only collective ones, wholly referable in its properties to those of its constituent species, populations and individuals.' In this view, nature presents a blooming, buzzing confusion of contingency that defies prediction and, indeed, theoretical understanding. At best, ecologists can identify the particular causes of particular events and, somewhat more generally, the empirical processes most significant with respect to a phenomenon of interest. Ecologist William Drury (1998, p. 23) inveighed against the 'strong tendency to accept the existence of self-organizing principles as inherent in natural systems'. He wrote, 'I feel that ecosystems are largely extemporaneous and that most species (in what we often call a community) are superfluous to the operation of those sets of species between which we can clearly identify important interactions . . . Once seen, most of the interactions are simple and direct. Complexity seems to be a figment of our imaginations driven by taking the "holistic" view' (p. 32).

In the absence of a consensus about how to define the concept ecosystem or even whether systems of the sort exist in nature, ecologists are likely to approach the identification of ecosystem services on a case-by-case basis. This makes sense. One does not need to define the concept ecosystem or say anything general or abstract about its structure or function to understand how in particular cases nature directly benefits humanity. For example, wild or naturally occurring insects often pollinate crops; when they do not, farmers often contract with beekeepers for commercial pollination services. Changes to ecosystems that threaten the population of wild pollinators – one thinks of pesticides, for example – would threaten the pollination services they provide.

The assumption that ecosystems or natural environments provide valuable services follows from the view that nature comprises everything – or at least everything not identified with human artifice. Yet, by the same token, nature is also the source of many disservices, such as earthquakes, floods, droughts, plagues, pests and all kinds of miseries and destructions, including universal death. Accordingly, while some changes to ecosystems lessen ecological services, others may prevent or mitigate ecological disservices and some changes may do both. Some balancing of nature's services with its disservices may be necessary to understand the value of natural environments. And there is little reason to believe that nature in an original state is on balance more supportive of than hostile to human well-being.

An insistence on the value of ecosystem services is surely helpful in those instances in which one can identify ecological processes that produce goods that markets can price. Obvious examples can be found in hunter-gatherer goods such as fish or trees captured from the wild. It would be feasible to value the natural system by comparing the price it might charge for its output relative to that of a competing industrial product. Indeed, wild and farmed salmon, for example, are often sold competitively in the same markets with the wild product fetching a premium.[15] But where, as is common, the assertion is that system survival is at stake, demonstration of a more comprehensive understanding of how ecosystems function seems critical. And yet there is no consensus among ecologists about how to define an ecosystem, delimit it or even whether the concept is a useful one (O'Neill, 2002).

This is where the dialogue can become sticky for the non-ecologist. Ecologists may argue that society should defer to their expertise as scientists, as much because of what ecological science does not know as what it does understand about complex systems. David Takacs (1996, p. 92) comments that ecologists 'tell us how little we know about the vanishing natural world and how severely we might pay for how little we know. They urge us to proceed with caution . . . therefore, we must preserve as much of the natural world as we can.' Takacs concludes that ecologists 'are thus masters of ignorance since what they do not know can hurt us and we ignore their ignorance at our peril'.

This observation brings us squarely back to the enormously important matter of prescriptions for dealing with profound ignorance and the resulting uncertainty about what any particular environmental action (or inaction) will produce by way of results in the natural world. As Samuel Hays (1985, p. 253) points out in his history of environmentalism, ecological science admittedly cannot provide even a small part of the information policy makers demand or need: 'Every advance in knowledge seemed to expand what was not known even more rapidly than it did what was known; society seemed to be faced with escalating ignorance.'

THE SAFE MINIMUM STANDARD: A BASIS FOR DIALOGUE?

To deal with risks or uncertainties associated with the possible collapse of ecosystems, ecologists and economists have discussed, separately and together, the possibility of establishing safe minimum standards (SMS) for environmental assets (that is, standards that would constrain or even supplant benefit-cost analysis when decisions involved sufficiently widespread or irreversible environmental effects).[16] The SMS concept, suggested by Ciriacy-Wantrup (1963), primarily concerns limits on the exploitation of the kinds of renewable resources that exhibit critical thresholds. Thus, emissions can overwhelm the capacity of sinks to absorb and recycle wastes. Species require minimum populations to survive. There may well be thresholds in the global climate system at which major changes in ocean currents, for example, could be triggered by modest additions to global average temperatures.

The appeal to SMS moves the burden of proof from those who would preserve resources to those who would exploit or develop them (Randall and Farmer, 1995). Given the depth and extent of our ignorance about the way ecosystems work – they are more complex than we can understand – ecologists argue we should err on the side of caution by allowing only those changes to natural systems that can be shown to be completely necessary. As ecologist Reed Noss (1991, p. 329) wrote, 'One key principle is prudence. In other words, we should err on the side of preservation when science is unable to provide clear answers.' The 'discipline of environmental science is fairly new,' opines historian of science Robert Nadeau, 'but with sufficient funding ecologists may one day be able to predict the behavior of complex systems'. Nadeau indicates that 'Minimizing negative environmental impacts of global activities is not a great price to pay in exchange for the opportunity to acquire knowledge that could allow our children and grandchildren to live out their lives on a hospitable planet' (Nadeau, 2003, pp. 146, 147).

Everyone is in favor of allowing future generations to live on a hospitable planet. However, it is knowledge, not ignorance that allows us to do this. Lord Jevons thought that his generation should save coal for the use of future generations – but this turned out to be unnecessary. We need to know what ecosystem services we should protect and what qualities or components are important to that protection. To save everything, in contrast, might require us to stop investing and thus giving form to new technologies. History suggests that this extreme measure might well do the opposite of conferring a benefit on future generations.

Consider the more or less ubiquitous problem of dealing with endangered species. Extinction is, for all currently practical purposes, irreversible. Since there are tens of millions of species – and given that genetic engineering may create millions more – which ones should we protect? The mere possibility that an endangered frog or frond harbors a valuable chemical is simply too slight to provide an incentive to any pharmaceutical company to bother to find out. Pharmaceutical companies develop 'drugs based on molecules that scientists can create for themselves in the laboratory, using new combinatorial chemistry techniques' (Macilwain, 1998, p. 535; Eisner, 2003). No industry sees any advantage in bioprospecting.[17] How, then, can we fashion an argument for saving endangered species on the basis of their instrumental value to future generations?

Ignorance provides only a shaky basis for preservation. Whether or not ignorance makes preservation the risk-averse strategy depends on what we are ignorant about when we have to decide. Thus, Ready and Bishop (1991) contrast two games played against nature. In one of these, when the decision must be made either to preserve the species or to develop its habitat and thus drive it to extinction, we know that the creature holds the cure to a dreaded disease, but we do not know whether the disease will actually become a human threat. In the other game we know that the disease will be a threat but not whether the species in question will hold a cure. Ready and Bishop show that in the first game preservation is the risk-averse (minimax loss) solution. But in the second ignorance at the time of the decision makes development and thus extinction the risk-averse choice. We cannot, in that game, penalize the development option for the loss of a cure that we do not know even exists. If this makes us uncomfortable, the obvious alternative, though not necessarily a realistic one, is to prevent all extinctions until we know all about each of tens or even hundreds of millions of species – at a significant research cost per species. We are back where we started with widespread and absolute prohibitions on development trumping the advance of technology that has so far improved the human lot. Technological advance may make our great grandchildren much better off than we – as it has made us better off than the pioneers who battled nature. Shall we act out of fear or hope?

A second problem involves what we must do to prevent extinction. Is preservation in zoological or botanical gardens sufficient? If not, how much of a host ecosystem must be protected? Of what must the protection consist? Can any human activity in the neighborhood be allowed? What sort and how much? For nearly every taxon in nearly every place – only a few tiny islands are exceptions – invasive or non-native species have vastly increased the available biodiversity, more than doubling the flora on large islands like New Zealand and Hawaii.[18] If the number and variety of plants and animals continues rapidly to increase in most ecosystems owing to invasions and if genetic engineering can produce nearly infinite amounts of additional genetic variety, if required, where are scarcities likely to arise?

In short, since ignorance means, in most of the major decision settings for which the SMS is advocated, inability to know when a standard produces safety, let alone what it is the minimally intrusive or costly way to achieve safety, the SMS concept may simply shift the terms of the ongoing discussion about cooperation, rather than producing agreement on how to cooperate. Would it not be preferable if, chastened by ignorance and uncertainty, ecologists and economists did not increase their reliance on their assumptions but questioned even whether what they know is true? Then they might work together in a framework of common sense and respect – honoring intellectual virtues such as clarity, openness to the opinions of others, the avoidance of righteousness and the recognition that one may be wrong.

RECOMMENDATIONS FOR BRIDGING THE GAP

After this diagnosis the question remains: what is the treatment plan? What, if anything, can be done to foster the ardently desired collaboration between ecologists and economists in the applied settings that could make so much difference to public policy dealing with the environment? Here are several suggestions, the first negative, the others positive.

Regulatory agencies should not expect blue ribbon committees to resolve philosophical problems that stand in the way of collaboration. While it is flattering to be chosen to participate in committees of 'the great and the good' and while the participants may actually learn something from the discussions, the chance that an ad hoc group will produce a consensus report that represents a breakthrough seems vanishingly small. Bromides, nostrums, invocations of the obvious and pabulum are more likely to emerge as committees find ways to pack squared circles into banalities and though these may trump a complete breakdown in communications, at least in expected value terms, the actual results are not likely to be much different in most experiments. And panels of interdisciplinary experts compete with other ways to spend limited budgets that seem more promising. An alternative strategy for agencies is to bring together teams of practitioners/researchers whose interests and experience are more applied than theoretical, with the understanding that each team will stay together for a significant period of time, several years at least, and will, during that time, be expected to provide useful input to actual decisions – input that reflects real collaboration, not one chapter of ecological truth followed by another of economic truth.[19] Parallel play cannot be the standard for interdisciplinary research. The agency should not specify in advance the approach or methodology to be taken, but there would be an explicit expectation that state-of-the-art predictive models, where these exist, would be applied where appropriate. And, given that benefit-cost analysis is unlikely to disappear from the government rulebook (unless the Obama administration issues an Executive Order on Regulatory Review that abandons it), techniques that produce summary measures in some units analogous to prices may be used. These techniques might be as familiar as benefit-cost and cost-effectiveness analysis or as challenging as multi-attribute utility analysis. Whatever is used, there would be no pretence that it represents the one true way to value ecosystem services.

When seeking to value or price ecosystem services, agencies should look as much at the cost as at the benefit side of the analysis. Economists and ecologists may agree on ways to describe (if not to value) damage to the environment and on ways to limit or minimize the costs involved in avoiding or mitigating that damage.[20] Collaborative efforts can also succeed if directed to cost-effective methods for analysing alternatives. For example, it would be useful to consider how such methods as complementary analysis for designing species survival strategies and habitat and wetland evaluation approaches (for example, Habitat Evaluation Procedure (HEP), Habitat Evaluation System (HES) and Wetlands Evaluation Techniques (WET)) can meaningfully be costed and used in predictive modeling settings to answer questions such as, 'What is the cheapest way to get a specified wildlife population effect?'

At a somewhat more abstract level, it would be interesting to see if sustained effort could put some empirical meat on the bones of the SMS idea. In particular, it should be possible to apply the SMS idea where there exist at least some working predictive ecological models. Could one show in a case study or two what would be given up in return for varying levels of safety (inevitably in probabilistic terms) in view of the prevailing levels of ignorance about model parameters and functional forms?

At a still greater level of generality, agencies might support explicitly cross-disciplinary research aimed at developing tools and techniques acceptable to both ecologists and economists. It would be arrogant to specify what these might look like or from what traditions they might spring, but there seems to be no lack of candidate ideas to examine in the world of decision analysis.[21]

Finally, recognizing that committees bringing together ecologists and economists to study environmental policy will continue to be created, they are bound to continue to confront the demand for ecological damage or production functions. While economists may acknowledge the infinite value of nature as a whole, decisions have to be made at the margin at least in the sense of the particular policy or project. Whether or not the natural world is worth at least $33 trillion because of its life-support functions will not make it any less important to operate in a way that attempts to do the best with what we have, including knowledge of that natural world as well as knowledge of human-created technologies. For the long haul, perhaps the best we can hope for is that the focus of discussion in such committees will slowly shift toward more ways to distinguish situations in which extreme risk aversion is the mind set of choice from situations in which we can, as an engineer might say, uncouple the parts of the system and deal with the one directly affected by whatever proposal is under discussion.

ACKNOWLEDGMENTS

We are grateful to Charles Howe and Duane Baumann for comments on an early outline version of this chapter and to A. Myrick Freeman, Charles Howe, Peter Loucks, Larry MacDonnell and V. Kerry Smith for extensive comments on a full draft. None of them, needless to say, bear any responsibility for this product. We also appreciate and have tried to reflect the comments obtained from assembled members of the Institute for Water Resources (IWR) during an authors' meeting at Fort Belvoir in June 2004.

NOTES

1. Judging by reactions to earlier versions of this chapter, it is important to emphasize here that this statement is not meant to sweep up all ecologists or all economists. It ignores not only variations across individuals but also across subdisciplines. In particular, it has been suggested that for the label ecologists we should substitute conservation biologists, as being more likely to capture what our label implies. But the experiences that triggered our interest in this problem involved people labeled as ecologists and we shall continue to use that broader name.
2. For a nearly contemporaneous look at the field of aquatic ecosystem modeling, see Russell (1975).
3. Status is here fudge for health. Some participants explicitly said that the goal was to be able to produce indices of ecosystem health analogous to the daily reports of the status of air quality routinely found in newspapers. Suter, however, then at ORNL had recently (1993) persuasively argued that this analogy to human health is neither meaningful nor helpful.
4. To anticipate the end of the story on the accomplishment dimension, the EPA went through a change of research leadership after about one-third of the project time and money had been expended. The new Assistant Administrator killed the project, leaving both groups high and dry with only a few focus group exercises as outputs. Nonetheless, an enterprising graduate student never gave up on publication and managed to place a multi-author paper in *Conservation Ecology*; Schiller et al. (2001).
5. Economists agreed with this characterization. Michael Toman, then at RFF, allowed that it amounted to a serious underestimate of infinity.
6. For example, Freeman wrote notes criticizing such gaffes as the use of gross landing values from fisheries instead of the net value or profit, the latter in many or perhaps most cases of open-access fisheries being zero or even negative. (personal communication; Freeman, 2003).
7. Marshall (1920) continues: 'Now, if the nation as a whole finds its stock of planing machines or ploughs inappropriately large or inappropriately small, it can redistribute its resources. It can obtain more of that in which it is deficient, while gradually lessening its stock of such things as are superabundant: but it cannot do that in regard to land; it can cultivate its land more intensively but it cannot get more . . . [F]rom the economic and from the ethical point of view, land must everywhere and always be classed as a thing by itself' (Appendix G, p. 28). For a good discussion of the views of Marshall with respect to land as a factor in production, see El Serafy (1991, pp. 168–75, esp. p. 169).
8. This statement glosses over a non-trivial problem: identifying the people concerned or, said a different way, the people whose preferences are to count in valuing the policy. As so-called non-use benefits or damages have become more respectable as components of analyses this question has loomed even larger than it was when Whittington and McRae raised it for the profession in their 1986 paper (Whittington and McRae, 1986).
9. It is necessary practically that the technical outputs of the ecological damage functions be translated into terms, indeed probably into pictures or images of some sort, to which lay people can relate. See the brief discussion of such an effort previously in war story 2.
10. The biggest breakthrough was almost certainly the acceptance of these methods, with some qualifications and cautions, by the National Oceanic and Atmospheric Administration (NOAA) Blue Ribbon Panel of Nobel Laureates (see NOAA 1993).
11. For anyone who wants to learn more about the indirect methods, there is no better place to begin than with Freeman (2003). Similarly, though it is somewhat more dated, an excellent initial source on the contingent or direct methods is Mitchell and Carson (1989).
12. For example, see Diener (1984), Erskine (1964), Guring (1960), Griffin (1986), Lane (1991, 1993), Diener et al. (1985), Argyle (1986) and Brickman et al. (1978). People in wealthier societies may be happier than those in impoverished ones, but this suggests

nothing about the relation between income or preference-satisfaction and happiness. In wealthier societies basic needs are more likely to be met; people have employment and health care, which do contribute to well-being.

13. For an excellent discussion of this major difficulty, as well as suggestions for improving matters, see Harris et al. (1989). Also highly relevant are the thoughts of Fischhoff (1991) and Slovic (1995).
14. It is important to note that this technique is the one recommended by the highly influential NOAA panel already mentioned.
15. John Krutilla's influential 1967 paper, 'Conservation reconsidered', made the case that naturalness or wildness of places would be of increasing value as the exploitation of the natural world for both marketed goods and mass recreation services grew in importance, the latter because of increasing population enjoying higher incomes and more leisure time, all the while learning by doing (Krutilla, 1967).
16. See, for example, Toman (1994), Ready and Bishop (1991) and Perrings and Pearce (1994).
17. When, in 1991, drug maker Merck responded to pleas of environmentalists by investing about $1 million – less than one-tenth of 1 per cent of its annual research budget – in taxonomic research in Costa Rica, it reaped a public relations bonanza because environmentalists never wearied of citing and celebrating the project as vindicating the economic value of wild rainforest species. Merck 'terminated its landmark bioprospecting contract with Costa Rica in 1999 after producing no commercially viable products' (Foreman, 2001, p. C1). Other efforts at bioprospecting in beautiful places have been similarly fruitless. Shaman Pharmaceuticals, which followed clues in ethnomedicine, filed for bankruptcy protection in 2001. Experts in drug discovery caution, 'The idea of exploiting the rain forests to find wonderful drugs is, quite frankly, not credible' (David Galas, President of Darwin Molecular, quoted in Macilwain, 1998, p. 537). Drug discoveries have been made (as by Merck) in toxic waste dumps and similarly disgusting and hazardous places where creatures evolve, for example, resistance to toxins. The toxic waste lagoons of Rahway, New Jersey might be preserved for bioprospecting more plausibly than the rainforests of the Amazon.
18. Sax et al. (2002, p. 774) have observed a 'highly consistent, approximately twofold, increase in the species richness of plants on oceanic islands' owing to plant introductions and invasions. Historically about 2000 plant species existed in the wild in New Zealand; fewer than ten are known to have become extinct. An additional 2000 exotic species have migrated to New Zealand, doubling plant biodiversity on that island (Sax et al., 2002, p. 768). Likewise, in Hawaii 'the native flora consists of about 1100 species – and an additional 4,600 exotic plants have been identified there . . .' (Vitousek, 1990, p. 8).
19. We have been told that such teams operate successfully within the water resources operations of the US Army Corps of Engineers.
20. An excellent study by Barnthouse et al. (1988) on the effect of cooling towers on fish recruitment, for example, finessed the problem of valuation by proposing an inexpensive and effective method of mitigation.
21. Since the late 1990s, the US Department of Energy, Office of Research has been supporting such work on a modest scale. For a product of one such collaboration, organized around ecosystem valuation in the face of uncertainty about catastrophic thresholds in ecosystem behavior, see Batabyal et al. (2003).

REFERENCES

Argyle, M. (1986), *The Psychology of Happiness*, New York, NY: Methuen and Co. Ltd.

Barbier, E.B., G.M. Brown, Jr, S. Dalmazzone, C. Folke, M. Gadgil, N. Hanley, C.S. Holling, W.H. Lesser, K.G. Mäler, P. Mason, T. Panayotou, C. Perrings, K. Turner and M. Wells (1995), 'The economic value of biodiversity', in H. Heywood (ed.), *Global Biodiversity Assessment*, Cambridge: Cambridge University Press, pp. 823–914.

Barnthouse, L.W., R.J. Klauda, D.S. Vaughan and R.L. Kendall (eds) (1988), *Science, Law and Hudson River Power Plants: A Case Study in Environmental Impact Assessment*, monograph no. 4, Bethesda, MD: American Fisheries Society.

Batabyal, A.A., J.R. Kahn and R.V. O'Neill (2003), 'On the scarcity value of ecosystem services', *Journal of Environmental Economics and Management*, **46** (2) (September), 334–52.

Blamey, R., M. Common and J. Quiggin (1995), 'Respondents to contingent valuation surveys: consumers or citizens', *Australian Journal of Agricultural Economics*, **39** (1), 263–85.

Bohm, P. (1994), 'CVM spells responses to hypothetical questions', *Natural Resources Journal*, **34** (1), 37–50.

Boyle, K.J., W. Desvousges, F.R. Johnson, R.W. Durnford and S.P. Huntington (1994), 'An investigation of part-whole bias in contingent valuation studies', *Journal of Environmental Economics and Management*, **27** (1) (July), 64–83.

Brickman P.D., D. Coates and R. Janoff-Bulman (1978), 'Lottery winners and accident victims: is happiness relative?', *Journal of Personality and Social Psychology*, **36** (8), 917–27.

Ciriacy-Wantrup, S. (1963), *Resource Conservation: Economics and Policies*, Berkeley, CA: University of California.

Costanza, R., R. d'Arge, R. de Groot, S. Farber, M. Grasso, B. Hannon, S. Naem, K. Limburg, J. Paruelo, R. O'Neill, R. Raskin, P. Sutton and M. van den Belt (1997), 'The value of the world's ecosystem services and natural capital', *Nature*, **387**, 253–60.

Daly, H.E. (1991), *Steady-state Economics*, 2nd edn with new essays, Washington, DC: Island Press.

Daly, H.E. (1996), *Beyond Growth: The Economics of Sustainable Development*, Boston, MA: Beacon Press.

Desvousges, W., A. Gable, R. Dunford and S. Hudson (1993), 'Contingent valuation: the wrong tool to measure passive use values', *Choices*, **8**, 9–11.

Diener, E. (1984), 'Subjective well-being', *Psychological Bulletin*, **95**, 542–75.

Diener, E., J. Horwitz and R.A. Emmons (1985), 'Happiness of the very wealthy', *Social Indicators Research*, **16**, 263–74.

Drury, W. (1998), *Chance and Change*, Berkeley, CA: University of California Press.

Eisner, T. (2003), 'Hard times for chemical prospecting', *Issues in Science and Technology*, **19** (4), 47–8.

El Serafy, S. (1991), 'The environment as capital', in R. Costanza (ed.), *Ecological Economics: The Science and Management of Sustainability*, New York: Columbia University Press, pp. 168–75.

Erskine, H. (1964), 'The polls: some thoughts about life and people', *Public Opinion Quarterly*, **28**, 517–28.

Fischhoff, B. (1991), 'Value elicitation: is there anything there?', *American Psychologist*, **46**, 835–47.

Foreman, J. (2001), 'Drug hunters can't see the rainforest for the medicines', *Boston Globe*, 27 March, p. C1.

Freeman, A.M. (2003), *The Measurement of Environmental and Resource Values: Theory and Methods*, Washington, DC: Resources for the Future.
Gilbert, F.S. and J. Owen (1990), 'Size, shape, competition and community structure in hoverflies', *Journal of Animal Ecology*, **59**, 21–39.
Griffin, J. (1986), *Well-Being*, Oxford: Oxford University Press.
Guring, G. (1960), *Americans View their Mental Health*. New York: Basic Books.
Hanemann, W.M. (1991), 'Willingness to pay and willingness to accept: how much can they differ?', *American Economic Review*, **81**, 635–47.
Harris, C.C., B.L. Driver and W.L. McLaughlin (1989), 'Improving the contingent valuation method: a psychological perspective', *Journal of Environmental Economics and Management*, **17** (3), 213–29.
Hays, S.P. (1985), *Beauty, Health and Permanence: Environmental Politics in the U.S. 1955–1985*, Cambridge, MA: Cambridge University Press.
Heilbroner, R.L. (1961), *The Worldly Philosophers*, New York: Simon and Schuster.
Johansson-Stenman, O. (1999), 'Environmental regulation in the presence of interdependent and context-dependent preferences and social norms', unpublished manuscript, Department of Economics, Göteborg University.
Knetsch, J.L. (1994), 'Environmental valuation: some problems of wrong questions and misleading answers', *Environmental Values*, **3**, 351–68.
Knetsch, J.L. and J.A. Sinden (1984), 'Willingness to pay and compensation demanded: experimental evidence of an unexpected disparity in measures of value', *Quarterly Journal of Economics*, **99**, 507–21.
Krutilla, J.V. (1967), 'Conservation reconsidered', *American Economic Review*, **LVII** (4), 777–86.
Lane, R.E. (1991), *The Market Experience*, New York: Cambridge University Press, Chapter 26.
Lane, R.E. (1993), 'Does money buy happiness?', *The Public Interest*, **113**, 56–65.
Locke, J. (1963), 'Second treatise of government', in P. Laslett (ed.), *Locke's Two Treatises of Government*, Cambridge: Cambridge University Press.
Macilwain, C. (1998), 'When rhetoric hits reality in debate on bioprospecting', *Nature*, **392**, 535–40.
Marshall, A. (1920), *Principles of Economics*, 8th edn, London: Macmillan, Chapter 4.
Marx, K. (1972), 'Grundrisse zur kritik der politischen okonomie', quoted in W. Leiss, *The Domination of Nature*, New York: Braziller.
McMichael, A.J., C.D. Butler and C. Folke (2003), 'New visions for addressing sustainability', *Science*, **302**, 1919–20.
Millennium Ecosystem Assessment (2003), *Ecosystem and Human Well-being: A Framework for Assessment*, Washington, DC: Island Press.
Mitchell, R.C. and R.C. Carson (1989), *Using Surveys to Value Public Goods*, Washington, DC: Resources for the Future.
Nadeau, R.L. (2003), *The Wealth of Nature: How Mainstream Economics has Failed the Environment*, New York: Columbia University Press.
National Oceanic and Atmospheric Administration (NOAA) (1993), 'Natural resource damage assessments under the Oil Pollution Act of 1990', *Federal Register*, **58** (10), 4601–14.
Noss, R.F. (1991), 'From endangered species to biodiversity', in K.A. Kolm (ed.), *Balancing on the Brink of Extinction*, Washington, DC: Island Press, pp. 227–46.

O'Neill, R.V. (2002), 'Is it time to bury the ecosystem concept? (with full military honors, of course!)', *Ecology*, **82** (12), 3275–84.
Perrings, C. and D.W. Pearce (1994), 'Threshold effects and incentives for the conservation of biodiversity', *Environmental and Resource Economics*, **4** (1), 13–28.
Randall, A. and M.C. Farmer (1995), 'Benefits, costs and the safe minimum standard of conservation', in D.W. Bromley (ed.), *Handbook of Environmental Economics*, Oxford and Cambridge: Basil Blackwell Ltd., pp. 26–44.
Ready, R.C. and R.C. Bishop (1991), 'Endangered species', *American Journal of Agricultural Economics*, **73** (2), 309–12.
Ready, R.C., J.C. Buzby and D. Hu (1996), 'Differences between continuous and discrete contingent value estimates', *Land Economics*, **72**, 397–411.
Rorty, R. (1991), 'Science as solidarity', in R. Rorty (ed.), *Objectivity, Relativism and Truth*, Cambridge: Cambridge University Press, pp. 35–45.
Russell, C.S. (ed.) (1975), *Ecological Modeling in a Resource Management Framework*, Washington, DC: Resources for the Future.
Sax, D.F., S.D. Gaines and J.H. Brown (2002), 'Species invasions exceed extinctions on islands worldwide: a comparative study of plants and birds', *American Naturalist*, **160**, 766–83.
Schiller, A., C. Hunsaker, V. Dale, C. Russell, M. Kane, G. Suter, G. Pion, M. Jensen and V. Konar (2001), 'Communicating ecological indicators to decision makers and the public', *Conservation Ecology*, **5**, 1.
Schkade, D.A. and J.W. Payne (1994), 'How people respond to contingent valuation questions: a verbal protocol analysis of willingness to pay for environmental regulation', *Journal of Environmental Economics and Management*, **26**, 88–109.
Sen, A. (1977), 'Rational fools: a critique of the behavioral foundations of economic theory', *Philosophy and Public Affairs*, **6**, 317–44.
Slovic, P. (1995), 'The construction of preferences', *American Psychologist*, **50**, 364–71.
Smith, A. (1991), *An Enquiry into the Nature and Causes of the Wealth of Nations*, first published 1776, Oxford: Oxford University Press.
Stern, P.C. and T. Dietz (1994), 'The value basis of environmental concern', *Journal of Social Issues*, **50**, 65–84.
Stern, P.C., T. Dietz and L. Kalof (1993), 'Value orientations, gender and environmental concern', *Environment and Behavior*, **25**, 322–48.
Suter, II., G.W. (1993), 'A critique of ecosystem health concepts and indexes', *Environmental Toxicology and Chemistry*, **12**, 1533–9.
Takacs, D. (1996), *The Idea of Biodiversity*, Baltimore, MD: Johns Hopkins University Press.
Toman, M. (1994), 'Economics and "sustainability": balancing trade-offs and imperatives', *Land Economics*, **70** (4), 399–413.
US Environmental Protection Agency (EPA) (2001), 'Review of the draft analytical plan for EPAs Second Prospective Analysis – benefits and costs of the Clean Air Act, 1990–2020', Science Advisory Board EPA-SAB-Council-ADV-01-004, Environmental Board (1400A) September, an advisory by a special panel of the Advisory Council on Clean Air Act Compliance Analysis, accessed 23 January, 2009 at www.epa.gov/science1/pdf/councila01004.pdf.
US Environmental Protection Agency (EPA) (2002), 'A framework for the economic assessment of ecological benefits', Science Policy Council, Social Science Discussion Workgroup, accessed at www.epa.gov/osa/spc/pdfs/feaeb3.pdf.
Vaughan, W.J., C.S. Russell, D.J. Rodriguez and A.C. Darling (1999), 'Willingness

to pay: referendum contingent valuation and uncertain project benefits', Inter-American Development Bank working paper ENV-130, Washington, DC.

Vaughan, W.J., C.S. Russell, D.J. Rodriguez and A.C. Darling (2000), 'Cost-benefit analysis based on referendum CV: dealing with uncertainty', *Journal of Water Resources Planning and Management*, **126**, 351–7.

Vitousek, P.M. (1990), 'Biological invasions and ecosystem processes; towards an integration of population biology and ecosystem studies', *Oikos*, **57**, 7–13.

Whittington, D. and D. MacRae (1986), 'The issue of standing in cost/benefit analysis', *Journal of Policy Analysis and Management*, **5**, 665–82.

6. Political decision making: real decisions in real political contexts

Peter Rogers, Lawrence MacDonnell and Peter Lydon

INTRODUCTION

> The problems posed by having to make explicit choices about the environment, something governments now have to do, have pointed up the weakness of modern government. So long as the problems faced by governments could be defined as technical problems, then they could be passed along to experts and the solutions arrived at by a nonpolitical or covertly political process. But environmental problems quickly became value judgment problems. Moreover, they obviously require governmental action. (Haefele, 1973, p. 15)

There seems to be great misunderstanding of what the concept of governance encompasses. In the eyes of many technical persons (engineers, economists and development planners, for example) governance is just about laws, regulations and institutions, typically exogenously given. The common belief is that there is an ideal set of laws, regulations and institutions that will result in good governance of water resources regardless of what else is happening in the country. Practical experiences, however, are quite different. In many settings apparently rational laws and regulations and appropriate institutional frameworks are in place, but the actual performance is nonetheless not good. Practical observers will note that unless textbook laws, institutions and regulations are based on a working political system, they cannot function as intended. In particular, one cannot consider a system of resource governance independent of the political system that is its framework, source and operating context. For an understanding of water governance, then, what is needed is an examination of the full interaction among politics, laws, regulations, institutions, civil society, water service producers and providers and the consumer-voter. The aim of this chapter is to take some initial steps down that road by:

- Broadening the set of political 'models' considered beyond the familiar, specifically by introducing the 'Polis' Model attributed to Stone (2002); and

- Exploring the complex history of the interstate sharing and management of the Colorado River in the light of alternative models of the political process.

The modest goal is to compare insights and thoughts about the future – both likely and desirable actions and outcomes in the light of the contrasting theories.

Political decision making for water has many levels of geographic scale and temporal sequencing. Often different levels, either spatial or chronological, overlap or are simultaneously in play, giving rise to subtle but real conflicts and incoherencies. For instance, in the case of scale, the need for an articulated federal water policy or for a national level water policy body can only be resolved at the highest political levels of government – leadership is required from the White House and, depending on the details of the arrangements contemplated, direct action by both houses of Congress may be necessary. In fact, the US Constitution is remarkably silent on water management. It implies by its silence that water management is a state subject only to be taken up at the national level when there is a compelling federal interest. Most water issues were then and still are local and can best be resolved at that level. But the federal government has become involved both in specific issue areas and in specific geographic contexts, in response to evolving concerns of interest groups and of the public more broadly, as has been described previously in Chapters 2 and 4 especially. The actual individuals involved in deliberations in Washington come to the capital from particular states, valleys and counties and are involved in their home districts' or states' very local projects, which will be heavily influenced by any changes in the current federal arrangements.

The time dimension has its own subtle rigidities that affect the political process and its sequencing. For example, often certain determinations or decisions, usually broad overarching ones, must be made before other actions can be taken. So the Carey Act of 1894 and the Reclamation Act of 1902 had to be formulated and in place before many consequential decisions for investments in western irrigated agriculture could be made. However, passage of even one such piece of legislation is the outcome of a major political struggle that cannot be undertaken lightly or repeated every year. Hence, huge inertia works against changing policies once they are in place and thus some flexibility and responsiveness are quickly lost from each new water management innovation. In addition to the inability of Congress to fight more than a limited number of serious battles in a given period, the major federal agencies involved do not wish to have their own programs disrupted by large policy changes and display a strong bureaucratic resistance to change. Finally, at any given moment, water stakeholders and interest groups have built their positions and developed their sources of political power by adapting to the status quo and they too can be relied upon to demonstrate large resistance to change. All of these inertial problems reside in any area of governance, such as energy or even military affairs, where there is always a powerful desire to fight the last war rather than change methods. Is inertia particularly oppressive in the water area? Is there something special in the nature of water or its provenance that makes its governance especially liable to the type of political rigidities and frictions just described? Given the difficulties described below experienced by federal agencies in achieving desired outcomes in the Colorado setting, the answer might seem to be yes. But it is perhaps more realistic, if less dramatic, to remain skeptical of such a claim. Sagoff (personal communication, 2004) points out by way of background that:

> In the nineteenth century, Congress encouraged settlers to migrate to the West. The Homestead Act of 1862 and the land acts that followed sought to build a Jeffersonian democracy of small freeholders throughout the arid West. These landowners were supposed to form associations patterned on town meetings to settle controversies that might arise among them. In refusing to allow public rangeland to be fenced, for example, Congress in 1885 perpetuated the hope that settlers could collaborate to manage common resources. The idea was not that Congress should deal with water and other resources – but that the people in those places would manage them. What happened, of course, was mayhem. By the end of the nineteenth century, timber, cattle, railroad and mining barons, after concentrating their economic and political hold on Western land and water, plundered places and ravaged resources that Congress had hoped associations of farmers, tradesmen and other small landowners would protect. Theodore Roosevelt spoke for the conservation movement by demanding that corporations 'be so supervised and regulated that they shall act for the interest of the community as a whole'. The sparse citizenry and the inchoate local government of the West, however, could hardly challenge mining, timbering and grazing barons. There had to be some form of governance – but no polity existed to take up the political reins.
>
> The result was that Congress asserted its power over Western resources basically to keep the robber barons from usurping them. Bureaucracies justified their power by appealing to expertise – the Progressive era faith that science could replace politics by identifying the public interest objectively. However, as the West became populated and the states gained power – and local interests became more diverse – layer upon layer of authority formed – leading to, gridlock, paralysis and a never-ending policy loop.

Nonetheless, under cover of apparent stasis, roles can change subtly – even if the final effect is to reinforce a status quo. Bonnor (2002), in recounting Buffalo Bill's important involvement with irrigation in

Wyoming at the turn of the twentieth century, describes the Reclamation Service's new professionals of that time as dedicated idealists who stood up for the public interest in opposing the local special interests. Compare this with Espelande's (1998) history of blocking the construction of the Orme Dam in the 1990s. Here the US Bureau of Reclamation (USBR), an 'Old Guard' of traditional engineering elites loyal to traditional interest groups, confronted a 'New Guard' of young experts who viewed their professional expertise to be 'rationality and planning, a commitment to commensuration and to making the bureaucracy more democratic' (p. 136). The New Guard aligned itself with a non-traditional interest group – the Yavapai Apaches – in stopping the dam and saving the Yavapai reservation. Over a long period of time we can expect, then, to see ebbs and flows of public interest behavior on the part of even the most hardened engineering agencies. In this example the USBR flipped from public interest idealism to rigid interest politics and then back to a more centered public interest posture.

Even for federal actions within the existing water non-policy, there is major resistance to any change in implementation. The political system has already adjusted to the status quo and each component knows how best to achieve its goals under the familiar circumstances. This makes the case of the Colorado River Basin, which we consider below, of particular salience because the federal government has played such an important role in promoting water development and, more recently, in facilitating needed changes. The federal role itself boils down to a set of water allocation and reallocation decisions smoothed by financial payoffs to some of the agricultural interests. What to most participants looks like a set of small decisions at the micro level, in reality could not have happened without strong action at the Congressional level.

There is a further long-term line of development we want to watch. The historical tradition is one of local interests in water for economic development, traditionally supported and argued over at the national level, but now we see a strong national interest in environmental protection beginning to assert itself at the local level. In the middle are all the beneficiaries of the traditional system. (See Chapters 2 and 4 for more on this theme.)

Politics is the arena in which changing values of water are fought. Water as a public resource is especially susceptible to political influences. Substantial individual and public entity rights to sue over water act as a brake on change, but this brake is eased a bit by the need for redefinition of rights to claim water. Such redefined rights must reflect the core private benefits which water rights carry, while separating out and preserving their public good dimensions. The tension between these divergent paths leads directly to political immobilization.

THE PRINCIPLES UNDERLYING WATER GOVERNANCE

The need for collective action, and hence the organization of government, stems from the realization that without collective enforcement of institutions, such as property rights, the anarchy which would likely result would only serve to render human life nasty, brutish and, ultimately, short. In a world inhabited by imperfect people, collective organization is required to prevent bad people from doing harm as much as enabling good people to do good (noting, of course, that the good and bad people may be the same people).

The principles underlying water governance are based upon theories of collective behavior. Unfortunately, no one, simple theory explains every situation and few universal rules exist. But under almost every circumstance in our day, the state plays an important role in defining property rights regimes, a major foundation of every society's collective behavior. Different property regimes, with their associated rights and obligations, include:

- Open access property – There is no defined group of users or owners and the benefits of the resource are available to anyone. Individuals have both privilege (the ability to act without regard to the interests of others) and no right (no capacity to affect the actions of others) with respect to use rates and maintenance of the asset. This applies particularly where the shared resource is so plentiful that the distributive and management issues of scarcity have not arisen.
- Common pool property – The management group (the owners) has a right to exclude non-members and non-members have a duty to abide by the exclusion. Individual members of the management group have both rights and duties with respect to use rates and maintenance of the property.
- Private property – Individuals have the right to undertake socially acceptable uses and a duty to refrain from socially unacceptable uses. Others (non-owners) have a duty to allow socially acceptable uses and a right to expect that only socially acceptable uses will occur.
- State (governmental) property – Individuals have a duty to observe use and access rules determined by the controlling agency of the state. The agency has the right to determine the access and use rules.

In most countries water is the 'property' of the state (i.e. the government),[1] but a typical history of water property rights follows a path starting as open access, which is appropriated by a group and becomes common pool property. The state ultimately appropriates these rights from the common pool resource ownership group, usually acknowledging existing uses, to create state property. The state is then faced with deciding how to deploy the unallocated resource to the national advantage. In various countries the state decided this can be best achieved by devolving licenses or property rights to some individuals and groups in civil society. How that can be done fairly and equitably without reducing incentives for efficient use of the resource is the central issue in governance in many nations at the beginning of the twenty-first century.

The USA is one of a handful of countries allowing private ownership of water rights, but even for the USA, most of the resource is still handled under state or common pool property relationships. That is, the uncaptured water is held in common subject to individual rights to make certain uses. The problems with water resources governance are due not only to the problems of water rights, but also to the unusual, contradictory nature of water itself. Water is a fugitive resource – it has public good dimensions, its supply engenders natural monopolies and its use both causes and suffers from major externalities.

MODELS OF GOVERNANCE

Traditional views of national water governance in the USA appear to be based upon a set of assumptions about how the federal system works. Those assumptions may no longer be relevant. The rational, top-down approach whereby laws and regulations are promulgated by Congress and implemented by the federal agencies in concert with the state and local authorities and with the concurrence of stakeholders and interest groups is only a sketch of what really happens. Nevertheless, there are several versions of that sketch. Here are three:

- Rational I The bureaucratic politics and process model – This model is based on political-bureaucratic bargaining in a federal system. Its focus is typically the Executive Branch, with the elected legislature hardly in the picture. Classic cases are drawn from US foreign policy problems, such as the 1962 Cuban missile crisis in which Congress was not a major player (Allison, 1971); but this is the opposite of the situation in water, where the Executive Branch until recently was largely excluded by Congress.[2]
- Rational II The Congressional behavior model – A second federal model concentrates on the elected Congress, where the explanatory principle for Congressional behavior is that Congressional representatives are single-minded seekers of re-election. It follows from this that representatives' goals are to improve the welfare

of their constituents in the shortest possible time frame. The realities of information processing are also important in describing Congressional behavior. With humanly limited capacities to absorb and judge, legislators are so overloaded with information that they have to be extremely selective in committing their attention. Legislators deal with this by specializing in a particular and limited subject area. In other areas they take cues from sources (colleagues, outside groups, committee reports) that they trust.

- Rational III The interest group model – When national legislators think about the constituency that elected them, they rarely, if ever, see an undifferentiated mass of individual voters. They instead see categories of interests; in some cases, only a few dominant interests. But there are literally thousands of active interest groups – environmental, agricultural, industrial, professional and academic, just to name some of the more obvious players – involved with water policy (734 000 public interest groups of all types were registered with the Internal Revenue Service in 1998 (Bornstein, 2004)). These groups often have overlapping concerns and overlapping memberships. They constitute vital channels for particular publics to participate in the federal governmental process. Pork barrel projects are the fodder for the well-known iron triangles (McCool, 1987) of legislators, bureaucrats and active interest groups that develop in specific issue fields. (The term 'pork barrel' was first used to describe the exchange of benefits in successive US Rivers and Harbors Acts.)

Useful developments of interest group theory are found in Dahl's (1961) regime theory and in Buchanan and Tullock's (1962) public choice theory. They attempt to predict which pattern of decision making will prevail based upon the concentration or diffusion of costs and benefits of public choices. For example, the problem with water governance in the USA is the tendency to privatize (concentrate) the benefits and socialize (diffuse) the costs. This refers to situations such as federally financed irrigation, where the benefits are concentrated in the hands of a few farmers and the costs are widely dispersed over society. This may be an acceptable situation, since regimes which distribute both benefits and costs widely tend to lead to inertia and underinvestment.

THE POLITICS OF WATER GOVERNANCE

The driving force in any area of governance is politics. As already noted, the conventional view of the relationship of politics to governance is based on one or another version of a rational model, in which actors at all levels of government and in the public at large seek to satisfy private desires through the use of government mechanisms. This is seldom a neat and tidy process and in 1936 Harold Lasswell may have said it best in the title of his book: *Politics: Who Gets What, When and How?* Governance is not seen as a simple linear process, but a discursive and a highly complex set of interactions between laws and institutions, and personal and group interests as well as the general interest. We shall look shortly at the new and expansive analysis of Deborah Stone (2002, p. 261) who brings in further non-ideal elements and says 'policy is more like an endless game of Monopoly than a bicycle repair'.

Approaches to Political Decision Making

The literature on politics and political theory and its implications for effective water governance in the USA contains two contrasting models of politics and public decision making. The first, which has come to be called 'public choice or rational choice' can be traced to Downs (1957), who attacked an earlier, naive economic theory of politics (ETP) by objecting to the notion that politicians were motivated to try to maximize something called a social welfare function. He hypothesized instead that political parties act in order to obtain income, prestige and power and that politicians are motivated by a vote-maximization objective rather than by altruistic or ideological objectives. Downs concluded that parties formulate policies in order to be elected rather than win elections to get the chance to formulate policies. The second model may be seen as growing out of earlier, institutional analysis such as that found in Maass (1951) who used it in his analysis of the relationship between Congress and the US Army Corps of Engineers (USACE) and in Maass et al. (1962). This approach examines the evolution, development and erosion of popular democratic institutions and sees subtle shifts away from popular democracy, which depends on group political mobilization in forms such as labor unions and fraternal associations, towards personal democracy, which involves increasing atomization of the electorate, with individuals and small groups exerting political influence through contributions of money, leaving the overtly political aspects to paid political professionals. This model is very much in keeping with the 'bowling alone' view of institutions of Putnam (2000). More recently, Crenson and Ginsberg (2002) presented a particularly pessimistic view of the general health of democratic institutions and democracy in the USA. They see citizens becoming customers of the government, not its controllers. For this phenomenon, they adduce many reasons, including the size of the population, the lack of direct taxation specifically to fund government's war-making ability and the role played by television and media in converting citizens to passive consumers.

The evolved version of this model replaces the highly rational version of policy analysis by Stone's (2002) insights, articulated in what she calls the Polis Model. By way of summary contrast, she characterizes the Rational Model taught in schools of public policy as:

- Identify objectives.
- Identify alternative courses of action for achieving the objectives.
- Predict the possible consequences of each alternative.
- Evaluate the possible consequences of each alternative.
- Select the alternative that maximizes the attainment of objectives.

These steps can be recognized as constituting most of what we do, or attempt to do, in our professional and intellectual lives. They are the foundation of documents like the US Water Resources Council's (WRC) *Principles and Guidelines* (P&G) of 1983 (WRC, 1983). So what could possibly be wrong with such an approach? Stone's response is that it misses the point of politics:

> A model of political reasoning ought to account for the possibilities of changing one's objectives, of pursuing contradictory objectives simultaneously, of winning by appearing to lose and turning loss into victory and, most unusual, of attaining objectives by portraying oneself as having attained them . . . Political reasoning is reasoning by metaphor and analogy. Stone, 2002, p. 9

In a series of tables Stone compares and contrasts the rational behavior models of politics with her Polis Model. For example, in the table reproduced here as Table 6.1 the comparison considers: what are the units of analysis, sources of conflicts, sources of ideas and nature of collective action. From this table, two radically different perceptions emerge on the nature of society. The unit of analysis shifts from the individual to the community. Self-interest is extended to include public interest. The conflicts now change from conflicts between self-interests to between self-interests and the public interest. The nature of collective activity moves from competition to a mixture of cooperation and competition. Most importantly, the criteria for decision making change from maximizing self-interest to the promotion of the public interest and loyalty to people and places. The nature of the information used tends to move from objective to ambiguous, interpretive, incomplete and strategically manipulated. Ultimately, what matters under the Polis Model are ideas, the pursuit of power and the maintenance of alliances.

Table 6.1 Comparison of Rational Model with Polis Model

Concepts of Society	Rational Behavior Model	Polis Model
Unit of analysis	Individual	Community
Motivations	Self-interest	Public interest (as well as self-interest), preserve community, develop alliances
Chief conflict	Self-interest vs public interest	Self-interest vs public interest (commons problems)
Source of people's ideas and preferences	Self-generation within the individual	Influences from outside
Nature of collective activity	Competition	Cooperation and competition
Criteria for individual decision making	Maximizing fulfillment of self-interest, minimizing cost	Loyalty (to people, places, organizations, products), maximizing self-interest, while promoting public interest
Building blocks of social action	Individuals	Groups and organizations
Nature of information	Accurate, complete, fully available	Ambiguous, interpretive, incomplete, strategically manipulated
How things work	Laws of matter (e.g. material resources are finite and diminish with use)	Laws of passion (e.g. human resources are renewable and expand with use)
Sources of change	Material exchange Quest to maximize own welfare	Ideas, persuasion, alliances, pursuit of power, pursuit of own welfare, pursuit of public interest

In a similar fashion, Stone examines each of the concepts of equity, efficiency, needs, liberty, interests, good rules and rights and shows why and how the Rational Model usually does not capture the complexity seen by the Polis Model. Stone essentially argues that, given what is known about political decision making, the questions need to be framed in a different way. To analyse water policy from Stone's point of view is a challenge to our conventional ways of thinking about water politics.

THE COLORADO RIVER BASIN: A CASE STUDY IN WATER GOVERNANCE

In this section we use the example of the Colorado River Basin to explore the nature of water governance. The water of the Colorado River and its tributaries is shared among seven states, 30 Native American tribes and two nations (MacDonnell, 2006). There are two interstate compacts and a US Supreme Court decision that apportion uses of the water among the states. There is a treaty fixing Mexico's share of the basin's water. With funding from Congress, the USBR has constructed and now operates major dams on the Colorado's main stream and several of its larger tributaries. There are large-scale, water-based environmental issues in the basin. In short, virtually the entire panoply of water issues requiring some form of governance exist in the basin.

Much has been written about the Colorado River Basin.[3] The following discussion is organized into three periods (pre-1929, 1929 to 1968 and 1969 to present) and is intended to highlight the development and use of governance mechanisms to address evolving basin concerns. Our purpose is to identify some of the major water-related issues that have arisen in the basin during the last century, to describe the primary institutional and governance responses and to consider the nature of these governance approaches.

Pre-1929: The Beginnings

The importance of the Colorado River is best understood not in terms of its length, the volume of its water, its utility as a highway of commerce or the fertility of its adjacent lands but in the remoteness and aridity of its setting. Modest in attributes when compared to other large US rivers, such as the Columbia, the Ohio or the Mississippi, the Colorado nevertheless stands out as the major river of the seven US and two Mexican states through which it or its tributaries pass. As a major source of water in a water-limited region, it has long been a focus of attention among competing interests.

Governance of water in the American West begins with the prior appropriation doctrine, state-level property rules for determining rights to use water that accord priority to the first to physically capture water and put it to beneficial use. It is a user-initiated system, under state supervision and following state requirements. Beneficial use vests the user with a property right that, among other things, ensures later users cannot interfere with previous users. Once the right is established, its owner is the primary decision maker respecting its use. All of the states with lands within the

Colorado River Basin follow some version of the rules of prior appropriation. State prior appropriation rules, however, only govern uses within that state, leaving unresolved the matter of how to share use of interstate waters such as the Colorado River.

To establish such interstate agreement, the basin states, with express Congressional authorization formed the Colorado River Commission in 1922. The Commission's purpose was to negotiate an interstate compact among the states to apportion use of the basin's water. The US Constitution authorizes states to enter into compacts, essentially contractual agreements that, once ratified by the state legislatures and Congress, take on the force of national law. Such compacts had been used to resolve interstate disputes such as boundary lines but never for apportionment of interstate rivers.

The Commission held 29 formal sessions over the course of the year 1922, as well as public hearings and field trips, in its ultimately successful quest for agreement. Unable to agree on specific state shares, the commissioners instead divided the basin into two parts and made apportionments to each sub-basin, Upper and Lower, divided at a point called Lee's Ferry in the canyons of northern Arizona. Users in each sub-basin were to have the right to consumptively use 7.5 million acre-feet annually of basin water. The faster growing Lower Basin states of Arizona, California and Nevada were authorized to consume an additional 1 million acre-feet before the states were to meet again and apportion the remaining unallocated water.[4] In addition, the Compact committed the Upper Basin states of Colorado, New Mexico, Utah and Wyoming to ensure delivery at Lee's Ferry of at least 75 million acre-feet of water in running ten year periods.

Use of the interstate compact authority was an innovative approach. The process brought the seven basin states together for the first time in direct negotiations respecting uses of the basin's water. It assured the slower growing states in the Upper portion of the basin that there would be water still available for their use whenever demanded by their economies. This assurance, in turn, enabled these states to go along with Congressional funding for federally constructed projects that would benefit California first. Apportionment of water to groups of states left open the question of individual state shares, however.

Irrigation interests in the fertile but desert-like Imperial Valley of California desperately needed Congressional funding to build a canal totally within the boundaries of the USA to replace the original system located largely in Mexico. The USBR had determined that a dam constructed in the vicinity of Boulder Canyon near Las Vegas would provide full regulation of this portion of the river, reduce the risk of downstream flooding, provide storage water that could be delivered to downstream users when needed and enable generation of hydroelectricity that could be sold to pay for project costs. And the basin states needed Congress to approve the Compact so it could go into effect.

These interests finally converged in the 1928 Boulder Canyon Project Act. This legislation authorized and funded construction of what became known as the Hoover Dam as well as the All American Canal. Storage water was to be provided to downstream users, under contract with the Secretary of the Interior. Because Arizona had refused to sign the 1922 Compact, Congress provided the Compact could go into effect with the ratification of only six states. Herbert Hoover, now president but previously the federal representative to the Colorado River Commission and its chair, declared the six-state Compact in effect in 1929.

With this Act, Congress expanded its vision of the Reclamation program well beyond modest projects to provide water for unsettled farm lands, putting the USBR into the business of large-scale, multi-purpose river development. The basin states got important experience in the use of Congress to help accomplish multiple interests in development of an interstate river. And water and power interests in California got access to resources needed for their growth without having to finance their enormous costs. The 'iron triangle' found success at a scale well beyond anything it had previously attempted.

1929 to 1968: The Era of Development

During this second period, Congress funded and the USBR constructed, six dams on the main Colorado, including Hoover and Glen Canyon, and three on major tributaries in the Upper Basin, as well as the All American Canal. In all, the federally constructed dams provided approximately 60 million acre-feet of water storage – four times the annual average flow of the river and its tributaries from the Upper Basin. It was an extraordinary period, in which the federal government took center stage in development of the Colorado River.

Initially, California was the major beneficiary. Arizona, still refusing to sign the Compact, took its fight to the US Supreme Court three times in the 1930s, with little to show. In 1944 Arizona yielded to the inevitable and signed the Compact. In that same year, the USA and Mexico entered into a treaty committing 1.5 million acre-feet of Colorado River water to Mexico annually. Even though no specific provision had been made for water deliveries to Mexico in the Colorado River Compact, this treaty obligation now superseded in priority all uses under that agreement.

Water development interests in the Upper Basin states grew increasingly uneasy as California moved steadily towards diversion and use of the 5.3

million acre-feet its users had under contract with the USA. With Arizona now signaling its intention to go after its unused share of Colorado River water by seeking Congressional funding for a Central Arizona Project (CAP) and with 1.5 million acre-feet committed to Mexico, it seemed time for the Upper Basin states to act. Upon a 'request' from Upper Basin Congressmen, the USBR took the initiative. In 1946 it completed a report describing the potential for water development in the Upper Basin and identifying the best locations for large-scale water storage projects (USBR, 1946). The report emphasized the need for the Upper Basin states to clarify individual state consumptive-use entitlements prior to moving ahead with additional federal water development. Upper Basin state representatives came together in 1948 and negotiated the Upper Colorado River Basin Compact, providing for percentage allocations to each of the states.

Recognizing the need to clarify its share of the Lower Basin Compact apportionment, Arizona filed suit against California in the US Supreme Court in 1952. More than ten years later, the Court held that Congress had already made a division of this apportionment in the Boulder Canyon Project Act: 4.4 million acre-feet to California, 2.8 million acre-feet to Arizona and 300 000 acre-feet to Nevada – all to come from the Colorado River main stream (*Arizona* vs *California*, 373 U.S. 546 [1963]). The Court placed the Secretary of the Interior in charge of allocating surpluses and shortages among the Lower Basin states. In addition, the Court ruled that reservations of Indian and federal lands along the main stream had rights to water under the Reserved Rights Doctrine.[5]

In 1956 Congress rewarded the Upper Basin states with the Colorado River Storage Project Act. Central to this bill was authorization of Glen Canyon Dam on the main stream just upstream of Lee's Ferry. Capable of storing 27 million acre-feet of water in its reservoir, Lake Powell, Glen Canyon provided assurance that the Lower Basin would receive at least its 7.5 million acre-feet annually while allowing upstream storage and use of water. Revenues from hydroelectric power generated at the dam would pay the bills. This act also authorized and funded Flaming Gorge Dam and Reservoir on the Green River (benefiting Wyoming and Utah), Navajo Dam on the San Juan (benefiting New Mexico) and the Aspinall Unit on the Gunnison (benefiting Colorado) – a wonderful example of distributive politics in action.

Arizona finally got the CAP, a complex system of facilities that would take Arizona's remaining unused entitlement of 1.4 million acre-feet of Colorado River water to Phoenix, Tucson and points between, with the 1968 Colorado River Basin Project Act. To get California's support, Arizona had to agree to subordinate its project rights to those of California.

Colorado, whose Congressman Wayne Aspinall chaired the critical House Interior and Insular Affairs committee, received authorization for five new projects, while Utah received authorization for the Central Utah Project that would enable movement of Colorado River water out of the basin and into the most populated portion of the state.

This was a period of extraordinary water and power development in the basin. Perhaps nowhere in the west was the role of the USBR more important. Its success in building the remarkable Hoover Dam vaulted the USBR to the forefront of government construction agencies. Its expansion during this period was a model for how to grow an agency. Basin state members of Congress could point to the federal funds they had successfully directed toward developing their local economies. The considerable hydroelectric power revenues from the 'cash register' dams satisfied most other members of Congress that federal funds would be returned to the USA and states were happy with access to cheap hydroelectric power and essentially free water.

Yet, as projects constructed by the USBR made more and more water available for human use, the number of claimants for that water grew even faster. Southern California continued its growth, boosted by an economy intensely stimulated by World War II. Indeed all of the basin states played important roles in support of the war economy and, following the war, continued to boom. Although the presumed need for additional irrigation water continued to drive river development, irrigation demand had peaked by the end of this period. Urban and industrial demands started playing increasingly important roles in motivating additional water development.

Perhaps most unsettling to those local interests watching closely over the basin's water matters in the 1960s was the uncertain extent of Indian claims to water. Tribal water claims had been acknowledged but left unresolved during the 1922 Compact negotiations. *Arizona* vs *California* not only reaffirmed tribal reserved rights, it established a standard for quantification far more expansive than previously imagined. It was not hard to create a scenario in which much or even all of the annual supply of Colorado River Basin water would be controlled by the tribes with reservations in the basin.

1969 to Present: The Emergence of Environmental Concerns and the Beginnings of Scarcity

Political support for federally directed river basin development peaked in the 1960s with the convergence of an Executive Branch strongly favoring an activist national government and a Congress still greatly enamored of getting federal funds to support local economies. Tellingly, however, no new water storage projects have been authorized in the basin since the 1968 Colorado River Project Act and not all the projects authorized were built. A combination of changing national politics and changes in the basin converged in this period, presenting new challenges and raising different problems of governance.

In the 1960s the US Fish and Wildlife Service (USFWS) listed two fish species found only in the Colorado River Basin as in danger of extinction – the humpback chub and the Colorado squawfish (now 'Colorado' pikeminnow). Barely noticed at the time, this action took on significance with passage of the Endangered Species Act (ESA) in 1973. One piece of an outpouring of Congressional legislation in the 1960s and 1970s aimed at providing nationwide protection for the environmental values of air, water, wildlife and land, the ESA prohibits federal agency actions likely to 'jeopardize' the continued existence of listed species.

In the review process for proposed new Reclamation dams in the Upper Basin in the late 1970s, the USFWS determined that additional depletions of water made possible by these dams would jeopardize the continued existence of the two listed species of fish. As a 'reasonable and prudent alternative' to the projects as proposed, the USFWS recommended altering project plans to provide for replacement of associated water depletions (for example, by dedicating a share of storage water to stream flow management) as well as providing funding support for the research to determine what was necessary to restore the fish populations to sustainable levels. For private projects subject to some federal permission, the USFWS proposed payment of a 'depletion charge' – a one-time payment assessed against expected water depletions, with the funds used for research and recovery activities. After many years of negotiations involving Upper Basin states, water users, electric power users and environmental organizations, a large-scale program was implemented, with actions intended to improve habitat conditions and increase fish populations to sustainable numbers (USFWS, 2003). A comparable program but covering more species has been put in place in the Lower Basin (Lower Colorado River Multi-Species Conservation Program, 2004). The effect of these programs is to allow continued water development so long as progress towards the recovery of the fish is made. Both of these programs depend on substantial federal funding, but the states and the water users make important contributions as well.

Salinity emerged as a significant new concern in the basin in the 1960s. Some geologic formations in the basin contain high concentrations of salts reflecting their origin as sediments in vast inland seas. Natural drainage from these formations carries salinity to the river. Irrigation return flows passing through saline soils add to this salinity (Miller et al., 1986).

The salinity content in the river's water increases measurably as it moves downstream.

Expanded irrigation in the 1960s at a Reclamation project near Yuma, Arizona produced a sharp spike in the levels of salinity in water passing into Mexico (Friedkin, 1988). In 1972, at the insistence of Mexican President Echeverria, the USA committed itself to fix the salinity problem. The basin states made it clear they considered this a federal, not a state, problem and demanded the solution not interfere with their consumptive uses. By Minute 242 to the 1944 Treaty the two nations agreed the salinity of water delivered to Mexico would be approximately the same as the salinity of water diverted into the All American Canal, the last diversion point on the river within the USA.

In 1974 Congress passed the Colorado River Basin Salinity Control Act, authorizing the USBR to construct and operate a desalting plant near Yuma. The plant would be capable of treating 145000 acre-feet of water annually prior to release into the Colorado for delivery to Mexico. In addition, Congress authorized the USBR to construct salinity control projects that would reduce loadings of salinity at locations upstream. Under the new Federal Water Pollution Control Act (FWPCA) Amendments, the US Environmental Protection Agency (EPA) agreed to establish only a single salinity water quality standard for the basin, at the diversion for the All American Canal. The basin states established the Colorado River Salinity Control Forum, the first and only truly basin-wide institution, to work with the USBR and EPA to implement the salinity control program.

By the 1980s, still another new problem had emerged: insufficient water to meet growing urban demands in Nevada and California. California users had exceeded that state's basic entitlement sometime in the 1950s. Expanding urban demands depended on using water legally allocated to other states but not yet used. Booming growth in Las Vegas promised to soon exhaust Nevada's modest 300000 acre-foot entitlement from the Colorado. New federal projects for Arizona and the Upper Basin states meant that much of the unused water soon would be taken.

The fast-growing city of San Diego started looking for ways to acquire rights to unused Upper Basin water. The Metropolitan Water District (MWD) of Southern California did the same. Las Vegas announced it would buy water from any source, at the right price. In response, several schemes for interbasin water marketing emerged. The Upper Basin states, however, resisted these efforts, arguing they were illegal under the so-called Law of the River and especially the 1922 Compact.[6] These states held fast to the deal made in 1922 by which they would some day be able to consumptively use up to 7.5 million acre-feet of basin water despite the fact that, after 80 years, they were only consuming about half that amount.

Perhaps now increasingly aware of growing Lower Basin dependence on the Upper Basin's unused apportionment, Upper Basin states asked the Secretary of the Interior in 1990 not to declare the availability of 'surplus' water for that year. Under *Arizona* vs *California,* the Secretary is authorized to deliver surplus water to users in the Lower Basin holding contracts. It was this water upon which many urban users in California depended. The Secretary declined to act without California's agreement and urged the states, including California, to come up with a common proposal respecting the availability of surplus water. Ultimately successful, the process required 12 years to complete (Lochhead, 2003).

From the standpoint of basin governance, this process is especially interesting. It was initiated by representatives of the basin states and included all seven states. Each state relied on trusted experts, either inside or outside state government, to carry on the negotiations. At times it included tribal representatives. The ultimate decision maker, of course, was the Secretary of the Interior.

Twelve years is a long time, but the stakes were high. To the Upper Basin states it was a test of the strength of the Law of the River. Was California truly limited to its basic entitlement? Or had decades of use beyond this amount established a new level of right? Was it now possible for California users to find a way to reduce uses of Colorado River water from roughly 5.2 million acre-feet to 4.4 million acre-feet? The best option appeared to be to shift a portion of the large volume of water going to irrigated agriculture in the state to the urban users who would otherwise face large cutbacks.

The first such water transfer involved lining portions of the All American Canal, paid for by MWD, with the saved water shifting from the Imperial Irrigation District (IID) to the MWD. Next, the MWD worked out a 'fallowing' program with the Palo Verde Irrigation District (PVID) by which water temporarily not diverted by PVID for irrigation use would be available to the MWD. In addition, San Diego negotiated a transfer agreement with IID primarily based on system and field efficiency improvements paid for by San Diego. Now assured it could meet demands within its basic entitlement, California signed on to a plan under which it would gradually reduce its reliance on surplus water. With minor modifications, the Secretary adopted the seven-state plan in 2002.

Just as agreement was finally reached about when to make available surplus water, however, the basin went into a multi-year drought. Now the Secretary was faced for the first time with what to do in the Lower Basin in the event of shortages. The Upper Basin states faced more clearly than ever the likelihood their users would never be able to consume 7.5 million acre-feet. And Colorado began discussing how it would reduce existing uses if necessary to meet Compact delivery obligations to the Lower Basin. To heighten uncertainty, climate change analyses indicated reliable supplies of water in the basin would decrease over time (NRC, 2007).

Thus, important changes are affecting the Colorado River Basin. The era of large-scale water development is over. The federal role has changed. Relatively recent national interests in water quality and endangered species, international relations with Mexico and obligations to Native American tribes have intruded into basin affairs, often placing the USA at odds with basin states. To date, Congress has been willing to provide the funding necessary to ensure these national interests are met without diminishing consumptive uses of water in the basin.

Perhaps the most fundamental challenge facing the basin at present is how to deal with overuse of its water (MacDonnell, 2005). According to the USBR's *Consumptive Uses and Losses Report* (USBR, 2004), annual average depletions of water in the basin including Mexico between 1996 and 2000 were about 19 million acre-feet – considerably more than the basin yields. Yet demands continue to increase.

Summarizing the Existing Governance System in the Colorado River Basin

There is no single organized, integrated system of governance, such as a river basin commission (RBC), for water-related decisions in the Colorado River Basin.[7] Rather there are multiple governance mechanisms, organized primarily around particular objectives, such as delivery of consumptive use rights, management of water facilities including hydroelectric power generation, salinity control, water for Mexico and endangered species protection. Many important governance issues have been addressed through processes developed specifically to reach agreement for that matter and then abandoned, leaving implementation to individual participants. Examples include the original Colorado RBC for basin water apportionment and the seven-state group for surplus and shortage water criteria. Arizona has resorted to the US Supreme Court four times in its disputes with California. Congress has been the institution through which many governance decisions have been made, a role greatly facilitated by the funding it has provided to implement the decisions. The US State Department decided how much basin water should go to Mexico. And, at times, the Secretary of the Interior has made critical decisions. At the heart of virtually all of these decisions are the states, usually operating through the governors and primarily representing the interests of their major Colorado River Basin water users.

From the outside, governance in the basin appears diffuse, poorly organized and even irrational. It certainly challenges the ordinary flowchart model of institutional decision making. From within, it can best be understood as evolutionary, adaptive and responsive to need. It is a mirror of the multiple and changing interests competing for the benefits provided in a large and complex water system. Mechanisms emerge as interests and needs change.

As we have suggested, this approach is consistent with the US preference for decentralized public and private decision making. It also reflects our federal system of government and our system of individual property rights in land and natural resources under which individual owners are the primary decision makers respecting their use, subject to governmentally imposed requirements. Governance serves to support these uses while seeking also to provide for politically determined interests that extend beyond these user-determined purposes. Holding this system together is the Law of the River.

With this extended case study in hand, we turn to a consideration of the Rational and Polis Models and apply them to what we have observed in the Colorado River Basin.

Comparison of the Rational and Polis Models

The highly diffuse nature of the multiple benefits available from a river and its water challenges the assumption of the Rational Model that it is possible to plan for and develop rivers like any other resource. Even multi-attribute analytical models have a limit on the number of considerations that can be included. Moreover, no single entity owns or controls a river. There are typically large numbers of riparian landowners who may own the underlying bed in non-navigable portions while the state likely owns the bed in navigable portions. In riparian law states, adjacent landowners hold some loosely defined legal rights to the use of the river and its water. In prior appropriation states, owners of water rights are regarded as having a property right to the use of certain water. Yet, while the water remains in the channel it is the property of no one and under the broad control of the state. Some rivers, such as the Colorado, pass through more than one state and may even pass into (or from) another country. In short, decisions respecting uses of rivers and their water are not made through a centralized, integrated process but through a large number of generally disconnected processes subject to certain national requirements (for example, regulation of pollutant discharges) but not subject to direct national control.

The USA has actively experimented with more centralized decision making respecting river basin development. The conservation era of the early 1900s ushered into governmental decision making a firm commitment to the importance of 'comprehensive' river basin development (Rogers, 1993). In this view rational and coordinated decision making would produce far greater economic benefits than the kind of piecemeal development of rivers and water occurring at that time. Subjecting private hydroelectric power development to federal licensing was a step in this direction. At the same time, Congress began expanding the planning and construction authority of the USACE beyond its traditional focus on navigation. Authorization of the Hoover Dam put the USBR in the business of river basin development. Creation of the Tennessee Valley Authority (TVA) explicitly rested on the belief that centralized river basin development supported by federal funding would bring greater economic benefits than unplanned development without government support. This movement toward federally planned, funded and implemented river basin development reached its peak with the 1965 Water Resources Planning Act, which authorized the creation of RBCs for all the major river basins. The last large-scale federally funded projects for the Colorado River Basin passed Congress in 1968. It was the end of an era.

It is tempting to suggest that the end of federal sponsorship of river basin development also ended our commitment to rational planning for rivers. But, in fact, it now seems clear that much of the apparently rationally based decision making respecting water development projects between 1930 and 1970 was better explained by the dynamic of the iron triangle in which politics generally trumped analysis. Whatever one's views about the efficacy of benefit-cost analysis and the manner in which it was used, the reality was that by 1970 the national economy had moved beyond the need for more large dams, for more subsidized hydroelectric power, for more inland seaports, for more water to irrigate farmlands. The rationale for directing large amounts of federal funding to such projects no longer stood up to scrutiny and political support eventually evaporated. Conversely, rational analysis of options for altering historical uses of rivers to meet contemporary interests remains essential. In many cases that analysis now turns more on science than on economics.

Events in the Colorado River Basin closely parallel this view. It suggests state water interests encouraged federal participation in river basin development to gain access to public funding that was otherwise unavailable. At the same time, settlement and development of the arid but arable lands of this region had become a national objective in 1902. The USBR's success in 'taming' the Colorado became a matter of national pride, especially with the depression era construction of the towering Hoover Dam between the walls of the Black Canyon. Sales of hydroelectric power, even at the reduced rates provided to public power purchasers, still generated revenues sufficient to justify the enormous commitment of federal funds involved. The basin states and the USBR showed great skill in obtaining Congressional support for a wide array of projects.

The decline of political interest at the national level for support of Reclamation projects after 1968 short-circuited several authorized but unbuilt projects in the Upper Basin. President Carter's famous 'hit list' met an angry response from these states, still hoping for federal funding. It is illustrative that the last Reclamation project Congress funded in the basin, the Animas La-Plata, is primarily intended to resolve tribal reserved right claims. On the other hand, Congress has continued to provide significant funding for projects to reduce salinity problems in the basin and, more recently, has provided the bulk of the funding needed to protect endangered species of fish so basin water use can continue essentially unimpaired. National political priorities have shifted.

It is perhaps more useful to understand that decisions about the uses of rivers and their water, as primarily publicly managed resources, are inherently political and that those who are most effective in convincing decision makers that their preferred use of the water best serves the interests most important to the decision makers will always drive political decisions. Traditionally, economic interests have been most effective at making their case for water development even while those interests have shifted over time. Indeed, some of these interests were even able to get public support to pay the costs of water development. As the nation has urbanized, urban water suppliers have become increasingly important in decisions respecting water. Such suppliers, with their taxing, bonding and rate charging authority, are fully capable of paying the costs of water development. More recently, recreation and environmental interests have begun to have more influence. Public funding related to water has accordingly shifted somewhat in the direction of river restoration. Water as an input to the production of market goods has declined in relative importance, while water for direct human use and for environmental purposes has become more important.

The highly rational quest for centralized river basin authorities that would be able to integrate all interests into a single, objectively based decision-making framework continues. That it has only found limited success in the USA and then primarily because federal funds were used to pay the costs, is not surprising in view of US culture and institutions. Yet there is something inherently integrative about rivers. Their uses are and, must be, shared. Upstream uses affect downstream uses. Private uses affect public uses. Human uses affect natural river functions.

The Colorado Basin study illustrates the inevitable need for communication among interests respecting uses of this shared resource. While no single forum exists for such communications and no single

Table 6.2 Major interest groups in Colorado River governance

Constitutionally required interest groups
The seven states and their agencies
The federal government and its agencies
The Native American tribes
The government of Mexico
Water user ad hoc groups
Irrigated farmers
Power producers
Municipalities (water and wastewater)
Industry
Tourism
Recreation groups
Colorado River Water Users Association
Other user groups
Environmental and sustainability groups
Sierra Club
American Rivers
Environmental Defense Fund
Pacific Institute
Other NGOs

decision-making authority exists to resolve conflicts, there are a variety of integrating mechanisms at work. They are messy in appearance but appear to work.

In discussing the pursuit of political action, it is imperative to identify every interest group and its aspirations and goals. In the case of Colorado River governance, the major interest groups are numerous, but can be broadly characterized as shown in Table 6.2.

How each of these groups and agencies will position themselves on the ongoing governance issues is hard to know precisely, but they each have fairly well-known core approaches and interests. For the sake of argument let us assume that the further regulation of water uses in the Colorado River Basin would divide the interest groups according to an intuitively plausible pattern. On one side would be the federal governments of the USA and Mexico aligned with the environment and sustainability groups; on the other side would be all the others (including the state governments). This gross assumption makes for lots of strange bedfellows, but may help advance our understanding of what political outcomes might be expected. In Table 6.3 we compare the Polis perspective to the Rational Model with regard to the next stages of political decision making in the Colorado River Basin.

Table 6.3 demonstrates that, even for a well-managed river basin, there are defensible alternate explanations likely to be useful in interpreting political maneuvering in the aid of favorable outcomes. The models implicit or explicit in these interpretations can conflict with the widely accepted Rational Model and demonstrate that there can be more than

Table 6.3 Comparison of the Rational and Polis Models

Rational Model	Polis Model
Context of the issues	
Social welfare is defined as the efficient national economic development (NED) goal, taken roughly to mean maximizing regional income to the citizens in the basin, constrained by environmental and ecosystem sustainability requirements and other social objectives, such as income distribution and fulfillment of treaties with the tribes and Mexico. Supporters of more and better governance in this framework tend to stress sustainability, protection of endangered species, economic efficiency and the public interest.	Powerful symbols have been invoked in the water conflicts in the Colorado River Basin. States' rights, public interest, autonomy for the tribes, environmental sustainability, extinction of species, freedom from government interference, saving the family farm, protecting a way of life, economic efficiency and water for the poor have all been widely advanced. Proponents stress freedom from federal interference and saving the family farm. The Colorado River Basin is rich in explanatory and persuasive stories. The lone farmer struggling against the aridity to settle the land produces crops and ultimately develops an independent and resilient society; settlers struggling against all odds to make a living in a hard and unforgiving landscape. We are all aware of the cowboy culture followed by the 'Chinatown' stories of exploitation by city slickers; also stories of noble savages, wise land management and a landscape managed for millennia. The issues boil down to the rights of individuals and states to decide upon their own local environments. The fight goes on through the courts and, hence, the need to find sympathetic plaintiffs (for example, small family farmers damaged by federal reduction in water supplies) and other controversial issues with federal laws, such as the ESA favoring animals rather than people.

Table 6.3 (continued)

Rational Model	Polis Model
Defining the problem	
The problem is largely a technical problem of making sure that the wise use of resources is achieved by careful assessment and calculations of all the benefits and costs. The role of a paternalistic father restraining headlong and irresponsible children from despoiling the patrimony is a strong story line. Finding an allocation of water that maximizes a measure of national economic welfare within constraints implied by environmental rules, redistributive goals and international treaty obligations.	Finding an acceptable water allocation taking into account the complex interaction of many elements: states' rights, tribal rights, inadequate water measured against potential demands, existence of threatened and endangered species. Powerful symbols are employed portraying a frontier society fighting for its way of life in a changing world.
Defining the goals	
The goals are to develop an equitable, efficient and environmentally sustainable management system for the basin. Straightforward application of the P&G would help articulate the goals. In the management of river basins efficiency has always been one of the beacons urging rational planning to produce the maximum national economic benefits. The Rational Model has the whole weight of traditional economic policy behind it here.	In any political struggle the goals need to be carefully articulated for each of the contending interest groups because the goals will certainly not be congruent among groups. Even within the coalitions that we have assumed for the Colorado Basin, there will be major disagreements on goals. The Polis Model's goals stress development of governance systems that emphasize local economic development and social controls. The Polis Model sees efficiency in terms of interests of the political players and these will largely be regional economic activities.
Finding the solutions	
The proponents of the Rational Model feel they have the full weight of federal and state laws, responsibilities and rights on their side. They believe they are the true protectors of the public interest against special interests.	As Stone says, 'The best way to fight against a proposed rule is to show that it will not work to accomplish a commonly held goal' (Stone, 2002, p. 405). Hence, the Polis Model emphasizes that the governance rules promulgated by the Rational Model will not necessarily lead to the desired outcomes. They will point

Table 6.3 (continued)

Rational Model	Polis Model
There is an attempt to apply federal rules to override local interests that impede overall goals. Incentives are aimed at making people change their behavior. In the Colorado River Basin no one single rational incentive will have greater impact than the free trading of water and water rights among users. The introduction of a variety of economic incentives including water trading will make the implementation of governance much easier. Incentives are part and parcel of the Rational Model. Use modern analytical methods, including: models of river hydrology, agricultural production and ecosystem functioning to examine alternative allocations that are feasible and to identify the allocation that maximizes national economic benefits minus costs.	to the problems that come with federal management in other river basins – for example, the conflicts between federal agencies' practices and the ESA in the Mississippi-Missouri. They will claim that local control will lead to better local and national outcomes. These attempt to reframe the issue not as a conflict between federal/public interest issues and local control, but rather as the search for the best set of rules for wise management of the river. The Polis Model will strenuously insist on maintaining the use and the users of water being kept as separate issues. Moreover, the allocation mechanisms are state or local issues not to be usurped by mechanistic federal rules or even worse the vagaries of the marketplace. The issue is reframed as a decentralized search for a consensus solution agreed to at the local and state levels. The courts are used to oppose federal government attempts to impose solutions.
Implementing solutions	
Clearly the federal interest will benefit more people rather than purely local concerns. The river basin is a national resource and should provide benefits to non-local as well as local people, particularly given the magnitude of federal tax monies spent in the past and in the present to maintain the system. Use decentralized incentives to encourage rational use at farm, factory or household level, including marginal cost pricing of water and the establishment of markets in water allocations.	In any policy dispute the actual magnitudes and the numbers of people benefiting from the proposed action are of major significance. The federal interest will benefit more people rather than purely local concerns, but the relative per capita magnitudes involved for the local people may be many times greater than the federally received, or perceived, benefits spread out over the entire US population. This leads to a mobilization of strong local economic interest groups. The 'continuous Monopoly game' cannot and should not be avoided by leaving important decisions in federal agency hands. Constant vigilance and interaction of interests will be required and the courts will continue to play a role as stages on which disputes are settled as they arise.

one way of understanding what has happened in the past and thus ultimately what is likely to happen in the future. While we are not necessarily advocating the wholehearted acceptance of the Polis Model, we do believe that it leads to a richer understanding of the 'game' being played over many decades by those interested in the use of Colorado River water. A synthesis of the approaches seems to be the next logical step in the development of our understanding.

SUMMARY AND CONCLUSIONS

We set out to understand political decision making about water development and management in the USA, in particular, the federal role and its relationship, if any, to achieving wise and sustainable use of the resource for the public interest. To this end the chapter explored two contrasting hypotheses in a fairly general way:

- That the process and outcomes of water decisions are controlled by considerations of rationality applied to the administration of the resource by the federal system paying attention to states' and citizens' rights and also to the needs and demands of stakeholders and interest groups.
- That the outcomes are the results of the contention and struggles among a wide variety of stakeholders and interest groups, only lightly controlled by the federal institutions except when egregious situations arise demanding federal intervention.

Our findings are not simple to characterize in terms of these two hypotheses. The first seems a sound approach and appears to be the one most dearly held by the federal agencies. However, the model of Congressional debate, approval of laws, the drafting of regulations for implementing the laws and implementation does not hold up when placed under examination. It turns out, for starters, that there are many conflicting definitions of rationality that depend upon one's point of view. For example, a federal official could look to the economically efficient use of investment dollars as the rational goal, but a Congressional representative could equally well look to maximizing the expenditure in their district and still claim a rational best decision. Or the USACE may believe that the economic goal of increasing (energy efficient) inland waterway traffic is a better and more sustainable one than the wetland restoration that environmental groups believe to be more efficient (because of wetlands' myriad life-support benefits) and sustainable. What we find in the literature and in the experience of practical people is that group decision making is bedeviled by such inconsistent preferences. And Arrow's Impossibility Theorem shows that a community of rational people may not be able to behave 'rationally' when the choice to be made is a collective one.[8]

In examining the long and rich history of the Colorado River Basin, we see during the extended period of river development a classic convergence of federal, state and local interests. Despite the famous fights between Arizona and California, all basin interests wanted the same thing: economic development. Coordinated river basin development suited these circumstances reasonably well.

In recent decades, however, we see a splintering of interests at all levels, greatly complicating the decision-making process and placing an increased premium on negotiation and compromise. As interests narrow, alliances occur at smaller scales. Moreover, these interests are apt to align in different ways dependent on the particular issue. We are no longer in a world well represented by the Rational Model.

In going from the Rational to the Polis Model, the Laws of Nature are replaced by Laws of Passion and the currency of exchange is seen to be ideas, persuasion and alliances rather than material exchanges (traditional log rolling trading of economic benefits). It is difficult to find in the literature a framework that functions at all levels and spheres in which water management decisions are made. For some sets of decisions, regime theory predicting outcomes according to whether costs and outcomes are concentrated or diffused, gives good insights. These predictions are more likely to be valid when the issue is investing tax dollars for economic returns, rather than allocating a mixed bag of quantitative (usually monetary) and qualitative goods, such as sustainability, water independence or high water quality. The latter situation is one of apples and oranges, in which comparing outcomes is difficult and the simple interest group model, or its regime variant, is not of great help.

Observing conflicting interest groups through the lens of the Polis Model is slightly more reassuring. Looking at a particular water issue in this way will encourage us to explore and apply different and broader concepts. If the federal agencies charged with defending the public interest can move to this more supple and responsive method of analysis, they may well judge that the process of discovery is more people-based than the mechanistic concepts used by other modeling approaches. Arrow tells us that group decision making can never be guaranteed to be rational where preferences for outcomes are not consistent. Stone tells us not to worry because rationality is only a small part of political decision making anyway.

Under both hypotheses the protection of a broad national public interest is viewed as the almost exclusive domain of the federal agencies. It is found, however, that all levels of government and all types of governance systems have concern for a public interest – albeit for smaller publics. Also, major unresolved conflicts remain among the leading federal agencies charged with stewardship of the water resources. The agencies have different mandates, different assessment methodologies, different regulations, different regional emphases and their own special publics. All of these add up to multiple and divergent conceptions of the public interest among federal officials and to less than entirely coherent behavior within the federal government itself. Coordination of the different federal policies essentially went out of the window with the demise of the WRC. Looking at the current state of affairs of water management in the USA, one would have to conclude that no one in the federal government is truly looking out for a distilled and integrated national public interest. There are instead guardians of separate 'national' interests that involve the manner in which water is used. The interests historically were promotion of commerce and economic development utilizing river resources. Today such interests are more fractionalized and are pursued by a mix of federal agencies.

This is not to say that all is lost. On the contrary, many progressive and strong governance systems in place today address the needs of many of the citizens. The short-lived RBCs mentioned earlier are good examples of conceptual models of regional water governance, administered through the states that should have led to protection of the public interests from overexploitation by private interests. However, they proved to be classic examples of the application of the Rational Model that simply do not appear to work in the real world and were subsequently discarded. These and other sub-national scale institutions need political and financial support if they are to continue to be relevant and useful. Comparing the very tenuous Apalachicola-Chattahoochee-Flint (ACF) and the Alabama-Coosa-Tallapoosa (ACT) River Basin Commission agreements among Georgia, Alabama and Florida with the more successful Colorado Compact points out the benefits of having the states come together (a bottom-up approach) and the insufficiencies of the federal initiative embodied in the southeastern case (a top-down approach). Of course, the Colorado Compact has been in existence since 1929 compared to a four or five year attempt in the ACT-ACF River Basins. Even so, there is little reason to expect the latter to ever be viewed as anything other than a federally imposed institution.

After giving the case history of the Colorado River Compact since its ratification in 1929, in Table 6.2 we analysed and compared what we loosely call the Rational Model with the Polis Model. We see how different the Polis Model is from the Rational Model. Powerful political symbols are manipulated to mobilize political support and bring its force to bear on decision makers. By and large, most people would like to believe they support goals larger than themselves. Therefore, big resonant issues more easily motivate citizens and interest groups than divvying up economic windfalls, even though contestants are willing to be hard-nosed about money. These themes repeat themselves throughout the table. Under the category of looking for solutions via incentives, the Rational Model says that: no single rational incentive will have greater impact than the free trading of water rights (evidently a way to advocate integrated river basin management). The Polis Model says: allocation mechanisms are a state and local issue not to be usurped by mechanistic federal rules or even worse the vagaries of the marketplace (evidently a claim to keep decision making closer to the benefited interests so they can (fight to) control the process). There seems to us little doubt that most readers of this chapter will be ardent supporters of the Rational Model statement, whereas it seems to us the contemporary world found in the Colorado River Basin can better be explained by the Polis Model. At least it helps those of us wedded to the Rational Model to broaden our perceptions of what rationality is all about and suggests unfamiliar, but potentially valuable, approaches.

Overtly using specifically political models of water governance, as suggested in this chapter, rather than clinging only to rational methods could have major implications for how the federal government does business in the water arena. Since it emerges that most of the control and protection of the public interest comes at all levels down the chain from federal to state to river basin to county, to city, two questions naturally arise: first, whether the federal bureaucracy currently employed is too often at cross purposes to its supposed missions; and, second, whether it is the right size.

If the analysis presented in this chapter is sound, then the role of federal bureaucracies in implementing water policy should be seriously reconsidered. Does this imply a need for a serious reduction in manpower at the federal level with the resources then reapplied at the state and local levels? It would certainly imply that staff re-education on a large scale, at all levels, be instituted by the agencies. This would force the professional staffs of the agencies to confront politics directly in their attempts to pursue water policy. Too often one hears the agency professionals complaining that they got a particular policy (or even project) right only to be subverted by political considerations. These political considerations should be central concerns in their attempts to guide policy.

NOTES

1. The legal fiction of governmental 'ownership' of water provides the basis for governmentally established rules governing its allocation and use.
2. Jimmy Carter ran into serious political problems with Congress after he vetoed a series of water projects in the 1978 Rivers and Harbors Omnibus bill based upon their poor benefit-cost ratios. Ronald Reagan made changes in federal water policy when he intervened in 1983 to enforce the cost sharing provisions for federal water resources expenditures.
3. The reader is referred to Hundley (1975), Fradkin (1968), Waters (1946), Reisner (1986), Weatherford and Brown (1986), Martin (1989), Pontius (1997), Meyers (1966), Getches (1985), Lochhead (2001) and MacDonnell (2006).
4. The USBR and the US Geological Survey (USGS) had been measuring flows of water in the river below Yuma, Arizona since the late 1800s. Based on these measurements, it was believed the basin produced an average yield of about 17.4 million acre-feet annually (not including already existing depletions both natural and man-made). Thus, the commissioners believed there would be additional water available for allocation at some time in the future. As it turned out, they were badly mistaken.
5. For an explanation of this doctrine see Appendix 6.1.
6. The Law of the River, includes the compacts, treaty, acts of Congress, US Supreme Court decisions, contracts, regulations and other agreements that collectively form the rules under which everything is supposed to operate. Few can claim to have a working knowledge of all these pieces. Most are able to function just within those pieces that directly affect their activities. Despite claims of its immutability, the Law of the River is far from clear, coherent and consistent. Yet, despite its complexity or, perhaps because of it, it has also been adaptable when necessary.
7. For a thoughtful argument for better-integrated basin-level decision making, see Getches (1997).
8. Arrow (1951). A common teaching example of the Arrow Paradox is where the participants have double-peaked (or 'inconsistent') preferences and try to decide among three alternatives using majority voting.

REFERENCES

Allison, G.T. (1971), *Essence of Decision: Explaining the Cuban Missile Crisis*, Boston, MA: Little Brown.

Arrow, K.E. (1951), *Social Choice and Individual Values*, New York: Wiley.

Bonnor, R.E. (2002), 'Buffalo Bill Cody and Wyoming water politics', *The Western Historical Quarterly*, **33** (4) 433–51.

Bornstein, D. (2004), *How to Change the World: Social Entrepreneurs and the Power of New Ideas*, Oxford: Oxford University Press.

Buchanan, J. and G. Tullock (1962), *Calculus of Consent: Logical Foundations of a Constitutional Democracy*, Ann Arbor, MI: University of Michigan Press.

Crenson, M.A. and B. Ginsberg (2002), *Downsizing Democracy: How America Sidelined its Citizens and Privatized its Public*, Baltimore, MD: Johns Hopkins University Press.

Dahl, R.A. (1961), *Who Governs?*, New Haven, CT: Yale University Press.

Downs, A. (1957), *An Economic Theory of Democracy*, New York: Harper and Row.

Espelande, W.N. (1998), *The Struggle for Water: Politics, Rationality and Identity in the American Southwest*, Chicago, IL: University of Chicago Press.

Fradkin, P.L. (1968), *A River No More: The Colorado River and the West*, Tucson, AZ: University of Arizona Press.
Friedkin, J.F. (1988), 'The international problem with Mexico over the salinity of the Lower Colorado River', in D.H. Getches (ed.), *Water and the American West*, Boulder, CO: Natural Resources Law Center, pp. 31–52.
Getches, D.H. (1985), 'Competing demands for the Colorado River', *University of Colorado Law Review*, **56**, 413–79.
Getches, D.H. (1997), 'Colorado River governance: sharing federal authorities as an incentive to create a new institution', *University of Colorado Law Review*, **68**, 573–658.
Haefele, E. (1973), *Representative Government and Environmental Management*, Baltimore, MD: Johns Hopkins University Press.
Hundley, Jr, N. (1975), *Water and the West: The Colorado River Compact and the Politics of Water in the American West*, Berkeley, CA: University of California Press.
Lasswell, H.D. (1936), *Politics: Who Gets What, When and How?*, New York: McGraw-Hill.
Lochhead, J.S. (2001), 'An upper basin perspective on California's claims to water from the Colorado River, Part I: the law of the river', *Water Law Review*, **4**, 290–330.
Lochhead, J.S. (2003), 'Part II: the development, implementation and collapse of California's plan to live within its basic apportionment', *Water Law Review*, **6**, 318–410.
Loucks, P. (2003), 'Managing America's rivers: who's doing it?', *International Journal of River Basin Management*, **1** (1), 21–31.
Lower Colorado River Multi-Species Conservation Program (2004), *Colorado River Multi-Species Conservation Program, Volume II: Habitat Conservation Plan*, J&S 00450.00, 17 December, Sacramento, CA: US Bureau of Reclamation.
Maass, A. (1951), *Muddy Waters: The Army Engineers and the Nation's Rivers*, Cambridge, MA: Harvard University Press.
Maass, A., M.M. Hufschmidt, R. Dorfman, H.A. Thomas, Jr, S.A. Marglin and G.M. Fair (1962), *The Design of Water-Resource Systems: New Techniques for Relating Economic Objectives, Engineering Analysis and Governmental Planning*, Cambridge, MA: Harvard University Press.
MacDonnell, L.J. (2005), 'The Colorado River: has it run out of water?', *The Water Report*, **16**, 1–6.
MacDonnell, L.J. (2006), 'Colorado River Basin', *Waters and Water Rights*, **7**, 3–62.
Martin, R. (1989), *A Story that Stands Like a Dam: Glen Canyon and the Struggle for the Soul of the West*, New York: Holt.
McCool, D. (1987), *Command of the Waters: Iron Triangles, Federal Water Development and Indian Water*, Berkeley, CA: University of California Press.
Meyers, C.J. (1966), 'The Colorado River', *Stanford Law Review*, **19**, 1–75.
Miller, T.O., G.D. Weatherford and J.E. Thorson (1986), *The Salty Colorado*, Washington, DC: The Conservation Foundation.
National Research Council (NRC) (2007), *Colorado River Basin Water Management: Evaluating and Adjusting to Hydroclimatic Variability*, Washington, DC: National Academies Press.
Pontius, D. (1997), 'Colorado River Basin study', report to the Western Water Policy Review Advisory Commission, Denver, CO.

Putnam, R.D. (2000), *Bowling Alone: The Collapse and Revival of American Community*, New York: Simon and Schuster.

Reisner, M. (1986), *Cadillac Desert: The American West and its Disappearing Water*, New York: Penguin Books.

Rogers, P. (1993), *America's Water: Federal Roles and Responsibilities*, Cambridge, MA: MIT Press.

Stone, D.A. (2002), *Policy Paradox: The Art of Political Decision Making*, revised edn, New York: Norton.

US Bureau of Reclamation (USBR) (1946), *The Colorado River: A Comprehensive Departmental Report on the Development of the Water Resources of the Colorado River Basin for Review Prior to Submission to Congress*, Washington, DC: Department of Interior.

US Bureau of Reclamation, Upper Colorado Region, Lower Colorado Region and Denver Office (USBR) (2004), *Colorado River System Consumptive Uses and Losses Report 1996–2000*, Denver, CO: US Bureau of Reclamation, Technical Service Center.

US Fish and Wildlife Service (USFWS) (2003), *Section 7 Consultation, Sufficient Progress, and Historic Projects Agreement and Recovery Implementation Program Recovery Action Plan (RIPRAP)*, Lakewood, CO: Upper Colorado River Recovery Program.

US Water Resources Council (WRC) (1983), *Economic and Environmental Principles and Guidelines for Water and Related Land Resources Implementation Studies*, Washington, DC: US Government Printing Office.

Waters, F. (1946), *The Colorado*, New York: Rinehart and Co.

Weatherford, G.D. and F.L. Brown (eds) (1986), *New Courses for the Colorado River: Major Issues for the Next Century*, Albuquerque, NM: University of New Mexico Press.

APPENDIX 6.1 OVERVIEW OF AMERICAN LAW FOR ALLOCATION OF WATER

Water, like other natural resources, is considered a form of property under American law and its use is governed primarily by state law. Because water is in motion and its benefits are widely shared, it is not readily susceptible to private ownership. Thus, state laws generally regard water as owned by the public or the state and establish rules providing for both its public and private use. Rights of use, which may themselves be treated as property rights or may be allocated by permit, are regarded as enabling enjoyment of the benefits of water and providing certain legal protections for those uses.

English law separated waterways that were navigable from those that were not. Non-navigable waterways were regarded as under the exclusive control of the riparian landowners and rights to the use of water were simply an extension of the rights of ownership of land. Riparian land ownership also carried rights to the use of water in navigable waterways, but public access for certain public uses such as travel, trade or fishing was to remain open. English law respecting use of groundwater similarly treated this resource as an extension of the ownership of land. Water extracted from underneath one's property was treated as owned in the same manner as any other physical aspect of one's property.

Riparian law principles were imported to the American colonies as part of English common law. Such principles worked reasonably well in this relatively humid part of the country in which coastal rivers were important primarily for travel and trade, there was adequate rainfall to support growing crops and direct human needs for drinking and other household uses could be readily met from streams and wells without interference with the uses of others.

Settlement of lands west of the 100th meridian raised new challenges. Here, rainfall is generally less than the amount necessary to grow crops. Rivers and streams are relatively few and largely unsuited for navigation. In this region a different approach for establishing rights to the use of water emerged, one based on taking physical possession of water and putting it to use. This approach came to be called prior appropriation. Under rules of prior appropriation, ownership of riparian land carries no special rights to the use of adjacent water. The principles of prior appropriation for use of surface water eventually became firmly embedded in the laws of all western states.

This appendix provides an introduction to the basic elements of riparian and prior appropriation law as systems for the allocation of rights to use surface water. It also includes a short introduction to the approaches states have taken to the use of groundwater. Finally, it briefly summarizes ways in which Federal laws affect allocation and use of water.[1]

Riparian Law

Originally a product of common law in which judges sought to define principles to sort out disputes between specific competing human uses of waterways and their water, the riparian doctrine seeks to accommodate the broad and varied range of interests implied by shared use. It provides special status to those who own lands immediately adjacent to a waterway, according such riparians undefined rights to the use and enjoyment of the waterway and its water. Among such riparians, it establishes a rough equality of right. As initially articulated in England and then in America, riparian water uses were to occur 'without diminishment in the quantity or quality of the water'. Under this view, riparian landowners could only use water on their lands, could not divert water away from other riparian landowners, could not take water out of the watershed and could not add pollution in amounts that would impair downstream uses. This strict rule of non-diminishment yielded inevitably in the 1800s, in the face of the realities of population increase and economic growth to one of 'reasonable' use. Under this rule, a court would assess the relative benefits of a use against the harm it caused to other riparians. In general, riparian principles seek an accommodation among competing uses of water along a shared waterway.

Regulated Riparian Systems

In recent years as out-of-stream uses of water have dramatically increased, most riparian law states have developed statutory systems governing allocation and use of water. Under such systems, removal of water from its source requires a permit issued by a state agency. Many states now also have special procedures applying to proposed uses of water out of the watershed in which the water originates. The right to obtain such a permit is not limited to those who own land riparian to the water source, but established riparian uses are protected from impairment by new uses. Permits are typically for a term of years. The amount of water that may be withdrawn is specified, as is the purpose and place of use. Conditions may be imposed, such as a requirement that withdrawals must be reduced in times of drought or that some minimum stream flow must be maintained. Permits may potentially be renewed as long as the use continues.

Prior Appropriation

Originally, rights to the use of water in states adopting the prior appropriation doctrine followed a rule of capture. The act of physically removing water from a stream and applying it to a use established the legal right. Beginning in the late 1800s, states moved to establish permit systems under which those wishing to make use of water must first apply to a state agency.

Prior appropriation follows a first-in-time, first-in right rule. The use established by the earliest appropriators (seniors) of water from a stream is protected from impairment by subsequent appropriators (juniors). Thus, for example, new diversions upstream from existing users cannot take water relied on by these users. In periods of high flows there may be enough water for all users. But, in low flow periods junior appropriators may have to stop diverting water to protect senior users. There is no sharing of shortages under this system.

The quantity of water that may be appropriated is limited to the amount that is beneficially used – that is, the amount reasonably necessary to accomplish the intended purpose of the use. Diversion of water in excess of that beneficially used is regarded as waste. The water right does not protect the diversion of wasted water.

Prior appropriation water rights are perpetual, so long as the use continues. Rights are fixed to a particular point of diversion, a maximum rate of diversion, a general type of use (such as irrigation or municipal) and a place of use. To change any of these elements of use requires a governmental review process to ensure there will be no injury to other water users. In most states water rights may be sold or otherwise transferred separate from the land on which they are used.

Groundwater Use

Groundwater law developed more slowly than the law for surface water. English law had allowed landowners essentially unlimited rights to withdraw water from underneath their land. Initially, American courts followed suit. With increasing evidence of the impacts of unlimited groundwater withdrawals, a reasonable use rule, similar to that for riparian users, emerged. In some states use of groundwater off the property from under which it was withdrawn is regarded as unreasonable. Conversely, any amount of use on the property is considered reasonable. There is a trend, however, toward seeking an accommodation of competing uses in recognition of the shared use of the resource. Certainly the effect of falling water elevations on other users can be a consideration.

Many western states follow the rules of prior appropriation for uses of groundwater. Rights of use thus are based on actual use, not on ownership of overlying land. Early rights enjoy certain protections against the effects of later uses.

Most states in which groundwater is heavily used have developed some kind of special management areas in which groundwater development is subject to public regulation. There may be spacing requirements to minimize well-to-well interference. There may be limits on pumping to regulate the drawdown of groundwater in a given aquifer (groundwater mining). There may be requirements to protect against pollution of the aquifer.

Traditionally, groundwater rules did not account for effects of groundwater withdrawals on surface water resources. In virtually all states the legal systems governing groundwater use developed separate from those governing surface water use. Increasingly, states are faced with the question of how to integrate competing groundwater and surface water uses. There is a clear trend toward regulating groundwater withdrawals to reduce adverse effects on surface water uses.

Effects of Federal Law on Water Allocation and Use

While state law governs the fundamental allocation of water uses, federal law plays an important role as well. The exercise of federal authority is based on powers found in the US Constitution. When the national government exercises constitutionally-derived authority, conflicting state laws must give way. The primary areas in which the USA has acted to affect the use of water are for protection of navigation and water quality, for the regulation of hydroelectric power generation, for water required for tribal and public land reservations and for uses of interstate rivers generally.

Navigation

The exclusive authority given to the US government to regulate matters of interstate commerce has been interpreted to give the USA power to protect the navigational uses of waterways. Thus, the USA, through the USACE, regulates actions of states or riparian landowners that might impede navigational uses. Congress also directs the USACE to implement projects improving navigational uses of coastal waterways and rivers, projects that now also include flood control functions.

Hydroelectric power generation

In 1920 Congress required any non-federal entity desiring to construct a dam for the purpose of generating hydroelectric power to obtain permission (embodied in a license) from a new federal agency, the Federal Power

Commission (FPC, now the Federal Energy Regulatory Commission). The primary motivation was to ensure that hydroelectric facilities were constructed in a manner that promoted comprehensive development of the river's economic values. In 1986 Congress included environmental considerations directly into licensing decisions.

Water quality

In 1972 Congress put in place a national program regulating the disposal of pollutants into the waters of the USA. Pollutant discharges into water are prohibited, except with a permit. Permit conditions set limits on allowable discharges of specific pollutants. The objective is to achieve and maintain water quality according to standards established to protect designated uses of water. Discharges of dredge and fill materials into waters of the USA are regulated as well, a requirement that has been extended by regulation to control filling wetlands.

Native American and federal reserved water rights

Federal courts have recognized the existence of 'reserved' water rights for areas of federal lands set aside for such purposes as Indian reservations or national forests. The effect is to reserve from state allocation some portion of water available within the land area encompassed by the reservation. Judicial recognition of such reserved rights originated with the 1908 US Supreme Court decision, *Winters* vs *U.S.* As subsequently extended and defined, reserved rights exist only as necessary to accomplish the primary purposes for which the reservation was established. Purposes are ascertained by reference to the treaty, Act of Congress or Executive Order establishing the reservation. To fit these rights into state law prior appropriation rules, reserved rights are given a priority as of the date the reservation was established. In contrast to prior appropriation, a reserved right need not have been used to exist. Federal law authorizes inclusion of federal claims to water, including reserved water rights, within state court adjudications of water rights.

Interstate rivers

Many waterways cross state lines. The primary mechanism for determining rights to use such rivers and their water between states is the interstate compact, authorized in the US Constitution. Compacts are negotiated by state representatives and may also include representatives of the federal government. The compact essentially serves as a contract between the parties. Congress must ratify interstate compacts and, once ratified, they are regarded as federal law.

Alternatively, disputes between states respecting uses of shared rivers may be settled through a special form of litigation filed directly with the US Supreme Court. Typically, the Court utilizes a 'special master' to serve as the finder of fact and to make recommendations respecting the law. The Court has developed the concept of 'equitable apportionment' to help guide its decisions, intended to fairly apportion uses on the basis of demonstrated need.

Note

1. For a succinct summary of basic American water law principles, see Getches (1997). A more complete treatment of water law is available in the one-volume treatise, Tarlock (1986) (with updates). For a multi-volume presentation, see Beck (1991) (with updates).

References

Beck, R.E. (1991), *Waters and Water Rights*, Charlottesville, VA: Michie.

Getches, D.H. (1997), *Water Law in a Nutshell*, 3rd edn, St Paul, MN: West Publishing Co.

Tarlock, A.D. (1988), *Law of Water Rights and Resources*, New York: C. Boardman.

7. Making the transition: moving water resources planning and management into the twenty-first century

Gerald E. Galloway

INTRODUCTION

The previous chapters discussed various aspects of water resources planning and management as they have evolved over roughly the last half-century and offered lessons learned, conclusions and recommendations for dealing with current or forecast problems. The purposes here are to provide a brief summary of these chapters, identify the significant issues now facing the nation in water resources planning and management, suggest future directions for the field in light of the challenges identified and offer recommendations for research that would support and complement efforts to deal with the challenges ahead.

WHERE HAVE WE BEEN?

In 1902, with passage of the Reclamation Act, the nation established a social and economic goal of drawing people to the more arid West by providing them with the water they would need to live and to support themselves by practicing irrigated agriculture. Hydropower, a major interest of conservationists seeking basin approaches to water planning during Teddy Roosevelt's presidency, became a regulatory mission of the federal government in the 1920s. During the Depression of the 1930s, the Administration of President Franklin Roosevelt took actions to prime the economic pump, including building large water projects, such as dams on the Missouri River and its tributaries. It also established the Tennessee Valley Authority (TVA) to bring economic development to that valley, while managing the river system as a system. Following on a series of legislative actions begun early in the century and designed to deal with regional flooding issues, in 1936 Congress passed legislation that made flood damage reduction a national program, proclaiming that floods 'constitute a menace to national welfare [and that] . . . it is the sense of Congress that flood control is a proper activity of the Federal government'. Further, the Act famously stated that the 'Federal government should improve or participate in improvements . . . for flood control purposes if the benefits to whomsoever they accrue are in excess of the estimated costs' (Flood Control Act of 1936, Section 1). In the second half of the 1940s a desire to provide jobs for servicemen returning from World War II led to more development for flood control, irrigation (reclamation), navigation and recreation. In short, the first half of the century was focused on development of water resources and protecting the populace from floods.

Over the next 50 years, as the nation became aware of the full impacts, physical and fiscal, of the developments of the previous decades, values shifted. The environment assumed a larger role in public policy and the public itself became more involved in decision making. The following sections are intended as a reminder for the reader of some of the landmarks in the world of water resources of the past five plus decades.

Seeing the Environment in a Different Light

Up to the 1950s, the task of water resources planners was to solve problems by designing projects. Little consideration had to be given to potential collateral impacts of the projects. Indeed, few recognized the need for such consideration. Multi-objective planning focused on bringing together the standard project purposes – water supply, hydropower, recreation, flood control and navigation – and few saw a need to deal with the environment, either social or ecological, that contained and surrounded the project.

The shift in values and the public's increase in willingness to become involved began in the late 1950s with the civil rights movement, broadened and accelerated with the Vietnam War protests and took on something of a green hue after the publication of Rachel Carson's *Silent Spring* (Carson, 1962). By the end of the 1960s it was clear the environment had to be given a place at the water resources development table, and President Nixon's signing of the National Environmental Policy Act (NEPA) on 1 January 1970 served as a statement that this had happened. Over the next four years, legislation protecting water and air, ensuring consideration of historic and archeological resources and protecting endangered and threatened flora and fauna were added or enhanced. Each law reflected a felt need to address a specific problem, but there existed no grand strategy for environmental protection. During the 1970s, federal rule making and court cases, primarily involving implementation of NEPA, ensured that federal actions, voluntarily or involuntarily, would attempt to take into consideration protection of the environment. Over the following two decades, engineers and economists, working with biologists, ecologists and other natural resources scientists, attempted to develop approaches that would permit water resources development to continue, albeit in a different fashion, with the natural environment protected and sometimes even enhanced.

Recognizing a Need to Cleanup a Legacy of Past Activity

Over the past four decades, billions of dollars have been spent in cleanup of toxic material and in addressing habitat and species issues. While progress has been made, there is still much to be accomplished. However, federal support of funding for cleanup is in decline (US Environmental Protection Agency (EPA), 2004a). For example, in 1987 in the Great Lakes the government identified 36 toxic hot spots for remediation. As this is written, work has not been completed at any of these locations but, as disturbing, because of the large number of separate but related programs, it is even difficult to determine how much has been spent or what the results of the partial remediations have been (International Joint Commission (IJC), 2003). More generally, there has been little effort to conduct formal *ex post* evaluations of these programs so the benefits cannot be accurately estimated. On the ecosystem restoration front, the $7.8 billion Everglades Restoration Project represents the first of several multi-billion dollar efforts that the federal government is being asked to fund, in whole or part. Standing in the wings are Louisiana Coastal Restoration; restoration of the mainstem Missouri, Upper Mississippi and Illinois Rivers; and many smaller projects. In fact, new water projects today typically include some aspect of restoration or mitigation. Unfortunately, while funds for the projects are appropriated in a manner that permits the construction and operation of the project, funding of mitigation frequently lags the construction by many years. These are deliberate choices made by either Congress or the Administration in establishing priorities for the budget year.

Moving the Center of Gravity of Politics and Public Interest

Federal funding for water projects has declined since the 1960s. Though there was a spike to support water and wastewater treatment facilities in the 1970s and 1980s, the spike has gone, and the current funding slope is downward. Certainly few large dams or other high cost projects have been started in the last two decades. Part of this decline stems from the obvious shift in societal values just discussed, but part also comes from increased involvement and decreased enthusiasm of the public where water projects are concerned. Cost sharing caused many who previously advocated water projects to back off once their money became a part of the solution. Buoyed by the report of the World Commission on Dams (2000), in the USA (as well as abroad), people are positing that the opportunity for big water projects has passed – that there is neither geographic opportunity nor political support for efforts that could cause environmental and social disruption. Federal budgeting has become more competitive, with tax cuts shrinking revenues, the specter of growing social security and medicare entitlements and the appearance of new claimants such as the Global War on Terrorism and homeland security pushing other needs aside. Decisions to rotate committee chairs in the House of Representatives reflect that even Congress is changing.

Changing the Guideposts for Water Planning and Management

The early 1960s saw considerable attention to the economics and engineering of water management. The Harvard Water Program (Maass et al., 1962) sparked a more sophisticated approach to planning water projects. President John Kennedy's interest in ensuring consideration of more than engineering and economics in project development and the publication of Senate Document 97 (US Senate, 1962) launched water planning in a new direction. The passage in 1965 of the Water Resources Planning Act codified a movement already underway and by the 1970s there was a Water Resources Council (WRC); seven basin planning commissions; the report of a National Water Commission (NWC) with its emphasis on a carefully coordinated and broad-based water program for the nation; legislative support for clean water and species protection; and a National Flood Insurance Program (NFIP) that proposed involving flood plain communities in controlling flood plain development and thereby reducing future flood damages. Perhaps the late 1970s represented the crest of this wave of attention to water planning and the establishment of broadly based guideposts for program execution. Actions in the early 1980s to: eliminate basin commissions; zero fund the WRC while leaving it in existence on paper; shift from *Principles and Standards* (P&S) (WRC, 1973) to *Principles and Guidelines* (P&G) (WRC, 1983); reduce money for water treatment, water monitoring and water projects; and institute federal–local cost sharing marked a dramatic shift in the importance of water on the national scene and began to reduce the utility of the guideposts. As Congressional committees and subcommittees reorganized to more closely align with where the money was going, formerly water-focused committees took on multiple responsibilities. Both the US Army Corps of Engineers (USACE) and the US Bureau of Reclamation (USBR) saw operations and maintenance becoming more critical elements in their portfolios and felt growing pressure to turn operations functions over to private contractors. The Great Mississippi Flood of 1993 and the 2002 drought that affected nearly half of the conterminous 48 states returned some attention to water resources issues and highlighted both the lack of coordination of federal programs and the problems with lack of harmony among the guideposts established earlier (US Executive Office of President (USEOP), 1994). As the twenty-first century began, the water programs of the federal government operated independently of each other and created problems for those at state and local levels. Environmental programs were organized along media-specific lines (see Chapter 4). Guideposts for planning activity were neither comprehensive nor inclusive. The accumulated laws, regulations and policies of successive Congresses and Administrations produced conflicts that threw new projects and operations into the courts or committed them to never ending analysis. Much of water resources planning has moved from a federally led, comprehensive approach towards a locally based, narrower, almost single purpose, focus.

Exploiting New Technology

In 1957, after the Soviet Union launched Sputnik, the USA went into high gear to improve its scientific base and 12 years later an American walked on the moon. The leaps in capabilities of information technology that accompanied the space race allowed scientists and engineers to acquire massive amounts of data about the natural and human environment, process and analyse these data, prepare more comprehensive and detailed plans for water programs and projects and operate them with a level of sophistication not considered possible only a few decades earlier. The pace in technical advance increased throughout the last half of the century and is expected to continue at a high level well into the current one.

Today, remote sensing platforms produce high quality imagery across a spectrum of wavelengths and provide highly accurate and timely information about all parts of the world, including bathymetric and topographic data and identification of even slight distinctions among land features, human made and natural. Global positioning systems now permit the identification of geographic location with an accuracy of less than a meter and along with massive computing power allow the development of Geospatial Information Systems (GIS) that permit the display and, more importantly, the analysis of a vast array of information in map format. This same computing power has allowed the development of hydrologic and hydraulic, climate and other water-related models whose complexity and realism exceed anything dreamed of a half-century earlier. The combination of GIS, mobile and advanced communication systems and miniaturization permit the almost real-time monitoring and assessment of data gathered from across the globe. A continuing challenge remains the ability to provide the resources required to fully exploit the capabilities of these technologies, to acquire and operate more sophisticated systems and to effectively use the data gathered by them.

CURRENT ISSUES

In the first decade of the twenty-first century the nation is already facing a number of significant conflicts over use of water. Major court cases have arisen over regulation of the flows of the Missouri River system by USACE; provision of minimum flows on the Rio Grande in New Mexico and in the Klamath Basin in Oregon; and allocation of the waters of the Apalachicola, Chattahoochee and Flint (ACF) Rivers in the Southeast USA and the Colorado River in the West. Public debate is underway over initiation of new lock construction on the Upper Mississippi River, over diversion of hydropower waters to provide flows to support salmon on the Columbia River, and even over the continued existence of dams that interfere with anadromous fish runs on both the east and west coasts. Much remains to be learned about how to deal with these challenges.

Implementing Adaptive Management

C.S. (Buzz) Holling and his colleague C.J. Walters brought adaptive management to the ecology scene in the 1980s (Walters and Holling, 1990). Defined as an approach that 'emphasizes the use of carefully designed and monitored experiments, based on input from scientists, managers and citizens, as opportunities to maintain or restore ecological resilience and to learn more about ecosystems' (National Research Council (NRC), 2002a, p. 5), it rapidly became an approach of choice for those engaged in the restoration of natural systems affected by development or the establishment of new habitat, such as in the development of new wetlands. Key components are choice of goals and objectives, creation of a plan, design of experiments that test the plan and strong stakeholder involvement not only in setting goals and objectives but also in vetting progress in achieving the goals and objectives. Adaptive management has been touted as a central component of projects such as the restoration of the Everglades and return to more natural flows on the Colorado River below Glen Canyon

Dam. It is being initiated in restoration efforts on the Missouri and Upper Mississippi systems. Going beyond ecology, USACE has even proposed its use in management of the decades-long effort to improve navigation on the Upper Mississippi. In this application it would involve not ecological intervention but modifying the construction schedule for a series of locks based on observed changes in demand for use of the navigation system.

While there is strong support for the concept in the ecology world, adaptive management is not without its critics. Those pressing for immediate action to deal with environmental issues, such as flow restoration, argue that the adaptive management framework does nothing more than delay execution of programs they know need to be done. Some would argue there is a need for more science behind the label to provide a better understanding of what adaptive management really means and how it can be carried out in a manner that will allow for appropriate fiscal controls. Indeed, politicians and program managers are starting to see approval of adaptive management as writing blank checks. When Congress authorized $7.8 billion for the Everglades restoration, it was supporting adaptive management. But Congress and the Administration, now being asked to consider the same sort of arrangement for other major projects such as Louisiana Coastal restoration ($12–$20 billion) and Upper Mississippi River restoration ($5 billion), are balking at granting approval for a series of high cost projects that will be defined over time.

Broadening Benefit-Cost Analysis

As indicated in previous chapters, especially Chapter 3, benefit-cost analysis has been part of the water resources project approval process for over 50 years. Only a few water projects leave USACE or other federal agencies without a favorable benefit-cost ratio and, to many, the higher the ratio, the better the project. But, in a 2004 report, a committee of the National Research Council reviewing USACE analytical methods and approaches recommended that 'benefit-cost analysis should not be used as the lone decision criterion in judging whether a proposed planning or management alternative in a USACE planning study should be approved' (NRC, 2004b, p. 6.) The committee further noted that reliance on a strict benefit-cost criterion fails as a decision rule for ranking investments because:

- Maximizing the difference between total benefits and total costs does not address equity considerations in the distribution of costs and benefits.
- There are likely to be significant uncertainties involved in development of the benefit and cost estimates.

In addition, all relevant costs and benefits may not be measurable in monetary terms. Certainly such aspects as changes in biodiversity and ethical concerns are not easily monetized (NRC, 2004b). As the authors of Chapter 3 noted, 'many project effects will never be monetized and some seem unlikely to be even quantified. The benefit-cost analysis must evolve to incorporate and to give appropriate weight to all project effects.'

In 2006 a House of Representatives Committee asked USACE to engage the National Academy of Public Administration (NAPA) in an evaluation of the USACE's budget process. In its report NAPA noted a 'current overemphasis on economic benefits' and a need to 'broaden the factors considered in selecting project alternatives . . .' and '. . . incorporate social equity considerations into the fundamental planning process' (NAPA, 2007, pp. 139, 140).

Understanding Ecological Restoration

Over the past 20 years, there has been a growing effort to restore ecosystems. Federally funded work, on both large and small scales, is underway across the USA. Some of the degradation targeted has been caused by activities such as farming and urban development; some has resulted directly from federal, state and local projects. But regardless of cause, a very large fraction of the problems being addressed involve water bodies or wetlands.

To have an effective, let alone an efficient restoration program or project, however, it is necessary to tie down the meaning of the word itself. The first and, in many ways, the hardest question is: restore to what condition? That inferred as existing prior to human occupancy? At a specific date, such as 1850? How do the restorers allow for or deal with human influences on the ecosystem that most likely will remain? For example, the nature of the run-off component of the hydrologic cycle has been profoundly altered by the uses to which we have put much of the landscape. Most of this will not change. Further, while we are considering removal of many small dams and other structures, many more are destined to remain and will continue to influence river ecosystems. The levees along the Mississippi River that allow millions to live and produce are almost certain to remain in place, even though we recognize that their presence has altered both the river's and the Gulf of Mexico's ecosystem.

Finally, as noted in Chapter 5, assessing the benefits and costs of ecological restoration will require economists and ecologists to learn to communicate and to develop tools and techniques acceptable to both groups.

Taking Climate Change into Account

The report, *Climate Change Impacts on the United States* (National Assessment Synthesis Team (NAST), 2000, p. 10), asserted that, with projected growth in greenhouse gas emissions, temperatures in the USA will likely rise 5 to 9 °F in the next 100 years and that the effects on water resources are likely to become issues in every region of the country. In February 2007 the Intergovernmental Panel on Climate Change (IPCC), established by the World Meteorological Organization (WMO) and the United Nations Environmental Program to assess scientific, technical and socioeconomic information relevant for the understanding of climate change, published *Climate Change 2007. The Physical Science Basis: Summary for Policymakers* (IPCC, 2007), indicating that:

- 'Global atmospheric concentrations of carbon dioxide, methane and nitrous oxide have increased markedly as a result of human activities since 1750 and now far exceed pre-industrial values determined from ice cores spanning many thousands of years.' (p. 2).
- 'Warming of the climate system is unequivocal, as is now evident from observations of increases in global average air and ocean temperatures, widespread melting of snow and ice, and rising global average sea level' (p. 5).
- 'At continental, regional and ocean basin scales, numerous long-term changes in climate have been observed. These include changes in Arctic temperatures and ice, widespread changes in precipitation amounts, ocean salinity, wind patterns and aspects of extreme weather including droughts, heavy precipitation, heat waves and the intensity of tropical cyclones' (p. 8).

The *Summary* (p. 9) highlighted that these recent '. . . trends, assessment of human influence on the trends and projections for extreme weather events for which there is an observed late Twentieth Century trend . . .' indicate that it is 'likely' that there will be increased incidence of extreme high sea levels, intense tropical cyclone activity and areas affected by droughts and that it is 'very likely' that there will be increases in heavy precipitation events (proportion of total rainfall from heavy falls) over most areas.

The NAST wrote in 2000:

> Today, while a political debate rages over the usefulness of the various climate change predictions, few, if any, planning efforts for water-resources projects and programs, even take into account the possible impacts of climate change. Neither Congress nor the Administration requires agencies to consider it. As

> a result, projects and actions are being undertaken today, with project lives extending over decades that may not be able to cope with tomorrow's conditions. What actions can and should be taken to mitigate the potential impacts of climate change? (NAST, 2000, p. 10)

Defining the Role of the Public

Stakeholders – local officials, project sponsors, non-governmental groups and the public at large – have always been involved in public decision making, but the nature of that involvement has evolved over the years (Galloway, 2003; Cortner, 1993). In the not-so-distant past, the government simply informed the public about a plan. Under pressure of environmental laws and lawsuits, the process was extended to include listening to feedback. The initiation of project cost sharing under the 1986 Water Resources Development Act (WRDA) further increased involvement by locals but also created tensions between the interests of local sponsors in pursuing a set of sponsor-favored alternatives and the federal need to address the broader public or national interest (NRC, 1999, 2004b). At the level expected today, public involvement programs are expensive and time consuming both for federal officials and the public. But, even the best planned project can be delayed or stopped by a failure to involve stakeholders in its early stages (Schkade et al., 1996). Much of the credit for the successful launch of the Everglades project belongs to the early development of a state-driven collaborative planning process in which 1600 stakeholders representing the competing users of the Everglades came together to develop a consensus on goals and objectives. Use of such a process requires the federal or state government to surrender some control and to accept the attendant challenges and the questions raised. What is the proper amount of local input and how is this to be balanced against the federal or broader public interest?

Extending Planning Objectives and Measuring Results

Planners today face a variety of technical challenges as they develop programs and projects. Factors only lightly considered in the past are moving to the forefront. For example, the Brundtland concept of sustainable development – meeting the needs of the present without compromising the ability of future generations to meet their own needs – is in effect a requirement for intergenerational equity (World Commission on Environment and Development (WCED), 1987) but offers little assistance in determining how this can be achieved. Contemporary social justice concerns are requiring planners not only to estimate the aggregate costs and benefits of projects but also the distribution of these costs and benefits among social classes and ethnic and racial groups.

Under the heading of measuring results the Government Performance and Results Act (GPRA) of 1993 requires government officials to develop metrics that permit the performance of programs to be assessed. Is water quality improving as predicted? Is the projected traffic on the waterway occurring? Is the ecosystem being restored? The GPRA also requires backward looks at projects and programs already completed though *ex post* audits of completed projects are seldom accomplished both because funds are generally not available and because there is a tendency to avoid reopening issues that seemingly have been settled. Yet, it will be difficult to fully comply with the GPRA, deal with the nuances of adaptive management and learn from past mistakes without a viable *ex post* audit program (NRC, 2004b).

Addressing Basin and Watershed Planning

The past decade has seen increased attention to watershed planning. In 1998 the US Department of Agriculture (USDA) and the EPA submitted to the White House a Clean Water Action Plan (CWAP) that included 'a new cooperative approach to watershed protection in which state, tribal, Federal and local governments and the public . . . work together to focus resources and implement effective strategies to solve [the most critical water quality] problems' (Browner and Glickman, 1998, one page letter). The Action Plan called on all federal agencies to participate in this process and sparked renewed efforts across the government to deal with water issues in a system or watershed context and, in 2003, the Secretary of the Interior issued Water 2025 with just such a focus (US Department of Interior (USDOI), 2003). In its *Civil Works Strategic Plan* (USACE, 2004, p. vii), USACE indicates that it intends to operate with an approach that emphasizes 'a holistic focus on water problems and opportunities; attention to the watershed as a logical geographic area for managing water resources; a systems approach for analyzing problems and solutions . . .'. Loucks (2003) argues for a collaborative intergovernmental effort to carry out integrated multi-state river basin planning and suggests that a single federal agency, the USACE, be designated to lead such efforts. Others suggest that the lead could be taken by agencies of the USDOI, the USDA or by EPA or that the lead evolve from the nature of activity in the particular basin. The 2007 NAPA report (p. 1) suggested that USACE 'must evolve from the role of "project engineer" to "watershed steward" and become a convener of intergovernmental prioritization processes'.

All of these government approaches emphasize watersheds and

collaboration among levels of government, though each program has its own focus. Some consideration of a list of so far unanswered questions, however, suggests there may be less here than the rhetorical flourishes might promise. For example:

- Where does the regional or basin coordination originate?
- Do federal or state agencies take the lead and which agency within that level of government?
- If the federal government is serious about this, why have the river basin commissions (RBCs) been zeroed out of the federal budget?
- What structure is needed to deal with multi-state basins?
- How does any basin plan contemplate organizing a decision process that can deal with disagreement and produce proposals that can be implemented?

One might even ask: are watersheds the proper vehicle for complex systems analysis? Today, human activity is increasingly responsive to economic influences. Growth follows the paths of highways and trade corridors. Metropolitan areas are not defined by watersheds but by the location of the population. Much of the wildlife population moves freely from watershed to watershed. Will the water planners of the twenty-first century have to deal with econosheds rather than watersheds and how will these econosheds relate to watersheds?

Where implementation, if not decision making, is concerned, advances in remote sensing and monitoring, information technology and GIS permit quicker and easier sharing of information and increasing connectivity among basin and watershed participants. For large basins, such as the Mississippi, it is even becoming possible to envision operating the many components of the basin as a collective whole so that, for example, the TVA releases on the Tennessee River do not conflict with releases on the Missouri and its tributaries by USACE and the USBR. For this sort of basin-wide operation, however, the question of who is in charge will be a very difficult one.

Integrating Groundwater into Planning and Management

In many parts of the country groundwater is out of sight and out of mind; in others it is recognized as a critical component of the overall water resources picture. Because aquifers are only partially mapped across the nation, the full extent of the resource can only be roughly estimated and a lack of monitoring makes it difficult to assess the quality even of the known sources. While groundwater specialists have appreciated the relationship between surface and groundwater, this relationship only came to the attention of water planners in the last decades. Thus, the value of wetlands as conduits for groundwater recharge was only recently considered in developing the benefits of wetland preservation. Recent studies of the Great Lakes Basin groundwater indicated that the contribution of groundwater to stream flow in the basin ranges from 79 per cent in the Lake Michigan Basin to 48 per cent in Lake Erie (Granneman et al., 2000). Over 40 million Americans rely on groundwater for their water supply and in many locations the aquifers on which they rely are being threatened. Yet, when water resources planning takes place, groundwater resources in the study areas are seldom treated in the same manner as surface water resources and integrated into the overall planning schema. Technology may offer some hope of faster and more accurate mapping and monitoring of groundwater, but progress will not be made until adequate funds are provided to carry out the needed work and groundwater is considered along with surface water in watershed planning.

Reducing Flood Damages

The devastating impacts of Hurricane Katrina on the Mississippi-Louisiana Gulf Coast refocused the nation's attention on the destructive power of floods. Over a million people lost their homes and jobs and more than 1300 people lost their lives. Entire neighborhoods in New Orleans were destroyed and may never be rebuilt. Although some residents have returned to the city, the majority of displaced and relocated citizens are still temporarily or permanently scattered in communities throughout the USA. Even prior to Katrina, national flood losses, after correction for inflation, have continued to grow in spite of decades of structural flood damage reduction efforts by the federal government and implementation of the NFIP, designed to provide some land use control over flood plain occupancy. While, from the 1930s through the 1980s, emphasis was placed on structural approaches to solving the flood damage problem – dams, levees, floodwalls and channel work – the failures of structures in New Orleans illustrated the fragility of such systems when they are not properly maintained and monitored and the residual risk that will always exist behind or below such structures. The movement toward non-structural approaches that began in the 1950s gained ground with the increased attention to the environment but has not been fully accepted by those at risk or in laws, regulations and rules. An inherent bias against non-structural approaches remains in the current P&G in that they establish national economic development (NED) as the objective of water project development and require projects based on non-quantifiable environmental and social benefits, typical of those found in non-structural efforts, to be approved on an exception-to-policy basis by the approving Federal Secretary governing development of flood damage reduction projects. Further, while current federal rules for communities participating in the NFIP permit the construction of buildings in the flood plain only as long as they do not raise the flood elevation by more than 0.1 feet, there is no required consideration of the cumulative impact of many such buildings. Similarly, the computation of the elevation of the flood with a 1 per cent chance of occurrence – the national standard for flood insurance – does not take into account future upstream conditions or climate change. The Association of State Floodplain Managers (ASFPM) argues that development in a floodway should only be permitted when it can be shown that it causes no adverse impact (ASFPM, 2004). A current evaluation of the NFIP is raising the question of the efficacy of the 1 per cent standard and the nature of the protection provided by levees and floodwalls (American Institutes for Research (AIR), 2005). Katrina also raised the issue of flood risk. A US Senate Committee constituted to examine Katrina response noted that, 'the inspection and maintenance regime in place to ensure that the levees, flood walls and other structures existing to protect the residents of the greater New Orleans area was in no way commensurate with the risk posed to these persons and their property' (US Senate, 2006, p. 17.)

Also following Katrina, an Interagency Performance Evaluation Task Force (IPET), chartered by USACE to examine the reasons for the failures in the New Orleans Hurricane Protection System, reported that:

> A risk-based planning and design approach would provide a more viable capability to inform decisions on complex infrastructure such as hurricane protection systems. The traditional approach, as used for the New Orleans protection measures, is component-performance-based, uses standards to define performance and relies on factors of safety to deal with uncertainty. It is difficult to examine the integrated performance of multiple components, and standards are usually limited to past experience. Risk-based planning is systems-based, requiring that the entire system be described in consistent terms and explicitly including uncertainty. Component performance is related to system performance as well as the consequences of that performance. The risk-based approach is well suited for consideration of a variety of measures of merit. Factors such as loss of life, environmental losses and cultural consequences can be included in decision making without reducing everything to one measure such as dollars. (USACE, 2006a, pp. 1–11)

According to a subsequent report by USACE addressing post-Katrina recovery planning,

> A new risk-based assessment methodology is being developed for use in preparing the LACPR [Louisiana Coastal Protection and Restoration] Final

> Technical Report. The methodology will assess alternative hurricane risk reduction plans and would produce valuation information describing the risks and consequences for various populations and assets along the coast under different planning scenarios and solution options . . . Ultimately, most risk reduction solutions are directed at altering the relative relationship between probability and consequence. This new risk assessment methodology is emerging from post-Katrina forensic efforts and is being proposed as an improved approach for future engineering work. (USACE, 2006b, p. 57)

A natural question is whether both the USACE risk-based approach and the NFIP standards-based approach should operate as official guides to flood plain decisions? Or is the cost of potential inconsistency too high?

Unjumbling Federal Water Policy, Legislation and Coordination

In Chapter 6 Rogers, MacDonnell and Lydon pointed out that 'looking at the current state of affairs of water management in the USA, one would have to conclude that no one in the federal government is truly looking out for a distilled and integrated national public interest'. Several recent NRC studies highlighted the conflicts existing among federal programs and the challenges faced by the agencies dealing with conflicting legislative mandates (NRC, 2004a, 2002a, 1999; NAPA, 2007). As noted in Chapter 2 and in federal reports (USEOP, 1994), since the demise of the WRC, little if any directed coordination exists among federal agencies engaged in water resources activities. The NAPA report found that USACE's budget process, which drives the bulk of water resources project development, '. . . is not completely of its own making and cannot be changed without the agreement of the Administration and Congress' (NAPA, 2007, p. xvi).

Within Congress there continues to be little coordination among the many committees with significant responsibilities for either authorizing or appropriating support of water resources. The Executive Branch's P&G apply to the development programs of four agencies (USACE, NRCS, USBR and TVA) but not to others with significant water-related funding (EPA, Federal Emergency Management Agency (FEMA) and Department of Housing and Urban Development (HUD)). Laws pertaining to FEMA's activity in the flood plain do not mesh with those governing the actions of the National Resources Conservation Service (NRCS) or USACE. Two recent administrations implemented programs that attempted to improve the situation. In 1991 President George H.W. Bush established Coastal America as a partnership among 13 sub-cabinet level agencies committed 'to work together and integrate their efforts with state, local and non-governmental activities'. The agencies 'pledged to coordinate their statutory responsibilities and combine their resources and expertise' (EPA, 2004b). In 1998 the Clinton Administration established the American Heritage Rivers Initiative, designating 14 rivers or segments thereof as Heritage rivers and appointing, for each, a River Navigator, whose mission was to bring together the efforts of the many federal agencies involved in dealing with the river (USEOP, 1997). Neither program provides a long-term substitute for federal coordination starting at the top.

Wisely Replacing Aging Infrastructure

The primary focus of water resources planning and management has historically been the development of new projects. Unfortunately, but perhaps understandably, considerably less attention has been given to ensuring that water projects, once built, are well maintained and rehabilitated or replaced when necessary. There is little evidence that the nation is aware either of the bill that eventually will be paid or the consequences of continuing to postpone its payment. This is despite efforts such as those of the American Society of Civil Engineers (ASCE) to bring the problem to public notice. An ASCE *Report Card for America's Infrastructure* has biennially assessed 12 areas of infrastructure vital to the nation's economic survival and quality of life. Four of the 12 areas involved water (ASCE, 2001, 2003, 2005). In its 2005 *Report Card* ASCE (2005) gave:

- Dams a grade of D, citing 3500 unsafe dams (those with deficiencies that leave them highly susceptible to failure). It was estimated that $10.1 billion will be needed to address all critical non-federal dams over the next 12 years.
- Both the nation's 54000 drinking water systems and its 16000 wastewater systems a D-. (Both water and wastewater systems were rated D in 2003.) To continue to comply with federal water regulations, drinking water systems face an annual shortfall of $11 billion needed to replace facilities nearing the end of their useful lives. The ASCE noted that the EPA estimated that $390 billion would be needed for infrastructure improvements to wastewater systems over the next 20 years. The EPA's own gap analysis, conducted in 2002, also noted similar shortfalls (EPA, 2002). The fiscal year (FY) 2008 EPA budget cuts nearly $400 million from the expected FY 2007 budget for wastewater infrastructure, the seventh straight year of declining budget requests for the Clean Water State Revolving Fund program (US House of Representatives, 2007).

- Navigable waterways a D- (downgraded from a D+ in 2003), with the ASCE noting that, of the 257 locks on the inland waterways, nearly half were obsolete in 2005 and that by 2020 that fraction is predicted to increase to four-fifths. It was estimated that the cost to replace the present system of locks would be more than $125 billion.
- Energy, which includes hydropower, a D (downgraded from a D+ in 2003). Attention was specifically drawn to the lack of investment in new transmission facilities.

In making decisions on funding of infrastructure maintenance and replacement, federal and state governments must establish priorities, consider alternative approaches and develop new funding mechanisms. Yet the existing governance structure does not provide a mechanism for evaluating and comparing the needs across all water sectors in order to establish a priority list. Some groups will advocate removal rather than replacement of some structures, especially dams, and much needs to be learned about the benefits and costs of such removals. The construction of major water and wastewater treatment facilities during the 1970s and 1980s was a one-time program that likely will not be repeated. Small communities were given the opportunity to build facilities that would accommodate hoped for growth and that required operation and maintenance funding beyond their means. Use of revolving funds to provide low interest loans to communities for rehabilitation or replacement has proven to be effective but adequate funding has not been forthcoming from the federal level. New attention is focusing on the feasibility and desirability of regional systems (NRC, 2005). Given current budget realities, infrastructure work will have to be phased over the next few decades.

Acknowledging the Private Sector

As just noted, with an aging infrastructure, both supply and treatment facilities will require modernization. When private organizations offer to design, build and operate new facilities and maintain costs at reasonable levels, the deal is appealing. While concerns about private control of water supply can be mitigated by sound contracts between local governments and private providers, a generalized suspicion is likely to remain. This is so even though in the USA there is and has been for many years a mix of public and private water utilities, so experience is not lacking (Boland, 2000). That the USA is not alone in its concerns may be seen from the debates on this subject at the Third and Fourth World Water Forums held in Kyoto, Japan in 2003 and Mexico City in 2006. At those events considerable attention focused on the negative aspects of privatization of water supply and water treatment, with forum participants expressing concern that turning management of water supply over to profit-making groups was pricing water out of the reach of the poor and giving private industry too much control. It is worth noting that similar concerns have been expressed over passing the missions of operating locks and dams and recreation facilities from the federal government to private contractors. The experiences of other agencies of the government with government to private conversions indicate that they can work if the proper contract mechanisms are put in place. For example, contractors have successfully taken over much of the information technology support for federal agencies. It seems unlikely, however, that the available evidence will be sufficient to quiet the suspicion.

FUTURE DIRECTIONS

The nation will have to develop new approaches to water resources planning and management to deal effectively with the water resources challenges of the twenty-first century. Here are suggestions for several fairly specific tasks that ought to be attacked sooner rather than later.

Deal with Organizational Structure and the Policy Void

The President, in coordination with the Congress, states and tribes, should facilitate joint development of national water policy and a national water vision. In revisiting national water resources policy and coordination, the role of states and regions should be accorded increased prominence, with the federal government playing a facilitative rather than directive role (Foster, 1998). Participants in the 2007 American Water Resources Association's (AWRA) National Water Policy Dialogue suggested that '. . . the center of gravity for national water actions should rest at the state level and be backed by appropriate support from the Federal government' (AWRA, 2007, p. 1).

The President also should reconstitute a water-focused coordinating organization within the US Executive Office of the President (EOP) to better coordinate federal water programs and agencies; assess the status of the nation's and the world's water environment; facilitate water research; and support state water resources planning and management programs. Such a coordinating body would also be expected to provide guidance on reconciling traditional water resources demands with those implied by environmental issues.

Assess the State of Water Resources in the Nation

The WRC conducted the last national water assessment in 1976. Frequent reports by the US Geological Survey (USGS) and the EPA provide insights into quantity and quality aspects of the nation's water resources but do not provide the breadth required of national assessment of the many uses of water and how water itself impacts human activity. In their 2004 report the Subcommittee on Water Availability and Quality (SWAQ, Committee on Environment and Natural Resources, National Science and Technology Council) outlined 'the need for coordinated science and technology efforts to better understand water supply and demand in the United States' (National Science and Technology Council, 2004). The 2007 SWAQ report focused on 'topics that would benefit from increased interagency and public/private collaboration and/or increased resources' (National Science and Technology Council, 2007).

What structure would best ensure frequent monitoring, assessment and reporting on the state of our water resources?

A 2002 National Water Policy Dialogue, conducted by the AWRA, noting a lack of coordination, spoke to the absence of a national water policy. The Dialogue called upon the President and Congress to:

- Develop a National Water Vision – to determine, in cooperation with the states and local governments, how the nation wants to deal with water.
- Formulate a National Water Policy that translates the vision into action. This is not a call for a federal water policy that directs the actions of federal, state and local governments, but for a policy that defines the shared responsibilities at each level of government.
- Provide for coordination and collaboration among federal agencies and with other agencies at state, regional and local levels in the conduct of their water resources activities.
- Address water issues on a holistic basis, using watersheds and basins as the setting for water resources projects and programs (AWRA, 2002).

Work in an Interdisciplinary Manner

We must demand that those who deal with water resources work in an interdisciplinary manner. Accomplishment of this goes far beyond the blending of programs in universities and the restructuring of organizations to provide for this balance. It will require that decision makers and legislators at the national, state and local level also take this approach. Single focus committees in legislative bodies need to be integrated to provide a more holistic approach to dealing with the challenges we face. Interdisciplinary groups that bring balanced and integrated alternative approaches to dealing with the challenges they face should support senior decision makers.

Adequately Support Water Resources Research

The road ahead in water resources and water resources planning and management is not clearly marked. Many unknowns and uncertainties will have to be dealt with as they are encountered. That said, there is strong reason to carry out broad efforts in research to work toward solving some of the challenges we do know must be faced. Research enabled us to address many of the issues encountered over the past half-century. It provided the technological advances already mentioned, along with new theories and methodologies that increased the available understanding of the water world in which humanity operates. But given the numerous challenges outlined in this text, the need for research will be at least as important in the decades ahead as it has been in the past. While the National Science Foundation continues to support basic research in water resources, it and other agencies could usefully pay more attention to applied research in planning and management. Decisions on water issues must be founded on unbiased, credible and current scientific data and information. There is a need to adequately support the water resources research programs outlined below.

General

Needs for water research have been investigated by any number of learned bodies including the NRC (NRC, 2004a, 2001; Baedecker and Friedman, 2000; Naiman et al., 1995). The authors of this book have also identified specific areas where water-related research will be valuable and where research needs can be focused or expanded.

In 2001 the Water Science and Technology Board of the NRC issued a report outlining important areas of water resources research needing to be addressed in the next 10 to 15 years in order to meet the challenges ahead (NRC, 2001). The report, which was initiated by the Board, reflected two years of Board discussions on the water issues facing the nation. It concluded that investments in water resources research can especially usefully be made in the following areas:

- Water availability – Investigations of surface water and groundwater availability should focus on the development of supply-enhancing technologies, on understanding the coupled hydrologic and biogeochemical cycles that control water quality and on developing means of preventing further declines in water quality. Data from networks of continuous ground-based and remote sensing instrumentation are critical for understanding responses to variable climates at different temporal and spatial scales. Monitoring is important not only for resource planning and regulatory activities, but also for assessing the effectiveness of water policies and management efforts.

- Water use – There is a great need to better understand the determinants of consumptive water use, the importance and scale of agricultural water use and the nature and impact of environmental uses of water. Research on the technologies and infrastructure for water conservation and recycling will be critical to meeting future water needs.
- Water institutions – Research leading to the development of improved water management institutions should receive much more emphasis in the research agenda of the twenty-first century than it has in the past. Efforts should focus on legal and economic institutions and should involve researchers from a broad array of social science disciplines (NRC 2001, p. 2).

Shortly after release of the 2001 NRC report, Congress requested that the NRC conduct a follow-on study to 'illuminate the state of the water resources research enterprise in the United States' (NRC, 2004a, p. 1). The resulting 2004 report refined and enhanced the findings of the 2001 report, examined investment patterns in research, addressed coordination of research and related institutional options for coordinating, prioritizing and implementing water research (NRC, 2004a). The 2004 report reviewed the research focus areas identified by the 2001 report and concluded that they reflected the current best statement of research needs. The need for research in these areas continues to exist.

Ecological restoration and environmental quality

Many restoration programs have been launched, but most of the available research has focused on small streams and rivers and only recently did work on large rivers become recognized for funding and other support. The social challenges of determining the ultimate objective of restoration may be equally large. The scientific challenges in understanding restoration ecology are immense and, as Moreau and Loucks indicate in Chapter 4, there has been insufficient focus on and a lack of progress in developing the broad concepts of ecosystem health and services that would support effective planning and management of ecosystem restoration. The critical questions needing answers cross the boundaries of the natural sciences and engineering. For example, the present level of understanding of the relationships between spatial and temporal patterns of flow and their effects on both aquatic and terrestrial ecosystems has progressed only a short distance beyond conceptual models. The research by Stanford (1994), Postel and Richter (2003) and others point out weaknesses in existing models. An emerging consensus among these investigators is that restoration of ecosystems to their natural conditions is most likely to occur if flow regimes are returned to their natural cycles and variability. That can be achieved only partially when streams are managed for multiple purposes. More comprehensive models of aquatic and riparian ecosystems and how they are affected by hydrologic variables are needed, not only to test the developing theories, but also to advance work on tools needed for ongoing projects. Substantial additional work is necessary to develop reliable predictive models to support environmental restoration programs and to assess potential impacts of projects designed primarily for other purposes. There is an equal need to examine the mechanism by which decisions concerning restoration are made and how, in multiple stakeholder environments, consensus on such issues can be reached. To add to the challenge, whether restoration efforts make a difference will not be fully understood for decades after programs are initiated. Achieving this understanding will require substantial investment in monitoring and assessment activities.

A second research need is to enhance understanding of the fate and transport of contaminants over long distances and extended time periods, especially for nutrients and toxic substances. Much of the current body of water quality models are based on phenomena that may occur over selected critical reaches of streams for selected periods of time. Nutrients, sediment and long-lived toxic substances frequently have their most significant effects hundreds of miles downstream and months or even years after the substances enter tributary streams.

Conduct of *Ex Post Facto* Research

As the previous chapters indicated, assessments of the performance of past water policies and projects can reveal many lessons. Methodologies for the conduct of such research exist but require special attention as they always rest on development of counter-factual scenarios. But the greatest challenge is likely to be to find the support for such applied research. Because it is not dealing with the search for theories or new methodologies it is generally not high on the funding priority lists of research agencies such as the National Science Foundation. Action agencies and their oversight committees in Congress will not be anxious to fund such activity because there may be bad news connected with the product. The Government Accountability Office (GAO) does report on specific elements of programs or projects but the products generally lack the rigor of more formally developed *ex post facto* evaluations. Further, with adaptive management being increasingly recognized as an appropriate strategy for addressing the uncertainty associated with environmental and other projects, the cost of continued monitoring and evaluation of projects should, in any case, be recognized as a valid project cost. If the monitoring is correctly designed, it can be useful both for real-time adaptation and for later evaluation.

New Paradigms for Stakeholder Involvement

If adaptive management is to be successful it is almost certainly true the stakeholders will have to be involved both in the monitoring and the choice of adaptations. Local sponsors of projects – be they states, municipalities or special organizations – are paying a cost share for many projects and expect to be at the decision table. But stakeholder involvement necessitates independent scientific support, perhaps in the form of a scientific panel available to them. There is considerable need for research in public involvement and especially in development of decision support systems (DSS). Efforts to further develop interactive, 'shared vision' simulation models could lead to improved stakeholder understanding and thereby enhance the decision process.

REFERENCES

American Institutes for Research (AIR) (September 2005), *Evaluation of the National Flood Insurance Program: Annual Report 2004–2005*, contract number 282-98-0029, task order 49, Washington, DC: American Institutes for Research.

American Society of Civil Engineers (ASCE) (2001), *2001 Report Card for America's Infrastructure*, Reston, VA: American Society of Civil Engineers.

American Society of Civil Engineers (ASCE) (2003), *2003 Report Card for America's Infrastructure*, Reston, VA: American Society of Civil Engineers.

American Society of Civil Engineers (ASCE) (2005), *2005 Report Card for America's Infrastructure*, Reston, VA: American Society of Civil Engineers accessed 15 April, 2008 at http://asce.org/reportcard/2005/index.cfm.

American Water Resources Association (AWRA) (31 December 2002), *Letter to the President of the United States, Speaker of the House of Representatives and Senate Majority Leader*, Middleburg, VA: American Water Resources Association.

American Water Resources Association (AWRA) (20 February 2007), *Letter to the President of the United States*, Middleburg, VA: American Water Resources Association.

Association of State Floodplain Managers (ASFPM) (2004), *No Adverse Impact: A Toolkit for Common Sense Flood Plain Management*, Madison, WI: Association of State Floodplain Managers.

Baedecker, M.J. and L.C. Friedman (eds) (2000), 'National research program in the hydrologic sciences', US Geological Survey, circular 1195, US Department of the Interior, US Geological Survey, Reston, Virginia.

Boland, J.J. (October 2000), 'The business of water: overview and context', *Water Resources Update*, **117**, 1–5.

Browner, C. and D. Glickman (14 February 1998), *Letter to the Vice President Accompanying the Clean Water Action Plan*, Washington, DC: US Environmental Protection Agency.

Carson, R. (1962), *Silent Spring*, New York: Houghton Miflin Co.

Cortner, H.J. (1993), *Reconciling Citizen, Analyst and Manager Roles in Democratic Governance: Public Involvement Challenges in the 1990s*, Alexandria VA: US Army Corps of Engineers, Institute for Water Resources.

Foster, C.H.W. (1998), 'The art and practice of living in place: lessons from New England regionalism', *Water Resources Update*, **111**, 18–22.

Galloway, G.E. (20 November 2003), *Report of the Coastal Louisiana Technical Summit*, Flood Control Act, 33 U.S.C. 701a, Reston, VA: American Society of Civil Engineers.

Granneman, N.G., R.J. Hunt, J.R. Nicholas, T.E. Reilly and T.C. Winter (2000), *The Importance of Groundwater in the Great Lakes Region*, investigation report 00-4008, Lansing, MI: US Geological Society.

Intergovernmental Panel on Climate Change (IPCC) (2007), *Climate Change 2007. The Physical Science Basis: Summary for Policymakers*, contribution of Working Group I to the Fourth Assessment Report of the Intergovernmental Panel on Climate Change, Geneva: Intergovernmental Panel on Climate Change.

International Joint Commission (IJC) (April 2003), *The Status of Restoration Activities in the Great Lakes Areas of Concern*, Washington, DC and Ottawa, ON: International Joint Commission.

Loucks, D.P. (2003), 'Managing America's rivers: who's doing it?', *International Journal of River Basin Management*, **1** (1), 21–31.

Maass, A., M.M. Hufschmidt, R. Dorfman, H.A. Thomas, Jr, S.A. Marglin and G.M. Fair (1962), *The Design of Water-Resource Systems: New Techniques for Relating Economic Objectives, Engineering Analysis and Governmental Planning*, Cambridge, MA: Harvard University Press.

Naiman, R.J., J.J. Magnuson, D.M. McKnight and J.A. Stanford (1995), *The Freshwater Imperative: A Research Agenda*, Covelo, CA: Island Press.

National Academy of Public Administration (NAPA) (2007), *Prioritizing America's Water Resources Investments: Budget Reform for Civil Works Construction Projects at the U.S. Army Corps of Engineers, Summary Report*, a report by a panel of the National Academy of Public Administration for the US Congress and the US Army Corps of Engineers, Washington, DC: National Academy of Public Administration.

National Assessment Synthesis Team (NAST) (2000), *Climate Change Impacts on the United States: The Potential Consequences of Climate Variability and Change*, Washington, DC: US Global Change Research Program.

National Research Council (NRC) (1999), *New Directions in Water-Resources Planning for the U.S. Army Corps of Engineers*, Washington, DC: National Academy Press.
National Research Council (NRC) (2001), *Envisioning the Agenda for Water-Resources Research in the Twenty-First Century*, Washington, DC: National Academy Press.
National Research Council (NRC) (2002), *The Missouri River Ecosystem: Exploring the Prospects for Recovery*, Washington, DC: National Academy Press.
National Research Council (NRC) (2004), *Confronting the Nation's Water Problems: The Role of Research*, Washington, DC: National Academy Press.
National Research Council (NRC) (2004b), *Analytical Methods and Approaches in Water Resources Project Planning*, Washington, DC: National Academy Press.
National Research Council (NRC) (2005), *Regional Cooperation for Water Quality Improvement in Southwestern Pennsylvania*, Washington, DC: National Academy Press.
National Science and Technology Council, Committee on Environment and Natural Resources, Subcommittee on Water Availability and Quality (2004), *Science and Technology to Support Fresh Water Availability in the United States*, Washington, DC: Executive Office of the President.
National Science and Technology Council, Committee on Environment and Natural Resources, Subcommittee on Water Availability and Quality (2007), *A Strategy for Federal Science and Technology to Support Water Availability and Quality in the United States*, Washington, DC: Executive Office of the President.
Postel, S. and B. Richter (2003), *Rivers for Life: Managing Water for People and Nature*, Washington, DC: Island Press.
Schkade, D.A., T.D. Feather and D.T. Capan (June 1996), *Environmental Valuation: the Role of Stakeholder Communication and Collaborative Planning*, report 96-R-17, Alexandria, VA: US Army Corps of Engineers, Institute for Water Resources.
Stanford, J.A. (July 1994), *Instream Flows to Assist the Recovery of Endangered Fishes of the Upper Colorado Basin*, biological report no. 24, Denver, CO: US Fish and Wildlife Service.
US Army Corps of Engineers (USACE) (March 2004), *Civil Works Strategic Plan*, Washington, DC: US Army Corps of Engineers.
US Army Corps of Engineers (USACE) (2006a), 'Performance evaluation of the New Orleans and Southeast Louisiana hurricane protection system', draft final report of the Interagency Performance Evaluation Task Force, vol. I, executive summary and overview 1 June, accessed 30 December, 2008 at https://ipet.wes.army.mil/.
US Army Corps of Engineers (USACE) (2006b), 'Louisiana coastal protection and restoration', preliminary technical report to the US Congress, US Army Corps of Engineers, New Orleans District, accessed 30 December, 2008 at www.lacpr.usace.army.mil.
US Department of Interior (USDOI) (2 May 2003), news release 'Water 2025: preventing crises and conflict in the west', US Department of Interior, Washington, DC.
US Environmental Protection Agency (EPA) (2002), *The Clean Water and Drinking Water Gap Analysis*, Washington, DC: Office of Water.

US Environmental Protection Agency (EPA) (7 October 2004a), 'Superfund budget history', accessed 30 December, 2008 at www.epa.gov/superfund/accomp/budgethistory.htm.

US Environmental Protection Agency (EPA) (7 October 2004b), 'Coastal America', accessed at www.epa.gov/owow/oceans/partnerships/coastam.html.

US Executive Office of the President (USEOP) (1994), *Sharing the Challenge: Floodplain Management into the 21st Century*, report of the Interagency Floodplain Management Review Committee, Washington, DC: Government Printing Office.

US Executive Office of the President (USEOP) (1997). *Federal Support of Community Efforts along American Heritage Rivers*, Executive Order 13061, Washington, DC: The White House.

US House of Representatives, Transportation and Infrastructure Committee (2007), 'FY 2008 President's budget request for the Department of Transportation and Environmental Protection Agency', press release accessed 30 December, 2008 at http://transportation.house.gov/hearings/hearingdetail.aspx?NewsID=28.

US Senate (1962), *Policies, Standards and Procedures in the Formulation, Evaluation and Review of Plans for Use and Development of Water and Related Land Resources*, Senate Document 97, 87th Congress, 2nd Session, Washington, DC: US Government Printing Office.

US Senate (2006), *Hurricane Katrina: A Nation Still Unprepared*, Committee on Homeland Security and Governmental Affairs, Washington, DC: US Government Printing Office.

US Water Resources Council (1973), 'Establishment of principles and standards for planning water and related land resources', *Federal Register*, **38** (174), 24778.

US Water Resources Council (1983), *Economic and Environmental Principles and Guidelines for Water and Related Land Resources Implementation*, Washington, DC: US Government Printing Office.

Walters, C.J. and C.S. Holling (1990), 'Large-scale management experiments and learning by doing', *Ecology*, **71** (6), 2060–68.

World Commission on Dams (November 2000), *Dams and Development: A New Framework for Decision Making*, London and Sterling, VA: Earthscan Publications Ltd.

World Commission on Environment and Development (WCED) (1987), *Our Common Future*, Oxford: Oxford University Press.

Index